Pitman Research Notes in Mat.
Series

Submission of proposals for consideration
Suggestions for publication, in the form of outlines and representative samples, are invited by the Editorial Board for assessment. Intending authors should approach one of the main editors or another member of the Editorial Board, citing the relevant AMS subject classifications. Alternatively, outlines may be sent directly to the publisher's offices. Refereeing is by members of the board and other mathematical authorities in the topic concerned, throughout the world.

Preparation of accepted manuscripts
On acceptance of a proposal, the publisher will supply full instructions for the preparation of manuscripts in a form suitable for direct photo-lithographic reproduction. Specially printed grid sheets are provided and a contribution is offered by the publisher towards the cost of typing. Word processor output, subject to the publisher's approval, is also acceptable.

Illustrations should be prepared by the authors, ready for direct reproduction without further improvement. The use of hand-drawn symbols should be avoided wherever possible, in order to maintain maximum clarity of the text.

The publisher will be pleased to give any guidance necessary during the preparation of a typescript, and will be happy to answer any queries.

Important note
In order to avoid later retyping, intending authors are strongly urged not to begin final preparation of a typescript before receiving the publisher's guidelines and special paper. In this way it is hoped to preserve the uniform appearance of the series.

Longman Scientific & Technical
Longman House
Burnt Mill
Harlow, Essex, UK
(tel (0279) 26721)

Titles in this series

1 Improperly posed boundary value problems
A Carasso and A P Stone
2 Lie algebras generated by finite dimensional ideals
I N Stewart
3 Bifurcation problems in nonlinear elasticity
R W Dickey
4 Partial differential equations in the complex domain
D L Colton
5 Quasilinear hyperbolic systems and waves
A Jeffrey
6 Solution of boundary value problems by the method of integral operators
D L Colton
7 Taylor expansions and catastrophes
T Poston and I N Stewart
8 Function theoretic methods in differential equations
R P Gilbert and R J Weinacht
9 Differential topology with a view to applications
D R J Chillingworth
10 Characteristic classes of foliations
H V Pittie
11 Stochastic integration and generalized martingales
A U Kussmaul
12 Zeta-functions: An introduction to algebraic geometry
A D Thomas
13 Explicit *a priori* inequalities with applications to boundary value problems
V G Sigillito
14 Nonlinear diffusion
W E Fitzgibbon III and H F Walker
15 Unsolved problems concerning lattice points
J Hammer
16 Edge-colourings of graphs
S Fiorini and R J Wilson
17 Nonlinear analysis and mechanics: Heriot-Watt Symposium Volume I
R J Knops
18 Actions of fine abelian groups
C Kosniowski
19 Closed graph theorems and webbed spaces
M De Wilde
20 Singular perturbation techniques applied to integro-differential equations
H Grabmüller
21 Retarded functional differential equations: A global point of view
S E A Mohammed
22 Multiparameter spectral theory in Hilbert space
B D Sleeman
24 Mathematical modelling techniques
R Aris
25 Singular points of smooth mappings
C G Gibson
26 Nonlinear evolution equations solvable by the spectral transform
F Calogero
27 Nonlinear analysis and mechanics: Heriot-Watt Symposium Volume II
R J Knops
28 Constructive functional analysis
D S Bridges
29 Elongational flows: Aspects of the behaviour of model elasticoviscous fluids
C J S Petrie
30 Nonlinear analysis and mechanics: Heriot-Watt Symposium Volume III
R J Knops
31 Fractional calculus and integral transforms of generalized functions
A C McBride
32 Complex manifold techniques in theoretical physics
D E Lerner and P D Sommers
33 Hilbert's third problem: scissors congruence
C-H Sah
34 Graph theory and combinatorics
R J Wilson
35 The Tricomi equation with applications to the theory of plane transonic flow
A R Manwell
36 Abstract differential equations
S D Zaidman
37 Advances in twistor theory
L P Hughston and R S Ward
38 Operator theory and functional analysis
I Erdelyi
39 Nonlinear analysis and mechanics: Heriot-Watt Symposium Volume IV
R J Knops
40 Singular systems of differential equations
S L Campbell
41 N-dimensional crystallography
R L E Schwarzenberger
42 Nonlinear partial differential equations in physical problems
D Graffi
43 Shifts and periodicity for right invertible operators
D Przeworska-Rolewicz
44 Rings with chain conditions
A W Chatters and C R Hajarnavis
45 Moduli, deformations and classifications of compact complex manifolds
D Sundararaman
46 Nonlinear problems of analysis in geometry and mechanics
M Atteia, D Bancel and I Gumowski
47 Algorithmic methods in optimal control
W A Gruver and E Sachs
48 Abstract Cauchy problems and functional differential equations
F Kappel and W Schappacher
49 Sequence spaces
W H Ruckle
50 Recent contributions to nonlinear partial differential equations
H Berestycki and H Brezis
51 Subnormal operators
J B Conway

52 Wave propagation in viscoelastic media
F Mainardi
53 Nonlinear partial differential equations and their applications: Collège de France Seminar. Volume I
H Brezis and J L Lions
54 Geometry of Coxeter groups
H Hiller
55 Cusps of Gauss mappings
T Banchoff, T Gaffney and C McCrory
56 An approach to algebraic K-theory
A J Berrick
57 Convex analysis and optimization
J-P Aubin and R B Vintner
58 Convex analysis with applications in the differentiation of convex functions
J R Giles
59 Weak and variational methods for moving boundary problems
C M Elliott and J R Ockendon
60 Nonlinear partial differential equations and their applications: Collège de France Seminar. Volume II
H Brezis and J L Lions
61 Singular systems of differential equations II
S L Campbell
62 Rates of convergence in the central limit theorem
Peter Hall
63 Solution of differential equations by means of one-parameter groups
J M Hill
64 Hankel operators on Hilbert space
S C Power
65 Schrödinger-type operators with continuous spectra
M S P Eastham and H Kalf
66 Recent applications of generalized inverses
S L Campbell
67 Riesz and Fredholm theory in Banach algebra
B A Barnes, G J Murphy, M R F Smyth and T T West
68 Evolution equations and their applications
F Kappel and W Schappacher
69 Generalized solutions of Hamilton-Jacobi equations
P L Lions
70 Nonlinear partial differential equations and their applications: Collège de France Seminar. Volume III
H Brezis and J L Lions
71 Spectral theory and wave operators for the Schrödinger equation
A M Berthier
72 Approximation of Hilbert space operators I
D A Herrero
73 Vector valued Nevanlinna Theory
H J W Ziegler
74 Instability, nonexistence and weighted energy methods in fluid dynamics and related theories
B Straughan
75 Local bifurcation and symmetry
A Vanderbauwhede
76 Clifford analysis
F Brackx, R Delanghe and F Sommen
77 Nonlinear equivalence, reduction of PDEs to ODEs and fast convergent numerical methods
E E Rosinger
78 Free boundary problems, theory and applications. Volume I
A Fasano and M Primicerio
79 Free boundary problems, theory and applications. Volume II
A Fasano and M Primicerio
80 Symplectic geometry
A Crumeyrolle and J Grifone
81 An algorithmic analysis of a communication model with retransmission of flawed messages
D M Lucantoni
82 Geometric games and their applications
W H Ruckle
83 Additive groups of rings
S Feigelstock
84 Nonlinear partial differential equations and their applications: Collège de France Seminar. Volume IV
H Brezis and J L Lions
85 Multiplicative functionals on topological algebras
T Husain
86 Hamilton-Jacobi equations in Hilbert spaces
V Barbu and G Da Prato
87 Harmonic maps with symmetry, harmonic morphisms and deformations of metrics
P Baird
88 Similarity solutions of nonlinear partial differential equations
L Dresner
89 Contributions to nonlinear partial differential equations
C Bardos, A Damlamian, J I Díaz and J Hernández
90 Banach and Hilbert spaces of vector-valued functions
J Burbea and P Masani
91 Control and observation of neutral systems
D Salamon
92 Banach bundles, Banach modules and automorphisms of C*-algebras
M J Dupré and R M Gillette
93 Nonlinear partial differential equations and their applications: Collège de France Seminar. Volume V
H Brezis and J L Lions
94 Computer algebra in applied mathematics: an introduction to MACSYMA
R H Rand
95 Advances in nonlinear waves. Volume I
L Debnath
96 FC-groups
M J Tomkinson
97 Topics in relaxation and ellipsoidal methods
M Akgül
98 Analogue of the group algebra for topological semigroups
H Dzinotyiweyi
99 Stochastic functional differential equations
S E A Mohammed

100 Optimal control of variational inequalities
V Barbu
101 Partial differential equations and dynamical systems
W E Fitzgibbon III
102 Approximation of Hilbert space operators. Volume II
C Apostol, L A Fialkow, D A Herrero and D Voiculescu
103 Nondiscrete induction and iterative processes
V Ptak and F-A Potra
104 Analytic functions – growth aspects
O P Juneja and G P Kapoor
105 Theory of Tikhonov regularization for Fredholm equations of the first kind
C W Groetsch
106 Nonlinear partial differential equations and free boundaries. Volume I
J I Díaz
107 Tight and taut immersions of manifolds
T E Cecil and P J Ryan
108 A layering method for viscous, incompressible L_p flows occupying R^n
A Douglis and E B Fabes
109 Nonlinear partial differential equations and their applications: Collège de France Seminar. Volume VI
H Brezis and J L Lions
110 Finite generalized quadrangles
S E Payne and J A Thas
111 Advances in nonlinear waves. Volume II
L Debnath
112 Topics in several complex variables
E Ramírez de Arellano and D Sundararaman
113 Differential equations, flow invariance and applications
N H Pavel
114 Geometrical combinatorics
F C Holroyd and R J Wilson
115 Generators of strongly continuous semigroups
J A van Casteren
116 Growth of algebras and Gelfand–Kirillov dimension
G R Krause and T H Lenagan
117 Theory of bases and cones
P K Kamthan and M Gupta
118 Linear groups and permutations
A R Camina and E A Whelan
119 General Wiener–Hopf factorization methods
F-O Speck
120 Free boundary problems: applications and theory, Volume III
A Bossavit, A Damlamian and M Fremond
121 Free boundary problems: applications and theory, Volume IV
A Bossavit, A Damlamian and M Fremond
122 Nonlinear partial differential equations and their applications: Collège de France Seminar. Volume VII
H Brezis and J L Lions
123 Geometric methods in operator algebras
H Araki and E G Effros
124 Infinite dimensional analysis–stochastic processes
S Albeverio
125 Ennio de Giorgi Colloquium
P Krée
126 Almost-periodic functions in abstract spaces
S Zaidman
127 Nonlinear variational problems
A Marino, L Modica, S Spagnolo and M Degiovanni
128 Second-order systems of partial differential equations in the plane
L K Hua, W Lin and C-Q Wu
129 Asymptotics of high-order ordinary differential equations
R B Paris and A D Wood
130 Stochastic differential equations
R Wu
131 Differential geometry
L A Cordero
132 Nonlinear differential equations
J K Hale and P Martinez-Amores
133 Approximation theory and applications
S P Singh
134 Near-rings and their links with groups
J D P Meldrum
135 Estimating eigenvalues with *a posteriori/a priori* inequalities
J R Kuttler and V G Sigillito
136 Regular semigroups as extensions
F J Pastijn and M Petrich
137 Representations of rank one Lie groups
D H Collingwood
138 Fractional calculus
G F Roach and A C McBride
139 Hamilton's principle in continuum mechanics
A Bedford
140 Numerical analysis
D F Griffiths and G A Watson
141 Semigroups, theory and applications. Volume I
H Brezis, M G Crandall and F Kappel
142 Distribution theorems of L-functions
D Joyner
143 Recent developments in structured continua
D De Kee and P Kaloni
144 Functional analysis and two-point differential operators
J Locker
145 Numerical methods for partial differential equations
S I Hariharan and T H Moulden
146 Completely bounded maps and dilations
V I Paulsen
147 Harmonic analysis on the Heisenberg nilpotent Lie group
W Schempp
148 Contributions to modern calculus of variations
L Cesari
149 Nonlinear parabolic equations: qualitative properties of solutions
L Boccardo and A Tesei
150 From local times to global geometry, control and physics
K D Elworthy

151 A stochastic maximum principle for optimal control of diffusions
U G Haussmann
152 Semigroups, theory and applications. Volume II
H Brezis, M G Crandall and F Kappel
153 A general theory of integration in function spaces
P Muldowney
154 Oakland Conference on partial differential equations and applied mathematics
L R Bragg and J W Dettman
155 Contributions to nonlinear partial differential equations. Volume II
J I Díaz and P L Lions
156 Semigroups of linear operators: an introduction
A C McBride
157 Ordinary and partial differential equations
B D Sleeman and R J Jarvis
158 Hyperbolic equations
F Colombini and M K V Murthy
159 Linear topologies on a ring: an overview
J S Golan
160 Dynamical systems and bifurcation theory
M I Camacho, M J Pacifico and F Takens
161 Branched coverings and algebraic functions
M Namba
162 Perturbation bounds for matrix eigenvalues
R Bhatia
163 Defect minimization in operator equations: theory and applications
R Reemtsen
164 Multidimensional Brownian excursions and potential theory
K Burdzy
165 Viscosity solutions and optimal control
R J Elliott
166 Nonlinear partial differential equations and their applications. Collège de France Seminar. Volume VIII
H Brezis and J L Lions
167 Theory and applications of inverse problems
H Haario
168 Energy stability and convection
G P Galdi and B Straughan
169 Additive groups of rings. Volume II
S Feigelstock
170 Numerical analysis 1987
D F Griffiths and G A Watson
171 Surveys of some recent results in operator theory. Volume I
J B Conway and B B Morrel
172 Amenable Banach algebras
J-P Pier
173 Pseudo-orbits of contact forms
A Bahri
174 Poisson algebras and Poisson manifolds
K H Bhaskara and K Viswanath
175 Maximum principles and eigenvalue problems in partial differential equations
P W Schaefer
176 Mathematical analysis of nonlinear, dynamic processes
K U Grusa
177 Cordes' two-parameter spectral representation theory
D F McGhee and R H Picard
178 Equivariant K-theory for proper actions
N C Phillips
179 Elliptic operators, topology and asymptotic methods
J Roe
180 Nonlinear evolution equations
J K Engelbrecht, V E Fridman and E N Pelinovski
181 Nonlinear partial differential equations and their applications. Collège de France Seminar. Volume IX
H Brezis and J L Lions
182 Critical points at infinity in some variational problems
A Bahri
183 Recent developments in hyperbolic equations
L Cattabriga, F Colombini, M K V Murthy and S Spagnolo
184 Optimization and identification of systems governed by evolution equations on Banach space
N U Ahmed
185 Free boundary problems: theory and applications. Volume I
K H Hoffmann and J Sprekels
186 Free boundary problems: theory and applications. Volume II
K H Hoffmann and J Sprekels
187 An introduction to intersection homology theory
F Kirwan
188 Derivatives, nuclei and dimensions on the frame of torsion theories
J S Golan and H Simmons
189 Theory of reproducing kernels and its applications
S Saitoh
190 Volterra integrodifferential equations in Banach spaces and applications
G Da Prato and M Iannelli
191 Nest algebras
K R Davidson
192 Surveys of some recent results in operator theory. Volume II
J B Conway and B B Morrel
193 Nonlinear variational problems. Volume II
A Marino and M K Murthy
194 Stochastic processes with multidimensional parameter
M E Dozzi
195 Prestressed bodies
D Iesan
196 Hilbert space approach to some classical transforms
R H Picard
197 Stochastic calculus in application
J R Norris
198 Radical theory
B J Gardner
199 The C* – algebras of a class of solvable Lie groups
X Wang

200 Stochastic analysis, path integration and dynamics
D Elworthy
201 Riemannian geometry and holonomy groups
S Salamon
202 Strong asymptotics for extremal errors and polynomials associated with Erdös type weights
D S Lubinsky
203 Optimal control of diffusion processes
V S Borkar
204 Rings, modules and radicals
B J Gardner
205 Numerical studies for nonlinear Schrödinger equations
B M Herbst and J A C Weideman
206 Distributions and analytic functions
R D Carmichael and D Mitrović
207 Semicontinuity, relaxation and integral representation in the calculus of variations
G Buttazzo
208 Recent advances in nonlinear elliptic and parabolic problems
P Bénilan, M Chipot, L Evans and M Pierre
209 Model completions, ring representations and the topology of the Pierce sheaf
A Carson

Volterra integrodifferential equations in Banach spaces and applications

G Da Prato & M Iannelli (Editors)

Scuola Normale Superiore, Pisa/Centro Internazionale per la Ricerca Matematica, Trento

Volterra integrodifferential equations in Banach spaces and applications

Longman Scientific & Technical

Copublished in the United States with John Wiley & Sons, Inc., New York

Longman Scientific & Technical,
Longman Group UK Limited,
Longman House, Burnt Mill, Harlow
Essex CM20 2JE, England
and Associated Companies throughout the world.

Copublished in the United States with
John Wiley & Sons, Inc., 605 Third Avenue, New York, NY 10158

First published 1989

AMS Subject Classification: (main) 45DO5, 45KO5, 45NO5
(subsidiary) 34A10, 45N1O, 73F99

ISSN 0269-3674

British Library Cataloguing in Publication Data
Da Prato, G.
Volterra integrodifferential equations in
Banach spaces and applications
1. Volterra integrodifferential equations
I. Title II. Iannelli, M. (Eds)
515.38, 515.3, QA431

ISBN 0-582-02882-5

Library of Congress Cataloging-in-Publication Data
Volterra integrodifferential equations in Banach spaces and applications
G. Da Prato & M. Iannelli, editors
p. cm.—(Pitman research notes in mathematics series, ISSN 0269-3674; 190)
Bibliography: p.
ISBN 0-470-21216-0
1. Volterra equations. 2. Integro-differential equations.
3. Banach spaces. I. Da Prato, Giuseppe. II. Iannelli, Mimmo
III. Series.
QA431.V64 1989 515.3'8—dc19 88–13277 CIP

Printed and bound in Great Britain
by Biddles Ltd, Guildford and King's Lynn

Contents

Foreword

This volume collects the contributions to the conference "Volterra Integrodifferential Equations in Banach Spaces and Applications" held in Trento (Italy) February 2-7, 1987. The meeting was part of the C.I.R.M. (Centro Internazionale per la Ricerca Matematica) program for the year 1987 and was sponsored by C.N.R. (Consiglio Nazionale delle Ricerche) and I.T.C. (Istituto Trentino di Cultura).

The contributions, by many experts in integrodifferential equations and abstract equations, constitute an up-to-date overview of the field. Many of these papers are devoted to abstract equations, and questions such as existence regularity and well posedness are considered. Other contributions are more concerned with applications problems and constitutive equations are also discussed. Most of the interest was concentrated on Volterra equations and applications to viscoelasticity theory.

We wish to thank all participants for their contributions and the secretary of the C.I.R.M., Mr. Augusto Micheletti, for his substantial help with organization.

Giuseppe Da Prato
(Scuola Normale Superiore
Pisa)

Mimmo Iannelli
(University of Trento)

List of participants

1) Paolo Acquistapace, Scuola Normale Superiore Pisa, Italy;
2) Wolfgang Arendt, Mathematisches Institut, Universität Tübingen, West Germany;
3) Flaviano Battelli, Istituto di Biomatematica, Università Urbino, Italy;
4) Claudio Bertoli, Istituto Motori del CNR, Napoli, Italy;
5) Rosanna Bressan Villella, Istituto di Analisi e Meccanica, Università Padova, Italy;
6) Maria Rosaria Capobianco, I.A.M. - C.N.R., Napoli, Italy;
7) Michel Chipot, Departement de Matématiques, Université de Metz, France;
8) Ph. Clément, Onderafdeling der Wiskunde, TH Delft, The Netherlands;
9) Constantin Corduneanu, Department of Mathematics, University of Texas at Arlington, U.S.A.;
10) Berardino D'Acunto, Dipartimento di Matematica e Applicazioni, Università Napoli, Italy;
11) Giuseppe Da Prato, Scuola Normale Superiore Pisa, Italy;
12) Domenica De Acutis, Dipartimento di Matematica, Università L'Aquila, Italy;
13) Odo Diekmann, Centrum voor Wiskunde en Informatica, Amsterdam, The Netherlands;
14) Hans Engler, Department of Mathematics, Georgetown University, Washington, U.S.A.;
15) Angelo Favini, Dipartimento di Matematica, Università Bologna, Italy;
16) Jerome A. Goldstein, Department of Mathematics, Tulane University, New Orleans, U.S.A.;
17) Ronald Grimmer, Mathematics Department, Southern Illinois University, Carbondale, U.S.A.;
18) Mats Gyllenberg, Institute of Mechanics, Helsinki University of Technology, Finland;
19) Kenneth B. Hannsgen, Mathematics Department, VPI & SU, Blacksburg, U.S.A.;
20) Melvin L. Heard, Department of Mathematics, The University of Illinois at Chicago, U.S.A.;

21) Mimmo Iannelli, Dipartimento di Matematica, Università di Trento, Italy.
22) V. Lakshmikantham, Department of Mathematics, University of Texas at Arlington, U.S.A.;
23) Günter Leugering, Fachbereich Mathematik, Techn. Hochschule Darmstadt, West Germany;
24) Alfredo Lorenzi, Dipartimento di Matematica, Università Milano, Italy;
25) Alessandra Lunardi, Dipartimento di Matematica, Università Pisa, Italy;
26) Richard C. MacCamy, Department of Mathematics, Carnegie Mellon University, Pittsburgh, U.S.A.;
27) Pierangelo Marcati, Dipartimento di Matematica, Università L'Aquila, Italy;
28) Jaroslav Milota, Department of Math. Analysis, Chalres University, Praha, Czechoslovakia;
29) Enzo Mitidieri, Dipartimento di Matematica, Università Trieste, Italy;
30) Jozef Myjak, Sissa Trieste, Italy;
31) John A. Nohel, Mathematics Research Center, University of Wisconsin-Madison, U.S.A.;
32) Enrico Obrecht, Dipartimento di Matematica, Università Bologna, Italy;
33) Jan Prüss, Fachbereich 17, UGHS Paderborn, West Germany;
34) Andrea Pugliese, Dipartimento di Matematica, Università Trento, Italy;
35) Michael Renardy, Department of Mathematics, VPI & SU, Blacksburg, U.S.A.;
36) Wilhelm Schappacher, Institut für Mathematik, Universität Graz, Austria;
37) Daniela Sforza, Dipartimento di Matematica, Università Pisa, Italy;
38) Horst R. Thieme, SFB 123, Universität Heidelberg, West Germany;
39) Mario Tosques, Dipartimento di Matematica, Università Pisa, Italy;
40) Luciano Tubaro, Dipartimento di Matematica, Università Trento, Italy;
41) Antonia Vecchio, I.A.M. - C.N.R., Napoli, Italy;
42) Sjoerd M. Verduyn Lunel, Centre for Mathematics, Amsterdam, The Netherlands;
43) Giorgio Vergara Caffarelli, Dipartimento di Matematica, Università Pisa, Italy;
44) Vincenzo Vespri, Dipartimento di Matematica, Università di Tor Vergata, Roma, Italy;
45) Augusto Visintin, Istituto di Analisi Numerica del CNR, Pavia, Italy;
46) Robert L. Wheeler, Department of Mathematics, VPI & SU, Blacksburg, U.S.A.;
47) Bing-gen Zhang, Department of Mathematics, Shandong Oceanography College, Qingdao, China.

P. ACQUISTAPACE & B. TERRENI

Quasilinear parabolic integrodifferential systems with fully nonlinear boundary conditions

0. INTRODUCTION

This paper concerns the study of local existence of continuously differentiable solutions $u: [0,T] \times \bar{\Omega} \to \mathbb{C}^N$ of quasilinear parabolic integrodifferential systems under fully nonlinear boundary conditions. We also prove some results on continuity of solutions with respect to the initial data. Although our method works in the general case of systems of order 2m, we just consider here, for the sake of simplicity, second order systems of the following kind:

$$\begin{aligned} &\frac{\partial u^h}{\partial t}(t,x) - \sum_{i,j=1}^{n} \sum_{k=1}^{n} A_{ij}^{hk}(t,x,u(t,x),Du(t,x)) \cdot D_j D_j u^k(t,x) = \\ &\qquad = F^h(t,x,u(t,x),Du(t,x)) + I^h(t,x,u) \text{ in } [0,T] \times \Omega,\ h = 1,\dots,N, \\ &u^h(0,x) = \phi^h(x) \text{ in } \Omega,\ h = 1,\dots,N, \\ &B^h(t,x,u(t,x),Du(t,x)) = J^h(t,x,u) \text{ in } [0,T] \times \partial\Omega,\ h = 1,\dots,N, \end{aligned} \tag{0.1}$$

where Ω is a bounded open set of $\mathbf{R}^n$ with boundary $\partial\Omega$ of class C^2 and $I(t,x,u)$, $J(t,x,u)$ are functionals of integral type:

$$\begin{aligned} I(t,x,u) := \int_0^t \int_\Omega & H(t,s,x,y;u(t,y),Du(t,y);u(s,x),Du(s,x),D^2u(s,x); \\ & u(s,y),Du(s,y),D^2u(s,y))\,dy\,ds \end{aligned} \tag{0.2}$$

$$\begin{aligned} J(t,x,u) := \int_0^t \int_{\partial\Omega} & K(t,s,x,y;u(t,y),Du(t,y);u(s,x),Du(s,x);u(s,y), \\ & Du(s,y))\,dy\,ds \end{aligned} \tag{0.3}$$

We list our assumptions.

(0.4) Regularity. The $N \times N$-matrices A_{ij} and the N-vectors B,F are sufficiently smooth: for instance, $A_{ij} \in C^2(\Lambda, \mathbb{C}^{N^2})$, $B \in C^3(\Lambda, \mathbb{C}^N)$, $F \in C^2(\Lambda, \mathbb{C}^N)$, where $\Lambda := [0,T] \times \bar{\Omega} \times \mathbb{C}^N \times \mathbb{C}^{nN}$. Concerning the N-vectors H,K, we require that $H \in C^2(\Lambda', \mathbb{C}^N)$, $K \in C^3(\Lambda'', \mathbb{C}^N)$, where

$$\Lambda' := \Delta \times \bar{\Omega}^2 \times \mathbb{C}^N \times \mathbb{C}^{nN} \times (\mathbb{C}^N \times \mathbb{C}^{nN} \times \mathbb{C}^{n^2N})^2,$$

$$\Lambda'' := \Delta \times \bar{\Omega} \times \partial\Omega \times (\mathbb{C}^N \times \mathbb{C}^{nN})^3 \quad \text{and } \Delta := \{(t,s) \in [0,T]^2;\ t > s\};$$

moreover, we need the following growth assumptions on the functions

$$D^\gamma H(t,s,x,y;u_1,p_1;u_2,p_2,q_2;u_3,p_3,q_3), D^\gamma K(t,s,x,y;u_1,p_1;u_2,p_2;u_3,p_3):$$

$$|D^\gamma H| \leq \begin{cases} c_0(M)\eta(t-s)\cdot(t-s)^{\alpha-1}[1+|q_2|+|q_3|] & \text{if there are no derivatives in } (t,q_2,q_3), \\ c_0(M)\eta(t-s)\cdot(t-s)^{\alpha-2}[1+|q_2|+|q_3|] & \text{if there are no derivatives in } (q_2,q_3) \text{ and just one derivative in } t, \\ c_0(M)\eta(t-s)\cdot(t-s)^{\alpha-1} & \text{if there are no derivatives in } t \text{ and just one in } (q_2,q_3), \\ c_0(M)\eta(t-s)\cdot(t-s)^{\alpha-2} & \text{if there are just one derivative in } t \text{ and just one in } (q_2,q_3), \\ \text{no assumptions} & \text{otherwise,} \end{cases}$$

provided $\sum_{i=1}^{3} [|u_i| + |p_i|] \leq M$ and $|\gamma| \leq 2$;

$$|D^\gamma K| \leq \begin{cases} c_0(M)\eta(t-s)\cdot(t-s)^{\alpha-1/2} & \text{if there are no derivatives in } t, \\ c_0(M)\eta(t-s)\cdot(t-s)^{\alpha-3/2} & \text{if there is just one derivative in } t, \\ \text{no assumptions} & \text{otherwise,} \end{cases}$$

provided $\sum_{i=1}^{3} [|u_i| + |p_i|] \leq M$ and $|\gamma| \leq 3$.

Here α is a fixed number in $]0,1/2[$ and $\eta(t)$, $t \geq 0$ is a continuous, non-decreasing function such that $\eta(0) = 0$.

(0.5) Ellipticity. The pair $\left(\sum_{j=1}^{n} A_{ij}(t,\cdot,u,p)D_iD_j,\ \sum_{i=1}^{n} \frac{\partial B}{\partial p_i}(t,\cdot,u,p)D_i\right)$, where $\frac{\partial B}{\partial p_i} := \{\frac{\partial B^h}{\partial p_i^k}\}_{h,k=1,\dots,N}$, is elliptic in the sense of [4,6], uniformly on bounded subsets of Λ. More precisely, consider the matrices

$$A(\theta;t,x,u,p;\xi,\rho) := \sum_{s,j=1}^{n} A_{sj}(t,x,u,p)\xi_s\xi_j + e^{i\theta}\rho^2 I,$$

$$\Gamma(t,x,u,p;\xi) := \sum_{j=1}^{n} \frac{\partial B}{\partial p_j}(t,x,u,p)\xi_j,$$

where $\theta \in \mathbf{R}$, $\xi \in \mathbf{R}^n$, $\rho \in \mathbf{R}$. Then we assume that for each $M > 0$ the following is true, provided $t \in [0,T]$, $x \in \bar{\Omega}$, $|u| + |p| \leq M$:

(i) there exist $\theta_M \in]\pi/2,\pi[$, $c_M > 0$ such that

$$|\det A(\theta;t,x,u,p;\xi,\rho)| \geq c_M(|\xi|^2 + \rho^2)^N \quad \forall\theta \in [-\theta_M,\theta_M],$$

$$\forall\xi \in \mathbf{R}^n, \quad \forall\rho \in \mathbf{R};$$

(ii) for each $\theta \in [-\theta_M,\theta_M]$, $\xi \in \mathbf{R}^n$, $\rho \in \mathbf{R}$ with $|\xi|^2 + \rho^2 > 0$ and $\xi\cdot\nu(x) = 0$ the polynomial $\tau \to \det A(\theta;t,x,u,p;\xi+\tau\nu(x),\rho)$ has precisely N roots $\tau_j^+(\theta;t,x,u,p;\xi,\rho)$ with positive imaginary part. Here $\nu(x)$ is the unit outward normal vector at x (which is supposed now to belong to $\partial\Omega$);

(iii) for each $\theta \in [-\theta_M,\theta_M]$, $\xi \in \mathbf{R}^n$, $\rho \in \mathbf{R}$ with $|\xi|^2 + \rho^2 > 0$ and $\xi\cdot\nu(x) = 0$ (here, again, $x \in \partial\Omega$) the rows of the matrix $\Gamma(t,x,u,p;\xi+\tau\nu(x))\cdot[A(\theta;t,x,u,p;\xi+\tau\nu(x),\rho)]^*$ are linearly independent modulo the polynomial $\tau \to \prod_{j=1}^{n} (\tau-\tau_j^+(\theta;t,x,u,p;\xi,\rho))$.

(0.6) Compatibility. The initial datum φ belongs to $C^1(\bar{\Omega},\mathbb{C}^N)$ and

$$B(0,x,\varphi(x),D\varphi(x)) = 0 \quad \forall x \in \partial\Omega .$$

Here are a few examples where the above hypotheses (neglecting the integro-differential terms) are fulfilled.

EXAMPLE 0.1. Let $A_{ij} \in C^1(\Lambda,\mathbb{C}^{N^2})$ and suppose that for each $M > 0$ there exists $\nu_M > 0$ such that if $|u| + |p| \leq M$

$$\mathrm{Re} \sum_{i,j=1}^{n} (A_{ij}(t,x,u,p)\cdot\eta,\eta)_{\mathbb{C}^N}\xi_i\xi_j \geqq \nu_M|\xi|^2|\eta|^2 \quad \forall\xi \in \mathbf{R}^n, \qquad \forall\eta \in \mathbf{R}^N,$$
$$\forall(t,x) \in [0,T] \times \bar{\Omega};$$

moreover take $B(t,x,u,p) := \{\sum_{i=1}^{n} b_i(t,x,u)p_i^h\}_{h=1,\ldots,N}$, with $b_i \in C^1([0,T] \times \bar{\Omega} \times \mathbb{C}^N, \mathbf{R})$ satisfying

$$\sum_{i=1}^{n} b_i(t,x,u)\nu_i(x) \neq 0 \quad \forall(t,x,u) \in [0,T] \times \partial\Omega \times \mathbb{C}^N.$$

This is the situation of [2].

EXAMPLE 0.2. Fix $N = 1$. For $i = 1,\ldots,n$ take $a_i \in C^3(\Lambda,\mathbf{R})$ such that for each $M > 0$

$$\sum_{i,j=1}^{n} \frac{\partial a_i}{\partial p_j}(t,x,u,p)\xi_i\xi_j \geqq \nu_M|\xi|^2 \quad \forall\xi \in \mathbf{R}^n, \quad (t,x) \in [0,T] \times \bar{\Omega},$$

provided $|u| + |p| \leq M$. Then, setting $A_{ij} := \frac{\partial a_i}{\partial p_j}$, $B := \sum_{i=1}^{n} a_i\nu_i$, we see that variational problems may be written in the form (0.1).

1. MAIN THEOREM

Let us fix any $p \in]n,\infty[$ and let φ_0 be a fixed element in $W^{2,p}(\Omega,\mathbb{C}^N)$. For each $r_0 > 0$, $N_0 > 0$ we set:

$$B(\varphi_0,N_0,r_0) := \{\varphi \in W^{2,p}(\Omega,\mathbb{C}^N) : \|\varphi-\varphi_0\|_{W^{2,p}} \leq r_0,\ B(0,\cdot,\varphi,D\varphi) = 0 \text{ on } \partial\Omega,$$
$$Q(0,\varphi) \in B_\infty^{2\alpha,p}(\Omega,\mathbb{C}^N) \text{ and } \|Q(0,\varphi)\|_{B_\infty^{2\alpha,p}} \leqq N_0\}, \qquad (1.1)$$

where $B_\infty^{2\alpha,p}$ is the Besov-Nikolskii space ([7, Definition 5.8]), and

$$Q(0,\varphi) := \sum_{i,j=1}^{n} A_{ij}(0,\cdot,\varphi,D\varphi)\cdot D_iD_j\varphi + F(0,\cdot,\varphi,D\varphi), \quad x \in \Omega; \qquad (1.2)$$

we also recall that the number $\alpha \in]0,1/2[$ was introduced in (0.4).

It is easy to check that $B(\varphi_0,N_0,r_0)$ is a closed subset of $W^{2,p}(\Omega,\mathbb{C}^N)$. Let us state our main result:

THEOREM 1.1 Assume (0.2),...,(0.5). There exists $\tau \in]0,T]$ such that for each $\varphi \in B(\varphi_0,N_0,r_0)$, problem (0.1) has a unique solution $u_\varphi \in C^{1+\alpha}([0,\tau], L^p(\Omega,\mathbb{C}^N)) \cap C^\alpha([0,\tau],W^{2,p}(\Omega,\mathbb{C}^N))$; moreover the map $\varphi \to u_\varphi$ is continuous, in the sense that there exists $c_2 > 0$, depending on $p,\alpha,\varphi_0,N_0,r_0$, such that

$$\|u_\varphi-u_\psi\|_{C^{1+\alpha}(L^p)} + \|u_\varphi-u_\psi\|_{C^\alpha(W^{2,p})} \leqq \tag{1.3}$$

$$\leqq c_2(p,\alpha,\varphi_0,N_0,r_0) \quad \{\|\varphi-\psi\|_{W^{2,p}} + \|Q(0,\varphi)-Q(0,\psi)\|_{B^{2\alpha,p}_\infty}\}.$$

If, in addition $\varphi \in C^2(\bar{\Omega},\mathbb{C}^N)$ and $Q(0,\varphi) \in C^{2,\alpha}(\bar{\Omega},\mathbb{C}^N)$, then

$$\frac{\partial u}{\partial t}, \sum_{i,j=1}^{n} A_{ij}(\cdot,\cdot,u,Du)\cdot D_iD_ju \in C^\delta([0,\tau],C(\bar{\Omega},\mathbb{C}^N)) \quad \forall\delta \in]0,\alpha[. \tag{1.4}$$

The proof will be outlined in the next sections; here we want to make a few remarks.

REMARK 1.2. We are forced to take $p \in]n,\infty[$ in order to guarantee that $u(t,\cdot),Du(t,\cdot)$ are continuous functions in $\bar{\Omega}$: this allows us to avoid any growth assumption on the nonlinearities A_{ij},f,B.

REMARK 1.3. The compatibility conditions $B(0,\cdot,\varphi,D\varphi) = 0$ on $\partial\Omega$ and $Q(0,\varphi) \in B^{2\alpha,p}_\infty(\Omega,\mathbb{C}^N)$ are necessary for the validity of the first assertion of Theorem 1.1, which is optimal in this sense (and we might also weaken (0.4) somewhat). On the other hand, we cannot replace, in (1.4), δ by α, because of the "bad" behaviour of the space $C(\bar{\Omega},\mathbb{C}^N)$ with respect to maximal time regularity in parabolic evolution problems: compare with [1, Remark 6.4].

REMARK 1.4. Theorem 1.1 is a local existence result; however, due to the fact that the compatibility conditions are preserved in time, it is clear that the local solution may be continued, by standard arguments, up to a maximal time interval $[0,T(\varphi)[$.

REMARK 1.5. We believe that the extension of Theorem 1.1 to the case of quasilinear parabolic systems of arbitrary order, with the elliptic part satisfying the assumptions of [4,6], is straightforward.

Our proof proceeds essentially as in [2], i.e. it uses two main ingredients: linearization of the problem, and the contraction principle. However, there are some differences with respect to [2], due to the presence of the integrodifferential terms and - which is more important - of the fully nonlinear boundary conditions; the latter generalization requires a slight refinement of the classical results of [3,4] from the point of view of the smoothness of the boundary coefficients. It is to be noted that in [2] we had very sharp assumptions from the point of view of smoothness, and consequently it was necessary there to apply a further regularization procedure of the solution, since the fixed point had been obtained in a space strictly larger than the space of data. The same regularization technique is not applicable here, because of the full nonlinearity at the boundary: this forces us to take stronger smoothness assumptions on the data, which allow us to get the fixed point directly in the optimal class.

2. AUXILIARY RESULTS

We collect here some propositions which shall be used later on. Consider the linear differential operators

$$A(x,D):=\sum_{i,j=1}^{n} a_{ij}(x)D_iD_j, \quad a_{ij}\in C(\bar{\Omega},\mathbb{C}^{N^2}), \tag{2.1}$$

$$B(x,D):=\sum_{i=1}^{n} b_i(x)D_i, \qquad b_i\in W^{1,p}(\Omega,\mathbb{C}^{N^2}),\ p>n, \tag{2.2}$$

and suppose that (0.5) is satisfied by the pair $(A(\cdot D),B(\cdot,D))$ uniformly in $\bar{\Omega}$. Then we have:

THEOREM 2.1. Under the above assumptions, there exist $\lambda_0>0$, $\theta_0\in\mathbb{R}$, $c_3>0$ (depending on p,A,B) such that for each $f\in L^p(\Omega,\mathbb{C}^N)$ $g\in W^{1,p}(\Omega,\mathbb{C}^N)$ and $\lambda\in S_{\lambda_0,\theta_0}:=\{z\in\mathbb{C}:|\arg(z-\lambda_0)|\le\theta_0\}\cup\{\lambda_0\}$, the problem

$$\lambda u - A(\cdot,D)u = f \text{ in } \Omega,$$
$$B(\cdot,D)u = g \quad \text{on } \partial\Omega, \tag{2.3}$$
$$u \in W^{2,p}(\Omega,\mathbb{C}^N)$$

has a unique solution, satisfying in addition, for each $\lambda \in S_{\lambda_0,\theta_0}$:

$$\sum_{r=0}^{2} [1+|\lambda-\lambda_0|]^{1-r/2} \, \|D^r u\|_{L^p} \leqq \tag{2.4}$$

$$\leqq c_3(p,A,B)[\ \|f\|_{L^p} + \inf_{\substack{\psi\in W^{1,p}(\Omega,\mathbb{C}^N) \\ \psi=g \text{ on } \partial\Omega}} \{[1+|\lambda-\lambda_0|]^{1/2}\|\psi\|_{L^p} + \|D\psi\|_{L^p}\}].$$

Proof. This result is proved in [6, Theorem 5.3] and [5, Theorems 12.2-13.1], under the stronger assumption $b_i \in C^1(\bar{\Omega},\mathbb{C}^{N^2})$.

Such assumption was necessary in order to apply the elliptic estimates of [3,4] which hold for all $p \in]1,\infty[$. But if we restrict p to be greater than n, it is easily seen that the argument of [3,4] still works when b_i is just in $W^{1,p}$: indeed, it is enough to make suitable modifications in [3, page 703] (see also [4, page 77]) as shown in the Appendix below.

With the generalized version of the elliptic estimates at our disposal, in order to get (2.4) and, consequently, uniqueness, we just need to repeat the argument of [5, Theorem 12.2]. Concerning existence, we first approximate the coefficients of $B(\cdot,D)$ in the $W^{1,p}(\Omega,\mathbb{C}^{N^2})$-norm by more regular ones, and next, we find (by [6, Theorem 5.3]) the corresponding solutions; finally, we note that, due to (2.4), such solutions converge in $W^{2,p}(\Omega,\mathbb{C}^N)$ to a solution of (2.3), as it is easily seen. □

Consider now the evolution problem

$$\frac{\partial u}{\partial t} - A(\cdot,D)u = f \text{ in } [0,T] \times \Omega,$$
$$B(\cdot,D)u = g \qquad \text{in } [0,T] \times \partial\Omega\ , \tag{2.5}$$
$$u(0,\cdot) = \varphi \qquad \text{in } \Omega$$

with A,B defined by (2.1), (2.2) and satisfying (0.5) uniformly in $\bar{\Omega}$. We have:

THEOREM 2.2. Fix $p > n$, $\alpha \in]0,1/2[$. If $f \in C^{\alpha}([0,T], L^{p}(\Omega,\mathbb{C}^{N}))$, $g \in C^{\alpha}([0,T],W^{1,p}(\Omega,\mathbb{C}^{N})) \cap C^{\alpha+1/2}([0,T],L^{p}(\Omega,\mathbb{C}^{N}))$, $\varphi \in W^{2,p}(\Omega,\mathbb{C}^{N})$ and in addition the compatibility conditions

$$B(\cdot,D)\varphi = g(0,\cdot) \text{ on } \partial\Omega, \quad A(\cdot,D)\varphi + f(0,\cdot) \in B^{2\alpha,p}_{\infty}(\Omega,C^{N}) \tag{2.6}$$

are fulfilled, then problem (2.5) has a unique solution $u \in C^{\alpha}([0,T],W^{2,p}(\Omega,\mathbb{C}^{N})) \cap C^{\alpha+1}([0,T],L^{p}(\Omega,\mathbb{C}^{N}))$; moreover there exists $c_4 > 0$, depending on p,α,A,B, such that:

$$\begin{aligned} \|u\|_{C^{\alpha}(L^p)} + \|u'\|_{C^{\alpha}(L^p)} &\leq c_4(p,\alpha,A,B)[\|\varphi\|_{W^{2,p}} + \\ &+ \|f(0,\cdot)\|_{L^p} + \|A(\cdot,D)\varphi + f(0)\|_{B^{2\alpha,p}_{\infty}} \\ &+ [f]_{C^{\alpha}(L^p)} + [g]_{C^{\alpha}(W^{1,p})} + [g]_{C^{\alpha+1/2}(L^p)}]. \end{aligned} \tag{2.7}$$

Proof. One can repeat the arguments of [8] using Theorem 2.1 instead of [8, Theorem 1.4]. □

Next, let us recall the following imbedding property:

PROPOSITION 2.3. Let $p > n$, $\alpha \in]0,1/2[$. Then we have the continuous imbeddings:

$$C^{\alpha}([0,T],W^{2,p}(\Omega,\mathbb{C}^{N}))\cap C^{\alpha+1}([0,T],L^{p}(\Omega,\mathbb{C}^{N})) \hookrightarrow \begin{cases} C^{\alpha+1/2}([0,T],W^{1,p}(\Omega,\mathbb{C}^{N})), \\ C^{\alpha+\sigma}([0,T],C^{1}(\bar{\Omega},\mathbb{C}^{N}))\ \forall\sigma\in]0,1/2-n/2p[, \\ C^{\alpha+\theta}([0,T],C(\bar{\Omega},\mathbb{C}^{N}))\ \forall\theta\in]0,1-n/2p[\end{cases}$$

Proof. The first inclusion follows by interpolation. Concerning the others, interpolation yields

$$C^{\alpha}([0,T],W^{2,p}(\Omega,\mathbb{C}^{N}))\cap C^{\alpha+1}([0,T],L^{p}(\Omega,\mathbb{C}^{N}))\hookrightarrow C^{\alpha+\gamma}([0,T],B^{2(1-\gamma),p}_{\infty}(\Omega,\mathbb{C}^{N}))$$

$$\forall\gamma \in]0,1[, \ \gamma \neq 1-\alpha,$$

and the results follow by Sobolev's Theorem. □

We conclude this section with an estimate of the growth of the nonlinearities appearing in problem (0.1) in terms of the growth of the unknown function u.

LEMMA 2.4. Assume (0.4). Let $v \in C^{\alpha}([0,T],W^{2,p}(\Omega,\mathbb{C}^N)) \cap C^{\alpha+1}([0,T], L^p(\Omega,\mathbb{C}^N))$ with $\|v\|_{C^{\alpha}(W^{2,p})} + \|v\|_{C^{\alpha+1}(L^p)} \leq M$, and for any function $f:\Lambda \to \mathbb{C}^h$ (where $h \in \mathbb{N}^+$ and Λ is defined in (0.4)) set

$$\tilde{f}(t,x) := f(t,x,v(t,x),Dv(t,x)),\ t \in [0,T],\ x \in \bar{\Omega}.$$

Then if D^{γ} is any derivative with respect to the variables $(t,x,u,p) \in \Lambda$, there exists $c_5 > 0$, depending on M, such that

$$\sum_{|\gamma|\leq 2} \|\widetilde{D^{\gamma}F}\|_{C([0,T]\times\bar{\Omega})} + \sum_{|\gamma|\leq 2} \sum_{i,j=1}^{n} \|\widetilde{D^{\gamma}A_{ij}}\|_{C([0,T]\times\bar{\Omega})} +$$
$$\sum_{|\gamma|\leq 3} \|\widetilde{D^{\gamma}B}\|_{C([0,T]\times\bar{\Omega})} + \sum_{|\gamma|\leq 1} \|\widetilde{D^{\gamma}B}\|_{C([0,T],W^{1,p})} \leq c_5(M).$$

Proof. Quite easy. □

A similar result holds for the integrands H,K of the functionals I,J defined in (0.2), (0.3):

LEMMA 2.5. Assume (0.4). Let $v \in C^{\alpha}([0,T],W^{2,p}(\Omega,\mathbb{C}^N)) \cap C^{\alpha+1}([0,T], L^p(\Omega,\mathbb{C}^N))$ with $\|v\|_{C^{\alpha}(W^{2,p})} + \|v\|_{C^{\alpha+1}(L^p)} \leq M$, and introduce the notations:

$$\tilde{H}(t,s,x,y):=H(t,s,x,y;v(t,y),Dv(t,y);v(s,x),Dv(s,x),D^2v(s,x);v(s,y),$$
$$Dv(s,y),D^2(s,y)),$$
$$\tilde{K}(t,s,x,y):=K(t,s,x,y;v(t,y),Dv(t,y);v(s,x),Dv(s,x);v(s,y),Dv(s,y)).$$

Then there exists $c_6 > 0$, depending on M,α, such that:

$$|D^{\gamma}H(t,s,x,y)| \leq \begin{cases} c_6(M,\alpha)\eta(t-s)\cdot(t-s)^{\alpha-1} & \text{if there are no derivatives in } t, \\ c_6(M,\alpha)\eta(t-s)\cdot(t-s)^{\alpha-2} & \text{if there is one derivative in } t \text{ at most.} \end{cases}$$

provided $|\gamma| \leqq 2$, and

$$|D^{\gamma}K(t,s,x,y)| \leqq \begin{cases} c_6(M,\alpha)\eta(t-s)\cdot(t-s)^{\alpha-1/2} & \text{if there are no derivatives in } t, \\ c_6(M,\alpha)\eta(t-s)\cdot(t-s)^{\alpha-3/2} & \text{if there is one derivative in } t \text{ at most,} \end{cases}$$

provided $|\gamma| \leqq 3$.

Proof. Completely straightforward. □

3. PROOF OF THEOREM 1.1

3.A. LOCAL EXISTENCE

Fix $p \in]n,\infty[$. Consider the Banach space

$$E_{\alpha,p}(T):=C^{\alpha}([0,T],W^{2,p}(\Omega,\mathbb{C}^N)) \cap C^{\alpha+1}([0,T],L^p(\Omega,\mathbb{C}^N)) \tag{3.1}$$

with its obvious norm; we also introduce

$$[u]_{E_{\alpha,p}(T)} := [u']_{C^{\alpha}(L^p)} + [D^2u]_{C^{\alpha}(L^p)}. \tag{3.2}$$

For each $\varphi \in B(\varphi_0,N_0,r_0)$ (see (1.1)) and $M > 0$ set:

$$B_{M,\alpha,p,T,\varphi} := \{v \in E_{\alpha,p}(T) : v(0,\cdot) = \varphi, \|v-\varphi\|_{E_{\alpha,p}(T)} \leqq M\}. \tag{3.3}$$

We linearize problem (0.1) by considering, for fixed $v \in B_{M,\alpha,p,T,\varphi}$, the problem:

$$\begin{cases} \frac{\partial u^h}{\partial t}(t,x) - \sum_{i,j=1}^{n}\sum_{k=1}^{N} A_{ij}^{hk}(0,x,\varphi(x),D\varphi(x))\cdot D_iD_ju^k(t,x)=F^h_{v,\varphi}(t,x) \text{ in} \\ \qquad [0,T]\times\Omega,\ h = 1,\dots,N, \\ u^h(0,x) = \varphi^h(x) \text{ in } \Omega,\ h = 1,\dots,N, \\ \sum_{i=1}^{n}\sum_{k=1}^{N} \frac{\partial B^h}{\partial p_i^k}(0,x,\varphi(x),D\varphi(x))D_iu^k(t,x)=G^h_{v,\varphi}(t,x) \text{ in } [0,T]\times\partial\Omega, h=1,\dots,N, \end{cases} \tag{3.4}$$

where:

$$F_{v,\varphi}(t,x) := F(t,x,v(t,x),Dv(t,x)) \tag{3.5}$$

$$- \sum_{i,j=1}^{n} [A_{ij}(0,x,\varphi(x),D\varphi(x))-A_{ij}(t,x,v(t,x),Dv(t,x))]\cdot D_iD_jv(t,x)$$

$$+ I(t,x,v),$$

$$G_{v,\varphi}(t,x) := \sum_{i=1}^{n} \frac{\partial B}{\partial p_i}(0,x,\varphi(x),D\varphi(x))\cdot D_iv(t,x)-B(t,x,v(t,x), \tag{3.6}$$

$$Dv(t,x)) + J(t,x,v).$$

In order to apply Theorem 2.2, we need the following lemmas, whose proofs will be given in Section 4.

<u>LEMMA 3.1</u>. Assume (0.4), (0.5), and let $\varphi \in B(\varphi_0,N_0,r_0)$. There exists a continuous, non-decreasing function $\omega_0(t)$, $t \geq 0$, with $\omega_0(0) = 0$, such that for each $K > 0$ and $v \in B_{K,\alpha,p,T,\varphi}$ with $\|v\|_{E_{\alpha,p}(T)} \leq K$ we have:

$$[F_{v,\varphi}]_{C^{\alpha}(L^p)} + [G_{v,\varphi}]_{C^{\alpha}(W^{1,p})} + [G_{v,\varphi}]_{C^{\alpha+1/2}(L^p)} \leq c_7(K)\omega_0(T).$$

<u>LEMMA 3.2</u> Assume (0.4), (0.5), and let φ, $\psi \in B(\varphi_0,N_0,r_0)$. There exists a continuous, non-decreasing function $\omega_1(t)$, $t \geq 0$, with $\omega_1(0) = 0$, such that for each $K > 0$ and $v \in B_{K,\alpha,p,T,\varphi}$, $w \in B_{K,\alpha,p,T,\psi}$ with $\|v\|_{E_{\alpha,p}(T)}$, $\|w\|_{E_{\alpha,p}(T)} \leq K$ we have

$$[F_{v,\varphi}-F_{w,\psi}]_{C^{\alpha}(L^p)} + [G_{v,\varphi}-G_{w,\psi}]_{C^{\alpha}(W^{1,p})} + [G_{v,\varphi}-G_{w,\psi}]_{C^{\alpha+1/2}(L^p)} \leq$$

$$\leq c_8(K)\{\|\varphi-\psi\|_{W^{2,p}} + \omega_1(T)\ [v-w]_{E_{\alpha,p}(T)}\}.$$

By Lemma 3.1, the data of problem (3.4) satisfy:

$$A_{ij}(0,\cdot,\varphi,D\varphi) \in C(\bar{\Omega},\mathbb{C}^{N^2}),\quad \frac{\partial B}{\partial p_i}(0,\cdot,\varphi,D\varphi) \in W^{1,p}(\Omega,\mathbb{C}^{N^2}),$$

$$F_{v,\varphi} \in C^{\alpha}([0,T],L^p(\Omega,\mathbb{C}^N)),\ G_{v,\varphi} \in C^{\alpha}([0,T],W^{1,p}(\Omega,\mathbb{C}^N)) \cap C^{\alpha+1/2}([0,T], L^p(\Omega,\mathbb{C}^N)),\ \varphi \in W^{2,p}(\Omega,\mathbb{C}^N),$$

and the compatibility conditions

$$B(0,\cdot,\varphi,D\varphi) = 0 \text{ on } \partial\Omega, \quad Q(0,\varphi) \in B^{2\alpha,p}(\Omega,\mathbb{C}^N) \tag{3.7}$$

hold. Hence, Theorem 2.2 implies the existence of a unique solution $u =: S(v)$ of problem (3.4), belonging to the space $E_{\alpha,p}(T)$. Consequently $w := u - \varphi$ is the unique solution of the problem

$$\begin{cases} \dfrac{\partial w}{\partial t} - \sum\limits_{i,j=1}^{n} A_{ij}(0,\cdot,\varphi,D\varphi) \cdot D_iD_jw = F_{v,\varphi} + \sum\limits_{i,j=1}^{n} A_{ij}(0,\cdot,\varphi,D\varphi)\cdot D_iD_j\varphi \text{ in } [0,T]\times\Omega, \\ w(0,\cdot) \text{ in } \Omega, \\ \sum\limits_{i=1}^{n} \dfrac{\partial B}{\partial p_i}(0,\cdot,\varphi,D\varphi)\cdot D_iw = G_{v,\varphi} - \sum\limits_{i=1}^{n} \dfrac{\partial B}{\partial p_i}(0,\cdot,\varphi,D\varphi)\cdot D_i\varphi \text{ in } [0,T]\times\partial\Omega, \end{cases} \tag{3.8}$$

and by (2.7), (3.5), (3.6), (1.2) the following estimate holds:

$$\|u-\varphi\|_{E_{\alpha,p}(T)} \leq c_7(p,\alpha, \|\varphi\|_{W^{2,p}}) \tag{3.9}$$

$$\{\|Q(0,\varphi)\|_{B_\infty^{2\alpha,p}} + [F_{v,\varphi}]_{C^\alpha(L^p)} + [G_{v,\varphi}]_{C^\alpha(W^{1,p})} + [G_{v,\varphi}]_{C^{\alpha+1/2}(L^p)}\}.$$

Hence, if we set $R := c_7(p,\alpha, \|\varphi\|_{W^{2,p}}) \|Q(0,\varphi)\|_{B_\infty^{2\alpha,p}}$ and choose $K > R$, we get by Lemma 3.1:

$$\|S(v)-\varphi\|_{E_{\alpha,p}(T)} \leq R + c_8(K)\omega_0(T), \tag{3.10}$$

where $\omega_0(T) \downarrow 0$ as $T \downarrow 0$.

Next, if $v,w \in B_{M,\alpha,p,\varphi,T}$ and $h := S(v)-S(w)$, then h solves the problem:

$$\begin{cases} \dfrac{\partial h}{\partial t} - \sum\limits_{i,j=1}^{n} A_{ij}(0,\cdot,\varphi,D\varphi)\cdot D_iD_jh = F_{v,\varphi}-F_{w,\varphi} \text{ in } [0,T]\times\Omega, \\ h(0,\cdot) = 0 \text{ in } \Omega, \\ \sum\limits_{i=1}^{n} \dfrac{\partial B}{\partial p_i}(0,\cdot,\varphi,D\varphi)\cdot D_iD_jh = G_{v,\varphi}-G_{w,\varphi} \text{ in } [0,T]\times\partial\Omega, \end{cases} \tag{3.11}$$

and again (2.7), (3.5), (3.6) yield:

$$\|S(v)-S(w)\|_{E_{\alpha,p}(T)} \leq c_9(p,\alpha,\|\varphi\|_{W^{2,p}}) \tag{3.12}$$

$$[F_{v,\varphi}-F_{w,\varphi}]_{C^\alpha(L^p)} + [G_{v,\varphi}-G_{w,\varphi}]_{C^\alpha(W^{1,p})} + [G_{v,\varphi}-G_{w,\varphi}]_{C^{\alpha+1/2}(L^p)}\};$$

thus by Lemma 3.2 we obtain:

$$\|S(v)-S(w)\|_{E_{\alpha,p}(T)} \leq c_{10}(K)\omega_1(T)\ \|v-w\|_{E_{\alpha,p}(T)} \tag{3.13}$$

where $\omega_1(T) \downarrow 0$ as $T \downarrow 0$.

Hence if $T =:\tau$ is small enough, the contraction principle is applicable and we get a unique solution u_φ of problem (0.1) in $[0,\tau]$, belonging to the class $E_{\alpha,p}(\tau)$. This proves the first assertion of Theorem 1.1; note that τ depends on $\varphi_0,N_0,r_0,p,\alpha$, but not on $\varphi \in B(\varphi_0,N_0,r_0)$.

3.B CONTINUOUS DEPENDENCE ON THE INITIAL DATUM

Next, fix φ, $\psi \in B(\varphi_0,N_0,r_0)$ and consider the difference $v := u_\varphi-u_\psi$. Redefine by F_φ, G_φ the functions (3.5), (3.6) (which depend in fact on u_φ and φ): then v solves:

$$\begin{cases} \dfrac{\partial v}{\partial t} - \displaystyle\sum_{i,j=1}^{n} A_{ij}(0,\cdot,\varphi,D\varphi)\cdot D_iD_jv = F^{\varphi,\psi} \text{ in } [0,\tau]\times\Omega,\ v(0,\cdot) = \varphi-\psi \text{ in } \Omega, \\ \displaystyle\sum_{i=1}^{n} \frac{\partial B}{\partial P_i}(0,\cdot,\varphi,D\varphi)\cdot D_iv = G^{\varphi,\psi} \text{ in } [0,\tau]\times\partial\Omega, \end{cases} \tag{3.14}$$

where

$$F^{\varphi,\psi} := F_\varphi-F_\psi + \sum_{i,j=1}^{n}[A_{ij}(0,\cdot,\varphi,D\varphi)-A_{ij}(0,\cdot,\psi,D\varphi)]\cdot D_iD_ju_\psi, \tag{3.15}$$

$$G^{\varphi,\psi} := G_\varphi-G_\psi - \sum_{i=1}^{n}[\frac{\partial B}{\partial p_i}(0,\cdot,\varphi,D\varphi) - \frac{\partial B}{\partial p_i}(0,\cdot,\psi,D\psi)]\cdot D_iu_\psi; \tag{3.16}$$

it is easy to verify that the compatibility conditions (3.7) hold. Concerning the regularity of $F^{\varphi,\psi}$, $G^{\varphi,\psi}$, we have:

<u>LEMMA 3.3.</u> Assume (0.4), (0.5) and let $\varphi,\psi \in B(\varphi_0,N_0,r_0)$. Then we have:

$$\|F^{\varphi,\psi}(0,\cdot)\|_{L^p} + \|G^{\varphi,\psi}(0,\ )\|_{W^{1,p}} \leq c_{11}(p,M,\varphi_0,r_0)\ \|\varphi-\psi\|_{W^{2,p}},$$

$$[F^{\varphi,\psi}]_{C^\alpha(L^p)} + [G^{\varphi,\psi}]_{C^\alpha(W^{1,p})} + [G^{\varphi,\psi}]_{C^{\alpha+1/2}(L^p)}$$

$$\leq c_{12}(p,\alpha,M,\varphi_0,r_0)\{\|\varphi-\psi\|_{W^{2,p}} + [u_\varphi-u_\psi]_{E_{\alpha,p}(\tau)}\cdot\omega_2(\tau)\},$$

where $\omega_2(t)$, $t \geq 0$, is a non-decreasing continuous function with $\omega_2(0) = 0$.

Proof. Easy consequences of (3.15), (3.16), (3.5), (3.6) and Lemma 3.2. □

By (2.7) and Lemma 3.3 we easily get:

$$\|u_\varphi-u_\psi\|_{E_{\alpha,p}(\tau)} \leq c_{12}(p,\alpha,M,\varphi_0,r_0)$$

$$\cdot\{\|\varphi-\psi\|_{W^{2,p}} + \|Q(0,\varphi)-Q(0,\psi)\|_{B^{2\alpha,p}_\infty} + \omega_2(\tau)[u_\varphi-u_\psi]_{E_{\alpha,p}(\tau)}\},$$

so that, possibly taking a smaller τ, we deduce (1.3). Thus we have shown continuous dependence on φ of the local solution u_φ of problem (0.1).

3.C HIGHER REGULARITY

Suppose finally that $\varphi \in C^2(\bar{\Omega},\mathbb{C}^N)$ and $Q(0,\varphi) \in C^{2\alpha}(\bar{\Omega},\mathbb{C}^N)$, and fix $\delta \in]0,\alpha[$. Then for each $p \in]\frac{n}{2(\alpha-\delta)},\infty[$ we can apply the preceding theory, obtaining a local solution $u := u_\varphi \in E_{\alpha,p}(\tau)$, where τ depends on p. Now we have by Proposition 2.3 (with $\sigma = 1/2 + \delta - \alpha$, $\theta = 1 + \delta - \alpha$):

$$u \in E_{\alpha,p}(\tau) \to C^{\delta+1/2}([0,\tau],C^1(\bar{\Omega},\mathbb{C}^N)) \cap C^{\delta+1}([0,\tau],C(\bar{\Omega},\mathbb{C}^N)).$$

Consequently we have $\frac{\partial u}{\partial t} \in C^\delta([0,T],C(\bar{\Omega},\mathbb{C}^N))$; moreover it is easy to see that

$$(t,x) \to F(t,x,u(t,x),Du(t,x)) + I(t,x,u) \in C^\delta([0,T],C(\bar{\Omega},\mathbb{C}^N)),$$

so that, by difference, we get $(t,x) \to \sum_{i,j=1}^{n} A_{ij}(t,x,u(t,x),Du(t,x))\cdot D_iD_ju(t,x)$

$\in C^{\delta}([0,T],C(\bar{\Omega},\mathbb{C}^N))$.

The proof of Theorem 1.1 is complete. □

4. PROOF OF LEMMAS 3.1-3.2.

Proof of Lemma 3.1. Fix $0 \leq \tau < t \leq T$ and set for $\xi \in]0,1]$:

$$P_\xi := (\tau+\xi(t-\tau),x,v(\tau,x)+\xi(v(t,x)-v(\tau,x)),Dv(\tau,x)+\xi(Dv(t,x)-Dv(\tau,x))),$$

$$\bar{P} := (0,x,\varphi(x),D\varphi(x))$$

$$Q_\xi := (\tau+\xi(t-\tau),s,x,y;v(\tau,y)+\xi(v(t,y)-v(\tau,y)),Dv(\xi,y)+ (Dv(t,y)-$$
$$Dv(\tau,y));\ v(s,x),Dv(s,x),D^2v(s,x);u(x,y),Dv(s,y),D^2v(s,y)).$$

Then recalling (3.5) we can write:

$$\|F_{v,\varphi}(t,\cdot)-F_{v,\varphi}(\tau,\cdot)\|_{L^p} \leq \|\int_0^1 \frac{d}{d\xi} F\ (P_\xi)d\xi\|_{L^p} \tag{4.1}$$
$$+ \|\int_0^1 \frac{d}{d\xi} [\sum_{i,j=1}^{n} (A_{ij}(\bar{P})-A_{ij}(P_\xi))\cdot(D_iD_jv(\tau,\cdot)+\xi(D_iD_jv(t,\cdot)-$$
$$- D_iD_jv(\tau,\cdot)))]d\xi\|_{L^p} + \|\int_\tau^t \int_\Omega H(Q_1)dyds\|_{L^p}$$
$$+ \|\int_0^\tau \int_\Omega \int_0^1 \frac{d}{d\xi} H(Q_\xi)d\xi dyds\|_{L^p} =: I_1 + I_2 + I_3 + I_4.$$

Now recalling Proposition 2.3 and Lemmas 2.4-2.5, we have (provided $\sigma \in]0,1/2-n/2p[$):

$$I_1 = \|\int_0^1 \{\frac{\partial F}{\partial t}(P_\xi)\cdot(t-\tau)+ \frac{\partial F}{\partial u}(P_\xi)\cdot(v(t,\cdot)-v(\tau,\cdot))+ \frac{\partial F}{\partial p}\ (P_\xi)\cdot(Dv(t,\cdot)\ -$$
$$- Dv(\tau,\cdot))\}d\xi\|_{L^p} \leq c(K)\ (t-\tau)^{\alpha+1/2}, \tag{4.2}$$

$$I_2 \leq \sum_{i,j=1}^{n} \Big\| \int_0^1 \Big\{ \frac{\partial A_{ij}}{\partial t}(P_\xi)(t-\tau) + \frac{\partial A_{ij}}{\partial u}(P_\xi)\cdot(v(t,\cdot)-v(\tau,\cdot)) + \frac{\partial A_{ij}}{\partial p}(P_\xi) \tag{4.3}$$

$$\cdot\,(Dv(t,\cdot)-Dv(\tau,\cdot))\Big\}\cdot(D_iD_jv(\tau,\cdot)+\xi(D_iD_jv(t,\cdot)-D_iD_jv(\tau,\cdot)))\Big\}d\xi\Big\|_{L^p}$$

$$+ \sum_{i,j=1}^{n} \Big\| \int_0^1 (A_{ij}(\bar{P})-A_{ij}(P_\xi))\cdot(D_iD_jv(t,\cdot)-D_iD_jv(\tau,\cdot))d\xi\Big\|_{L^p}$$

$$\leqq c(K)\,[(t-\tau)^{\alpha+\sigma} + (t-\tau)^{\alpha}\,\omega_0(T)],$$

$$I_3 \leqq c(K) \int_\tau^t \int_\Omega \|\eta(t-s)(t-s)^{\alpha-1}[1+|D^2v(s,\cdot)|+|D^2v(s,y)|]\|_{L^p}\, dyds$$

$$\leqq c(K)(t-\tau)^{\alpha}\,\omega_0(T); \tag{4.4}$$

finally concerning I_4 we write

$$I_4 = \int_0^\tau \int_\Omega \int_0^1 \Big\{\frac{\partial H}{\partial t}(Q_\xi)(t-\tau) + \frac{\partial H}{\partial u_1}(Q_\xi)\cdot(v(t,y)-v(\tau,y)) + \frac{\partial H}{\partial p_1}(Q_\xi)$$

$$\cdot\,(Dv(t,y)-Dv(\tau,y))\Big\}d\xi dyds\Big\|_{L^p} \tag{4.5}$$

and all integrals but the first are easily estimated by $c(K)(t-\tau)^{\alpha+1/2}$. The first one needs some care: it does not exceed

$$\eta(T)c(K) \int_0^\tau \int_0^1 (\tau-s+\xi(t-\tau))^{\alpha-2}(t-\tau)d\xi ds = \eta(T)\,c(K) \int_0^\tau \int_0^{t-\tau} (\beta+\sigma)^{\alpha-2}d\beta d\sigma$$

$$\leqq \eta(T)\,c(K)(t-\tau)^{\alpha}. \tag{4.6}$$

Hence recalling (4.1),...,(4.5) we get the estimate for $[F_{v,\varphi}]_{C^\alpha(L^p)}$.
The estimates for G_v can be obtained similarly: indeed, setting

$$R_\xi := (\tau+\xi(t-\tau),s,x,y;v(\tau,y)+\xi(v(t,y)-v(\tau,y));Dv(\tau,y)+\xi(Dv(t,y) -$$

$$- Dv(\tau,y));\; v(s,x),Dv(s,x);v(s,y),Dv(s,y)),$$

we have:

$$G_{v,\varphi}(t,x)-G_{v,\varphi}(\tau,x) = \sum_{i=1}^{n} \frac{\partial B}{\partial p_i}(\bar{P})\cdot(D_i v(t,x)-D_i v(\tau,x)) \tag{4.7}$$

$$- \int_0^1 \{\frac{\partial B}{\partial t}(P_\xi)(t-\tau) + \frac{\partial B}{\partial u}(P_\xi)\cdot(v(t,x)-v(\tau,x)) + \frac{\partial B}{\partial p}(P_\xi)\cdot(Dv(t,x)-Dv(\tau,x))\}d\xi$$

$$+ \int_\tau^t \int_{\partial\Omega} K(R_1)dyds$$

$$+ \int_0^\tau \int_{\partial\Omega} \int_0^1 \{\frac{\partial K}{\partial t}(R_\xi)(t-\tau) + \frac{\partial K}{\partial u_1}(R_\xi)\cdot(v(t,y)-v(\tau,y)) + \frac{\partial K}{\partial p_1}(R_\xi)\cdot(Dv(t,y) - Dv(\tau,y))\}d\xi dyds$$

$$= T_1 + T_2 + T_3 + T_4.$$

Now we need to rewrite $T_1 + T_2$ as:

$$T_1 + T_2 = \int_0^1 (\frac{\partial B}{\partial p}(\bar{P}) - \frac{\partial B}{\partial p}(P_\xi))\cdot(Dv(t,x)-Dv(\tau,x))d\xi \tag{4.8}$$

$$- \int_0^1 \{\frac{\partial B}{\partial t}(P_\xi)(t-\tau) + \frac{\partial B}{\partial u}(P_\xi)\cdot(v(t,x)-v(\tau,x))\}d\xi,$$

and it is easy to see that

$$\|T_1 + T_2\|_{L^p} + \|T_3\|_{L^p} \leqq c(K)\ \omega_0(T)(t-\tau)^{\alpha+1/2}; \tag{4.9}$$

the same holds for the integrals appearing in T_4, with the only exception of the first, whose L^p-norm does not exceed

$$\eta(T)c(K) \int_0^\tau \int_0^1 (\tau-s+\xi(t-\tau))^{\alpha-3/2}(t-\tau)d\xi ds \leqq \eta(T)\ c(K)(t-\tau)^{\alpha+1/2}. \tag{4.10}$$

Thus we get the estimate for $[G_{v,\varphi}]_{C^{\alpha+1/2}(L^p)}$.

Finally, in order to estimate $[G_{v,\varphi}]_{C^\alpha(W^{1,p})}$, we evaluate the gradient (with respect to x) of (4.7), and treat each term as before: more precisely, the first integral in T_4 generates some terms needing an estimate like (4.6), whereas the two terms containing $D^2v(t,x)-D^2v(\tau,x)$, which arise from T_1 and from the last integral in T_2, have to be coupled together; the result is

$$\int_0^1 [\frac{\partial B}{\partial p}(\bar{P}) - \frac{\partial B}{\partial p}(P_\xi)]\cdot(D^2v(t,x)-D^2v(\tau,x))d\xi,$$

whose L^p-norm can be easily estimated by $c(K)\omega_0(T)(t-\tau)^\alpha$. The remaining terms can also be estimated, tediously but easily, by the same quantity. This shows that the estimate for $[G_{v,\varphi}]_{C^\alpha(W^{1,p})}$ is true. The proof of Lemma 3.1 is now complete.

Proof of Lemma 3.2 Fix $0 \leq \tau < t \leq T$ and set for $\xi, \eta \in]0,1]$:

$$v^\eta(x):=v(\tau,x)+\eta[v(t,x)-v(\tau,x)], \quad w^\eta(x):=w(\tau,x)+\eta[w(t,x)-w(\tau,x)],$$

$$z_{\eta\xi}(x):=w^\eta(x)+\xi[v^\eta(x)-w^\eta(x)]$$

$$\beta_\xi(s,x):=w(s,x)+\xi[v(s,x)-w(s,x)],$$

$$\bar{P}_\xi:=(0,x,\psi(x)+\xi(\varphi(x)-\psi(x)),D\psi(x)+\xi(D\varphi(x)-D\psi(x))),$$

$$P_{\eta\xi}:=(\tau+\eta(t-\tau),x,z_{\eta\xi}(x),Dz_{\eta\xi}(x))$$

$$Q_{\eta\xi}:=(\tau+\eta(t-\tau),s,x,y;\beta_\xi(\tau+\eta(t-\tau),y),D\beta_\xi(\tau+\eta(t-\tau),y);\beta_\xi(s,x),$$
$$D\beta_\xi(s,x),D^2\beta_\xi(s,x);\beta_\xi(s,y),D\beta_\xi(s,y),D^2\beta_\xi(s,y)).$$

Then a very tedious calculation shows that:

$$\|F_{v,\varphi}(t,\cdot)-F_{w,\psi}(t,\cdot)-F_{v,\varphi}(\tau,\cdot)+F_{w,\psi}(\tau,\cdot)\|_{L^p} \tag{4.11}$$

$$= \|\int_0^1 (\frac{d}{d\xi} F_{w+\xi(v-w),\psi+\xi(\varphi-\psi)}(t,\cdot) - \frac{d}{d\xi} F_{w+\xi(v-w),\psi+\xi(\varphi-\psi)}(\tau,\cdot))d\xi\|_{L^p}$$

$$\leq \|\int_0^1 \frac{d}{d\eta} \int_0^1 \{\frac{\partial F}{\partial u}(P_{\eta\xi})\cdot(v^\eta-w^\eta) + \frac{\partial F}{\partial p}(P_{\eta\xi})\cdot(Dv^\eta-Dw^\eta)\}d\xi d\eta\|_{L^p}$$

$$+ \|\int_0^1 \frac{d}{d\eta} \int_0^1 \sum_{i,j=1}^n \{\frac{\partial A_{ij}}{\partial u}(\bar{P}_\xi)\cdot(\varphi-\psi) - \frac{\partial A_{ij}}{\partial u}(P_{\eta\xi})(v^\eta-w^\eta)$$

$$+ \frac{\partial A_{ij}}{\partial p}(\bar{P}_{\xi})\cdot(D\varphi-D\psi) - \frac{\partial A_{ij}}{\partial p}(P_{\eta\xi})\cdot(Dv^{\eta}-Dw^{\eta})\}\cdot D_iD_jz_{\eta\xi}\, d\xi d\eta\|_{L^p}$$

$$+ \left\|\int_0^1 \frac{d}{d\eta}\int_0^1 \sum_{i,j=1}^{n} (A_{ij}(\bar{P}_{\xi})-A_{ij}(P_{\eta\xi}))\cdot(D_iD_jv^{\eta}-D_iD_jw^{\eta})d\xi d\eta\right\|_{L^p}$$

$$+ \int_0^1\int_{\tau}^{t}\int_{\Omega} \frac{d}{d\xi}H(Q_{1\xi})dyds d\xi\|_{L^p}$$

$$+ \left\|\int_0^1 \frac{d}{d\eta}\int_0^1\int_0^{\tau}\int_{\Omega} \frac{d}{d\xi}H(Q_{\eta\xi})dyds d\xi d\eta\right\|_{L^p};$$

no particular problems (except for length and boredom!) arise in estimating each term; we just need some care for the last one, since it generates, among others, several terms containing second order derivatives of H, with one derivative with respect to t: for instance, the first of such terms is

$$\int_0^1\int_0^1\int_0^{\tau}\int_{\Omega} \frac{\partial^2 H}{\partial t\partial u_1}(Q_{\eta\xi})\cdot(t-\tau)(v^{\eta}-w^{\eta})dyds d\xi d\eta;$$

all such terms can be estimated as in (4.6).

Thus we easily arrive at the estimate for $[F_{v,\varphi}-F_{w,\psi}]_{C^{\alpha}(L^p)}$.

The estimates for $G_{v,\varphi}-G_{w,\psi}$ are similar in nature and we omit their proof for the sake of brevity. We only remark that, analogously to the proofs above, we need to couple together suitably the terms which contain derivatives of v,w "of maximal order" (i.e. terms containing $Dv(t,x)-Dw(t,x)-Dv(\tau,x)+Dw(\tau,x)$ for the estimate in the $W^{1,p}$-norm, and terms containing $D^2v(t,x)-D^2w(t,x)-D^2v(\tau,x)+D^2w(\tau,x)$ for the estimate in the L^p-norm); moreover some other terms of integrodifferential type, containing one time derivative of the integrand K, have to be treated as in (4.6), (4.10).

This concludes the proof of Lemma 3.2. □

APPENDIX

This short appendix is devoted to the proof of the slight refinement of [3], which we referred to in the proof of Theorem 2.1.

We follow the notations of [3, pages 702-703]. Let $p > n$ and assume that the coefficients of the boundary operators B_j belong to $W^{\ell-m_j,p}(\Sigma)$, with norms bounded by k. Due to Sobolev's Theorem, the coefficients of B_j also belong to $C^{\ell-m_j-1}(\bar{\Sigma})$, so that the oscillation in Σ_r of their $(\ell-m_j-1)$-th order derivatives does not exceed $\omega(r)$, where $\omega(R) \downarrow 0$ as $R \downarrow 0$. Our goal is the proof of [3, Theorem 15.1] under the above weaker assumptions.

It is sufficient to remark that

$$\|(B_j'(0;D)-B_j'(x;D)-B_j''(x,;D))u(x,t)\|_{W^{\ell-m_j,p}} \leq c\{K\ \|u\|_{W^{\ell-1,p}}+\omega(r)\ \|u\|_{W^{\ell,p}}\};$$

hence we obtain that [3, Formula (15,3)'] still holds with r replaced by $\omega(r)$. The remaining part of the proof does not need any change. □

References

[1] P. Acquistapace, Evolution operators and strong solutions of abstract linear parabolic equations, preprint Sc. Norm. Sup. Pisa (1987).

[2] P. Acquistapace, B. Terreni, On quaslinear parabolic systems, preprint Sc. Norm. Sup. Pisa (1987).

[3] S. Agmon, A. Douglis, L. Nirenberg, Estimates near the boundary of solutions of elliptic partial differential equations satisfying general boundary conditions, Comm. Pure Appl. Math. 12 (1959) 623-727.

[4] S. Agmon, A. Douglis, L. Nirenberg, Estimates near the boundary for solutions of elliptic partial differential equations satisfying general boundary conditions, Comm. Pure Appl. Math. 17 (1964) 35-92.

[5] H. Amann, Existence and regularity for semilinear parabolic evolution equations, Ann. Sc. Norm. Sup. Pisa (4) 11 (1984) 593-676.

[6] G. Geymonat, P. Grisvard, Alcuni risultati di teoria spettrale per i problemi ai limiti lineari ellittici, Rend. Sem. Mat. Univ. Padova 38 (1967) 121-173.

[7] P. Grisvard, Equations différentielles abstraites, Ann. Sci. Ec. Norm. Sup. (4) 2 (1969) 311-395.

[8] B. Terreni, Non-homogeneous initial-boundary value problems for linear parabolic systems, preprint Dip. Mat. Univ. Pisa (1987).

Paolo Acquistapace
Scuola Normale Superiore
I-56100 Pisa
Italy

Brunello Terreni
Dipartimento di Matematica
Università di Pisa
I-56100 Pisa
Italy.

W. ARENDT & H. KELLERMANN

Integrated solutions of Volterra integrodifferential equations and applications

This paper is supported by the Deutsche Forschungsgemeinschaft.

In order to treat an abstract Cauchy Problem

$$u'(t) = Au(t) \quad (t \geqq 0);\ u(0) = u_0 \tag{0.1}$$

for a linear operator A on a Banach space X, the theory of one-parameter semigroups is available and in fact highly developed. From the point of view of the theory of Laplace transforms it can be described as follows: A linear operator A is the generator of a C_0-semigroup if and only if there exists $w \geqq 0$ such that $(w,\infty) \subset \rho(A)$ (the resolvent set of A) and the function $\lambda \to (\lambda - A)^{-1}$ is a Laplace transform; i.e. there exists a strongly continuous mapping $T : [0,\infty) \to L(X)$ satisfying $\|T(t)\| \leqq Me^{wt}$ $(t \geqq 0)$ for some $M \geqq 0$ such that

$$(\lambda - A)^{-1} = \int_0^\infty e^{-\lambda t} T(t)dt \quad (\lambda > w). \tag{0.2}$$

Of course, $(T(t))_{t\geq 0}$ is then the C_0-semigroup generated by A and the solutions of (0.1) are given by $u(t) = T(t)x$ $(x \in D(A))$.

Starting from this observation, the theory of integrated semigroups has recently been initiated (see [1], [2], [10], [13]). Let $n \in \mathbb{N} \cup \{0\}$. A linear operator A on a Banach space X is called the generator of an n-times integrated semigroup, if there exists $w \geqq 0$ such that $(\lambda - A)$ is invertible for $\lambda > w$ and $\lambda \to (\lambda - A)^{-1}/\lambda^n$ is a Laplace transform of a strongly continuous function $S : [0,\infty) \to L(X)$ (and then $(S(t))_{t\geq 0}$ is called the n-times integrated semigroup generated by A and the solutions of (0.1) are given by $u(t) = d^n/dt^n\, S(t)x$ (whenever this derivative exists)).

The purpose of this article is to generalize this concept to Volterra integrodifferential equations on a Banach space

$$u'(t) = \int_0^t Au(t-s)\, d\eta(s) \tag{0.3}$$

where $\eta: [0,\infty) \to \mathbf{R}$ is exponentially bounded and of bounded variation on each finite interval. Thus we develop a theory of <u>n-times integrated solution</u> families of (0.3). This concept coincides with the theory of integrated semigroups in the case when $\eta = 1_{(0,\infty)}$, and it contains the theory for the Cauchy problem of second order (i.e. the theory of integrated cosine functions) by taking $\eta(t) = t$ $(t \geqq 0)$. We obtain a characterization theorem as an easy consequence of a vector-valued version of Widder's theorem proved in [2]. It coincides with the Hille-Yosida type theorem proved by Da Prato-Iannelli [5] in the case n = 0. Our approach has the advantage that also operators whose domain is not dense are admitted.

Applications to differential operators for the Cauchy problem of first and second order are given. We briefly describe the results we obtain for $A = i\Delta$. It is well known that A is the generator of a C_0-semigroup on $L^p(\mathbf{R}^N)$, $1 \leqq p \leqq \infty$, if and only if p = 2. However, we show that A generates, for suitable n, an n-times integrated semigroup on the spaces $L^p(\mathbf{R}^N)$ $(1 \leqq p \leqq \infty)$, $C_0(\mathbf{R}^N)$ and $C_b(\mathbf{R}^N)$. Since L^2-solutions of the associated Cauchy problem always exist, this can be interpreted as a regularity result for the solutions of $u'(t) = i\Delta u(t)$.

Acknowledgement: We would like to thank Mikhael Balabane, Frank Neubrander and Jan Prüss for discussions, comments and suggestions.

SECTION 1. VOLTERRA INTEGRODIFFERENTIAL EQUATIONS GOVERNED BY AN n-TIMES INTEGRATED SOLUTION FAMILY

Let $A : D(A) \to X$ be a closed linear operator on a Banach space X and $\eta : [0,\infty) \to \mathbf{R}$ be a function which is of bounded variation on each compact interval [0,T] (T > 0).
We consider the integrodifferential problem

$$P(A,\eta) \begin{cases} u'(t) = \int_0^t Au(t-s)d\eta(s) \\ u(0) = x \end{cases}$$

We assume throughout that η is exponentially bounded; i.e. there exist $N \geqq 0$, $\beta \geqq 0$, such that

$$|\eta(t)| \leqq N\, e^{\beta t} \qquad (t \geqq 0). \tag{1.1}$$

Then we can define the Laplace transform $\hat{\eta}$ of η by

$$\hat{\eta}(\lambda) = \int_0^\infty e^{-\lambda t}\, d\eta(t) \qquad (\lambda > \beta).$$

We assume further that $\hat{\eta}(\lambda) \neq 0$ $(\lambda > \beta)$.

DEFINITION 1.1. Let $n \in \mathbb{N} \cup \{0\}$. A strongly continuous family $(S(t))_{t \geqq 0}$ of bounded linear operators on X is called an n-times integrated solution family of $P(A,\eta)$ if there exist $M \geqq 0$, $w \geqq \beta$ such that

(i) $\|S(t)\| \leqq Me^{wt}$ $(t \geqq 0)$,

(ii) $S(0) = I$ if $n = 0$; $S(0) = 0$ if $n > 0$,

(iii) $(\lambda - \hat{\eta}(\lambda)A) : D(A) \to X$ is bijective $(\lambda > w)$ and

$$(\lambda - \hat{\eta}(\lambda)A)^{-1}/\lambda^n = \int_0^\infty e^{-\lambda t} S(t)\,dt \qquad (\lambda > w). \tag{1.2}$$

A 0-times integrated solution family is also called solution family for short.
We say $P(A,\eta)$ is governed by an n-times integrated solution family if such a family exists.

If η is the characteristic function $1_{(0,\infty)}$, then $P(A,\eta)$ is the usual abstract Cauchy problem (0.1). In that case a solution family is the same as the C_0-semigroup generated by A; an n-times integated solution family is the same as the n-times integrated semigroup generated by A (see [2]).

If $\eta(t) = t$ $(t \geqq 0)$, then $P(A,\eta)$ is equivalent to the Cauchy problem of second order

$$
p_2(A) \begin{cases} u''(t) = Au(t) & (t \geqq 0) \\ u(0) = x \\ u'(0) = 0, \end{cases}
$$

and the solution family of $P(A,\eta)$ is the same as the cosine family generated by A.

Throughout this section we assume now that $(S(t))_{t\geqq 0}$ *is an n-times integrated solution family of* $P(A,\eta)$*, where* $n \in \mathbb{N} \cup \{0\}$. We establish existence and uniqueness results for the inhomogeneous problem

$$
P(A,\eta,f) \begin{cases} u'(t) = \int_0^t Au(t-s)d\ (s) + f(t) & (t \in [0,T]) \\ u(0) = x, \end{cases}
$$

where $f \in C([0,T],X)$ is given.

By a *solution* of $P(A,\eta,f)$ we understand a function $u \in C^1([0,T],X)$ satisfying

$$
\int_0^t u(t-s)d\eta(s) \in D(A),\ u'(t) = A\int_0^t u(t-s)d\eta(s) \quad (t \in [0,T]),\ u(0) = x.
$$

REMARK. Using this notion of solution and the results of this section one can easily show that, for $n = 0$, the solution families of Definition 1.1 essentially coincide with the "resolvent families" of Da Prato-Iannelli [5]). In fact, the following holds. Suppose that A is densely defined. A strongly continuous family of bounded operators $(S(t))_{t\geqq 0}$ is a solution family of $P(A,\eta)$ if and only if

(a) $S(t)x \in D(A)$ and $AS(t)x = S(t)Ax$ for all $x \in D(A)$, $t \geqq 0$,

(b) there exist $M \geqq 0$, $w \in \mathbb{R}$ such that $\|S(t)\| \leqq Me^{wt}$ $(t \geqq 0)$,

(c) for each $x \in D(A)$ the function $u(t) := S(t)x$ defines a solution of $P(A,\eta)$.

Let $x \in X$. We define $v \in C([0,T], X)$ by

$$v(t) = S(t)x + \int_0^t S(t-s)f(s)ds.$$

The following theorem shows now that solving $P(A,\eta,f)$ is reduced to verifying regularity of v.

THEOREM 1.2. Suppose $P(A,\eta)$ is governed by an n-times integrated solution family.
Then the problem $P(A,\eta,f)$ has a solution if and only if $v \in C^{n+1}([0,T],X)$.
In that case the solution u is given by $u = v^{(n)}$.

For the proof we need a series of auxiliary results.

LEMMA 1.3. Let $g \in C([0,T],X)$. Then

$$u(t) = \int_0^t g(t-s)d\eta(s)$$

defines a function $u \in C([0,T],X)$. If $g \in C^1([0,T];X)$ and $g(0) = 0$, then $u \in C^1([0,T];X)$ and $u'(t) = \int_0^t g'(t-s)d\eta(s)$.

Proof. The first assertion is immediate. If $g \in C^1([0,T],X)$ such that $g(0) = 0$, then by Fubini's theorem

$$u(t) = \int_0^t \int_0^{t-s} g'(r)drd\eta(s) = \int_0^t \int_s^t g'(r-s)drd\eta(s)$$

$= \int_0^t \int_0^r g'(r-s)d\eta(s)dr$. Since $v(r) = \int_0^r g'(r-s)d\eta(s)$ defines a continuous function, the claim follows. □

LEMMA 1.4. Let $x \in D(A)$. Then $S(t)x \in D(A)$ and $AS(t)x = S(t)Ax \quad (t \geqq 0)$.

Proof. Let $\mu > w$. Then for all $y \in X$, $\lambda > w$ one has by definition of $(S(t))_{t \geqq 0}$

$$\int_0^\infty e^{-\lambda t}S(t)(\mu-\hat{\eta}(\mu)A)^{-1}ydt = (\lambda-\hat{\eta}(\lambda)A)^{-1}/\lambda^n\ (\mu-\hat{\eta}(\mu)A^{-1}y$$

$$= (\mu-\hat{\eta}(\mu)A)^{-1}(\lambda-\hat{\eta}(\lambda)A)^{-1}y/\lambda^n = (\mu-\hat{\eta}(\mu)A)^{-1}\int_0^\infty e^{-\lambda t}S(t)ydt$$

$$= \int_0^\infty e^{-\lambda t}(\mu-\hat{\eta}(\mu)A)^{-1}S(t)ydt.$$

By the uniqueness theorem for Laplace transforms it follows that

$$S(t)\ (\mu-\hat{\eta}(\mu)A)^{-1} = (\mu-\hat{\eta}(\mu)A)^{-1}S(t). \tag{1.3}$$

Now let $x \in D(A)$ and $y = (\mu-\hat{\eta}(\mu)A)x$. Then by (1.3), $S(t)x = (\mu-\hat{\eta}(\mu)A)^{-1}S(t)y \in D(A)$ and $(\mu-\hat{\eta}(\mu)A)S(t)x = S(t)y = S(t)(\mu-\hat{\eta}(\mu)A)x$. This implies that $AS(t)x = S(t)Ax$ (since $\hat{\eta}(\mu) \neq 0$). □

<u>LEMMA 1.5</u>. For all $x \in X$, $t \geq 0$ we have

$$\int_0^t \int_0^s S(s-r)xd\eta(r)ds \in D(A) \qquad \text{and}$$

$$A\int_0^t \int_0^s S(s-r)xd\eta(r)ds = S(t)x - (t^n/n!)x. \tag{1.4}$$

<u>Proof</u>. For $y \in D(A)$ and $\lambda > w$ we have

$$\int_0^\infty \lambda^{n+1}\ e^{-\lambda t}\ t^n/n!\ ydt = y$$

$$= \lambda\int_0^\infty \lambda^n\ e^{-\lambda t}\ S(t)ydt - \hat{\eta}(\lambda)\int_0^\infty \lambda^n e^{-\lambda t}S(t)Aydt$$

$$= \int_0^\infty \lambda^{n+1}\ e^{-\lambda t}\ S(t)ydt - \int_0^\infty \lambda^n e^{-\lambda t}\int_0^t S(t-s)Ayd\eta(s)dt$$

$$= \int_0^\infty \lambda^{n+1}\ e^{-\lambda t}\ S(t)ydt - \int_0^\infty \lambda^{n+1}e^{-\lambda t}\int_0^t\int_0^s S(s-r)Ayd\eta(r)dsdt.$$

Cancelling by λ^{n+1} one obtains from the uniqueness theorem that

$$\int_0^t \int_0^s S(s-r)Ay\,d\eta(r)ds = S(t)y - t^n/n!\; y \quad (y \in D(A),\ t \geq 0). \tag{1.5}$$

Now let $x \in X$. Define $y = (\lambda-\hat{\eta}(\lambda)A)^{-1}x$ where $\lambda > w$ is fixed. Let $z = \int_0^t \int_0^s S(s-r)x\,d\eta(r)ds$, where $t > 0$. We have to show that $z \in D(A)$ and

$$Az = S(t)x - t^n/n!\; x. \tag{1.6}$$

Using Lemma 1.4 and (1.5) we obtain

$$z = (\lambda-\hat{\eta}(\lambda)A) \int_0^t \int_0^s S(s-r)y\,d\eta(r)ds$$

$$= \lambda\int_0^t \int_0^s S(s-r)y\,d\eta(r)ds - \hat{\eta}(\lambda)[S(t)y - t^n/n!\; y] \in D(A) \text{ and}$$

$(\lambda-\hat{\eta}(\lambda)A)z = \lambda z - \hat{\eta}(\lambda)[S(t)x-t^n/n!\; x]$, which gives (1.6). □

COROLLARY 1.6. If $y \in D(A)$, then $S(\cdot)y \in C^1([0,T],X)$ and

$$d/dt\; S(t)\; y = \begin{cases} t^{n-1}/(n-1)!\; y + \int_0^t S(t-r)Ay\,d\eta(r) & (n > 0) \\ \int_0^t S(t-r)Ay\,d\eta(r) & (n = 0). \end{cases} \tag{1.7}$$

The following lemma gives us the key to prove Theorem 1.2. We show that the function v (given by $v(t) = S(t)x + \int_0^t S(t-s)f(s)ds$) solves some integrated version of the problem $P(A,\eta,f)$.

LEMMA 1.7. For all $t \in [0,T]$ one has

$$\int_0^t \int_0^{t-s} v(r)dr\,d\eta(s) \in D(A) \quad \text{and}$$

$$v(t) = t^n/n!\; x + A\int_0^t \int_0^{t-s} v(r)dr\,d\eta(s) + \int_0^t (t-s)^n/n!\; f(s)ds. \tag{1.8}$$

Proof. Introducing the expression for $S(t)$ given by (1.4) into the following function v_0 yields:

$$v_0(t) := \int_0^t S(t-s)f(s)ds$$

$$= A\int_0^t\int_0^{t-s}\int_0^w S(w-r)f(s)d\eta(r)dwds + \int_0^t (t-s)^n/n!f(s)ds$$

$$= A\int_0^t\int_0^{t-s}\int_r^{t-s} S(w-r)f(s)dwd\eta(r)ds + \int_0^t (t-s)^n/n!f(s)ds$$

$$= A\int_0^t\int_0^{t-s}\int_0^{t-s-r} S(w)f(s)dwd\eta(r)ds + \int_0^t (t-s)^n/n!f(s)ds$$

$$= A\int_0^t\int_0^{t-r}\int_0^{t-s-r} S(w)f(s)dwdsd\eta(r) + \int_0^t (t-s)^n/n!f(s)ds$$

$$= A\int_0^t\int_0^{t-r}\int_s^{t-r} S(w-s)f(s)dwdsd\eta(r) + \int_0^t (t-s)^n/n!f(s)ds$$

$$= A\int_0^t\int_0^{t-r}\int_0^w S(w-s)f(s)dsdwd\eta(r) + \int_0^t (t-s)^n/n!f(s)ds$$

$$= A\int_0^t\int_0^{t-r} v_0(w)dwd\eta(r) + \int_0^t (t-s)^n/n!f(s)ds.$$

Moreover by (1.4),

$$\int_0^t\int_0^{t-r} S(w)xdwd\eta(r) = \int_0^t\int_r^t S(w-r)xdwd\eta(r) = \int_0^t\int_0^w S(w-r)xd\eta(r)dw \in D(A)$$

and

$$A\int_0^t\int_0^{t-r} S(w)xdwd\eta(r) = S(t)x - t^n/n!x.$$

Since $v(t) = v_0(t) + S(t)x$ we obtain (1.8) as a consequence of this and the above identity for v_0. □

PROOF OF THEOREM 1.2. Let u be a solution of $P(A,\eta,f)$. Then by (1.7)

$$\int_0^t d/ds[S(t-s)u(s)]ds = -\int_0^t S'(t-s)u(s)ds + \int_0^t S(t-s)u'(s)ds$$

$$= -\int_0^t (t-s)^{n-1}/(n-1)!\ u(s)ds - \int_0^t\int_0^{t-s} S(t-s-r)Au(s)d\eta(r)ds$$

$$+ \int_0^t S(t-s) \int_0^s Au(s-r)d\eta(r)ds + \int_0^t S(t-s)f(s)ds$$

in the case when $n > 0$, whereas the first term has to be replaced by 0 if $n = 0$. Now by Fubini's theorem the second term equals

$$- \int_0^t \int_0^{t-r} S(t-s-r)Au(s)dsd\eta(r) = - \int_0^t \int_r^t S(t-s)Au(s-r)dsd\eta(r)$$

$$= - \int_0^t \int_0^s S(t-s)Au(s-r)d\eta(r)ds,$$ and so it cancels with the third term.

Now

$$\int_0^t d/ds[S(t-s)u(s)]ds =$$

$$= S(t-s)u(s)\Big|_0^t = \begin{cases} -S(t)x & \text{if } n > 0 \\ u(t)-S(t)x & \text{if } n = 0. \end{cases}$$

Hence $v(t) = S(t)x + \int_0^t S(t-s)f(s)ds$

$$= \begin{cases} \int_0^t (t-s)^{n-1}/(n-1)!\ u(s)ds & \text{if } n > 0 \\ u(t) & \text{if } n = 0. \end{cases}$$

This shows one implication to hold.

In order to prove the other assume that $v \in C^{n+1}([0,T];X)$. Using Lemma 1.3 and the fact that A is closed one obtains by differentiating (1.8) that $\int_0^t v(t-s)d\eta(s) \in D(A)$ and

$$v'(t) = t^{n-1}/(n-1)!\ x + A \int_0^t v(t-s)d\eta(s) + \int_0^t (t-s)^{n-1}/(n-1)!\ f(s)ds \quad \text{if } n > 0 \tag{1.9}$$

and $v'(t) = A \int_0^t v(t-s)d\eta(s) + f(t)$ if $n = 0$.

This shows that v is a solution if $n = 0$. If $n > 0$, since A is closed, one can differentiate (1.9) n-times and obtains that

$$\int_0^t v^{(n)}(t-s)d\eta(s) \in D(A) \quad \text{and}$$

$$v^{(n+1)}(t) = A\int_0^t v^{(n)}(t-s)d\eta(s) + f(t).$$

Since $v^{(n)}(0) = x$, this shows that $v^{(n)}$ is a solution of $P(A,\eta,f)$. □

Next we give sufficient conditions on the initial value x and the inhomogeneity f for $P(A,\eta,f)$ to have a solution.

THEOREM 1.8. Let $x \in D(A^{n+1})$ and $f \in C^{n+1}([0,T],X)$ such that $f^{(k)}(0) \in D(A^{n-k})$ $(k = (0,1,...,n-1)$. Then $P(A,\eta,f)$ has a unique solution with initial value x.

Proof. We let $v(t) = S(t)x + v_0(t)$ where $v_0(t) = \int_0^t S(s)f(t-s)ds$. First we show that $S(.)x \in C^{n+1}[0,\infty),X)$. In fact, we claim that $S(.)y \in C^k([0,\infty),X)$ for all $y \in D(A^k)$ and

$$d^k/dt^k\, S(t)y = \begin{cases} t^{n-k}/(n-k)!\, y + \int_0^t S(t-s)^{(k-1)}Ayd\eta(s) \\ \qquad \text{if } k = 1,2,...,n \qquad \text{and} \\ \int_0^t S(t-s)^{(k-1)}Ayd\eta(s) \quad \text{if } k > n. \end{cases} \tag{1.10}$$

For k = 1 this follows from Corollary 1.6. Assume now that the assertion is true for k. Let $y \in D(A^{k+1})$. Then $S(.)^{(k-1)}Ay \in C^1([0,\infty),X)$ by the inductive assumption. So by Lemma 1.3,

$$d/dt \int_0^t S(t-s)^{(k-1)}Ayd\eta(s) = \int_0^t S(t-s)^{(k)}Ayd\eta(s).$$

This shows (1.10) to hold for k+1.
Next we show that $v_0 \in C^{n+1}([o,T],X)$. In fact, we show by induction that $v_0 \in C^k([0,T],X)$ and

$$v_0^{(k)}(t) = S(t)f^{(k-1)}(0) + d/dt\,S(t)f^{(k-2)}(0) + \ldots +$$

$$d^{k-1}/dt^{k-1}\,S(t)f(0) + \int_0^t S(s)f^{(k)}(t-s)ds \qquad (1.11)$$

$$(k = 1,\ldots,n+1).$$

For $k = 1$ this is proved by differentiating v_0 by the usual rule. Assume that (1.11) holds for $k \leq n+1$. Since by (1.10) $S(.)y \in C^k$ if $y \in D(A^k)$, it follows that $v_0^{(k)} \in C^1$ and we obtain (1.11) for $k+1$ by differentiating.

We have shown that $v \in C^{n+1}([0,T];X)$ which implies the claim by Theorem 1.2. □

COROLLARY 1.9. If $x \in D(A^{n+1})$, then there exists a unique function $u \in C^1[0,\infty),X)$ such that $u(t) \in D(A)$ $(t \geq 0)$ and

$$u'(t) = \int_0^t An(t-s)d\eta(s) \qquad (t \geq 0)$$

$$u(0) = x.$$

Proof. One deduces from (1.10) that $u(t) = v^{(n)}(t) \in D(A)$ $(t \geq 0)$. □

SECTION 2. REAL CHARACTERIZATION

Throughout this section we assume that A is a closed linear operator and $\eta:[0,\infty) \to \mathbb{R}$ a function satisfying the assumptions of Section 1.

THEOREM 2.1. Let $n \in \mathbb{N} \cup \{0\}$, $w \geq \beta$, $M \geq 0$. The problem $P(A,\eta)$ is governed by an (n+1)-times integrated solution family $(S(t))_{t\geq 0}$ satisfying

$$\limsup_{h\downarrow 0} \|S(t+h) - S(t)\| \leq Me^{wt} \qquad (t \geq 0) \qquad (2.1)$$

if and only if $(\lambda-\hat{\eta}(\lambda)A)$ is bijective whenever $\lambda > w$ and

$$\|(\lambda-w)^{k+1}\, d^k/d\lambda^k((\lambda-\hat{\eta}(\lambda)A)^{-1}/\lambda^n)/k!\| \leq M \; (\lambda > w,\; k = 0,1,\ldots). \qquad (2.2)$$

Note that (2.2) makes sense since the function $\lambda \to (\lambda - \hat{\eta}(\lambda)A)^{-1}$ is C^∞ on (w,∞).

Theorem 2.1 is an immediate consequence of the vector-valued version of Widder's theorem given in [2, Theorem 4.1].

THEOREM 2.2. Suppose that A is densely defined. Let $n \in \mathbb{N} \cup \{0\}$. Then the following assertions are equivalent.

(i) Problem $P(A,\eta)$ is governed by an n-times integrated solution family.

(ii) There exist $w \geqq \beta$, $M \geqq 0$ such that $(\lambda-\hat{\eta}(\lambda)A)$ is bijective whenever $\lambda > w$ and

$$\|(\lambda-w)^{k+1}\, d^k/dt^k((\lambda-\hat{\eta}(\lambda)A)^{-1}/\lambda^n)k!\| \leqq M \;(\lambda > w,\ k = 0,1,2,\ldots).\quad(2.3)$$

Proof. (i) → (ii). Let $(T(t))_{t\geqq 0}$ be an n-times integrated solution family satisfying $\|T(t)\| \leqq Me^{wt}$ $(t \geqq 0)$, where $w \geqq \beta$. Then for $k = 0,1,2,\ldots,\lambda > w$,

$$\|(\lambda-w)^{k+1}\, d^k/d\lambda^k[(\lambda-\hat{\eta}(\lambda)A)^{-1}/\lambda^n]/k!\|$$

$$= \|(\lambda-w)^{k+1}\, d^k/d\lambda^k \int_0^\infty e^{-\lambda t}\, T(t)dt\|/k!$$

$$= \|(\lambda-w)^{k+1} \int_0^\infty (-t)^k e^{-\lambda t}\, T(t)dt\|/k!$$

$$\leqq (\lambda-w)^{k+1}\, M \int_0^\infty t^k e^{-\lambda t}\, e^{wt} dt/k! = M.$$

(ii) → (i). By Theorem 2.1 it follows from condition (ii) that $P(A,\eta)$ is governed by an (n+1)-times integrated solution family $((S(t))_{t\geqq 0}$ satisfying (2.1). It follows from Corollary 1.6 that the function $S(.)y$ is continuously differentiable for all $y \in D(A)$. Because of (2.1) it follows from this that $S(.)x$ is continuously differentiable for all $x \in \overline{D(A)} = X$. Define $T:[0,\infty) \to L(X)$ by $T(t)x = d/dt\, S(t)x$ $(t \geqq 0,\ x \in X)$. Then T is strongly continuous and satisfies $\|T(t)\| \leqq Me^{wt}$ $(t \geqq 0)$. Hence for $\lambda > w$ one obtains by integrating by parts,

$$(\lambda-\hat{\eta}(\lambda)A)^{-1}x/\lambda^n = \lambda \int_0^\infty e^{-\lambda t} S(t)x dt = \int_0^\infty e^{-\lambda t} T(t)x dt \;(x \in X).$$

Thus $(T(t))_{t\geqq 0}$ is an n-times integrated solution family of $P(A,\eta)$. □

REMARK. In the case when n = 0, Theorem 2.2 is known. In fact, if η is the characteristic function of $(0,\infty)$, then it is just the Hille-Yosida theorem; if $\eta(t) = t$, then it is the characterization theorem for generators of cosine functions due to Sova; for arbitrary η finally, it is essentially the generation theorem of Da Prato-Ianelli [5].

SECTION 3. A COMPLEX CONDITION

The real condition discussed in the preceding section has the advantage of involving merely real values λ. Moreover, at least when the operator is densely defined, it gives the best value of the integration order n.

However, the disadvantage of the condition is that all derivatives of $(\lambda-\hat{\eta}(\lambda)A)^{-1}$ have to be estimated, which might be impossible in concrete cases.

Now we discuss the condition that $(\lambda-\hat{\eta}(\lambda)A)^{-1}$ is polynomially bounded in a right half plane. In general, this condition is much easier to verify. However, usually not the best integration order n is obtained (cf. Remark 3.3).

We start with the underlying result for vector-valued Laplace transforms.

PROPOSITION 3.1. Let Y be a Banach space, $M \geqq 0$, $w \geqq 0$, $\alpha \in \mathbb{R}$ and let $r : [\mathrm{Re}\,\lambda > w] \to Y$ be an analytic function satisfying

$$\|r(\lambda)\| \leqq M|\lambda|^{\alpha} \qquad (\mathrm{Re}\,\lambda > w). \tag{3.1}$$

Then there exists a locally Hölder continuous function $f : [0,\infty) \to Y$ such that

(a) for all $w' > w$ there exists $M' \geqq 0$ such that

$$\|f(t)\| \leqq M'e^{w't} \qquad (t \geqq 0),$$

(b) $f(0) = 0$,

(c) $r(\lambda) = \lambda^{[\alpha]+2} \displaystyle\int_0^\infty e^{-\lambda t} f(t)\,dt \quad (\mathrm{Re}\,\lambda > w)$.

Here a function $f : [0,\infty) \to Y$ is said to be locally Hölder continuous if there exist $\varepsilon > 0$ and $c : [0,\infty) \to [0,\infty)$ such that

$$\|f(s) - f(t)\| \leqq c(T)\ |s - t|^{\varepsilon} \quad (s,t \in [0,T])$$

for all $T > 0$.

REMARK. We are indebted to Jan Prüss for the argument in the following proof which leads to the Hölder continuity of f.

Proof. Let $w' > w$ and $\gamma := w' + i\mathbb{R}$. Define

$$f(t) = 1/2\pi i \int_{\gamma} e^{\mu t}\ (1/\mu)^{[\alpha]+2}\ r(\mu)d\mu \qquad (t \geqq 0).$$

The integral converges since

$$\|e^{\mu t}\ (1/\mu)^{[\alpha]+2}\ r(\mu)\| \leqq Me^{w't}\ (1/|\mu|)^{2+[\alpha]-\alpha} \quad (\mu \in \gamma) \tag{3.2}$$

This implies (a).

We show that f is Hölder continuous. Let $\varepsilon > 0$ such that $[\alpha]-\alpha+2-\varepsilon > 1$. For $\lambda \in \gamma$ we have

$$|e^{\lambda s} - e^{\lambda t}| \leqq |\lambda|\ e^{w'T}\ |t-s| \qquad \text{and}$$

$$|e^{\lambda s} - e^{\lambda t}| \leqq 2e^{w'T} \quad \text{for all } s,\ t \in [0,T].$$

Consequently, $|e^{\lambda s} - e^{\lambda t}| \leqq |e^{\lambda s} - e^{\lambda t}|^{\varepsilon} \cdot |e^{\lambda s} - e^{\lambda t}|^{1-\varepsilon}$

$$\leqq |\lambda|^{\varepsilon}\ e^{\varepsilon w'T}|t - s|^{\varepsilon}\ \ 2^{1-\varepsilon}\ e^{w'(1-\varepsilon)T}$$

$$\leqq 2\ e^{w'T}\ |\lambda|^{\varepsilon}|t-s|^{\varepsilon} \quad (s,\ t \in [0,T],\ \lambda \in \gamma).$$

Now $f(t) - f(s) = 1/2\pi i \int_{\gamma} (e^{\mu t} - e^{\mu s})\ (1/\mu)^{[\alpha]+2}\ r(\mu)d\mu.$

The integrand is bounded by

$$|e^{\mu t} - e^{\mu s}| \cdot (1/|\mu|)^{[\alpha]+2}\ \|r(\mu)\|$$

$$\leqq\ 2e^{w'T}|\mu|^{\varepsilon} \cdot |t-s|^{\varepsilon} \cdot (1/|\mu|)^{[\alpha]+2} \cdot M \cdot |\mu|^{\alpha}$$

$$\leq 2e^{w'T}.M.(1/|\mu|)^{[\alpha]-\alpha+2-\varepsilon}.|t-s|^{\varepsilon}$$

which is integrable over γ.

Hence $\|f(s) - f(t)\| \leq ce^{w'T}.|t-s|^{\varepsilon}$ $(s,t \in [0,T])$. Next we show (c) to hold. Let $\lambda > w'$. Then

$$\lambda^{[\alpha]+2} \int_0^\infty e^{-\lambda t} f(t)dt$$
$$= \lambda^{[\alpha]+2} \int_0^\infty e^{-\lambda t}\ 1/2\pi i \int_\gamma e^{\mu t}\ (1/\mu)^{[\alpha]+2}\ r(\mu)d\mu dt$$
$$= \lambda^{[\alpha]+2}\ 1/2\pi i \int_\gamma \int_0^\infty e^{-(\lambda-\mu)t}dt\ (1/\mu)^{[\alpha]+2}\ r(\mu)d\mu$$
$$= \lambda^{[\alpha]+2}\ 1/2\pi i \int_\gamma 1/(\lambda-\mu)\cdot(1/\mu)^{[\alpha]+2}\ r(\mu)d\mu$$
$$= r(\lambda) \text{ by residue calculus.}$$

Finally, to see (b), we observe that $\mu \to (1/\mu)^{[\alpha]+2}\cdot r(\mu)$ is analytic in $[\mathrm{Re} > w]$ and $\|(1/\mu)^{[\alpha]+2}\cdot r(\mu)\| \leq M(1/|\mu|)^{2+[\alpha]-\alpha}$. So

$$f(0) = \lim_{R\to\infty}\ 1/2\pi i \int_{\gamma(R)} (1/\mu)^{[\alpha]+2}\cdot r(\mu)d\mu = 0,$$

where $\gamma(R)$ denotes the closed path consisting of the line $\{is : -R \leqq s \leqq R\}$ and the semicircle $\{Re^{ith} : -\pi/2 \leqq \xi \leqq \pi/2\}$. □

<u>THEOREM 3.2.</u> Assume that there exist $w > \beta$, $M \geqq 0$, and $\alpha \geqq -1$ such that

(a) $(\lambda-\hat{\eta}(\lambda)A)$ is bijective for $\mathrm{Re}\ \lambda > w$ and

(b) $\|(\lambda-\hat{\eta}(\lambda)A)^{-1}\| \leq M|\lambda|^{\alpha}$ $(\mathrm{Re}\ \lambda > w)$.

Then $P(A,\eta)$ is governed by an $[\alpha]+2$ - times integrated solution family $(S(t))_{t\geqq 0}$.

Moreover, S is Hölder continuous in the following sense: There exist $\varepsilon > 0$ and $c : [0,\infty) \to [0,\infty)$ such that

$$\|S(t) - S(s)\| \leq c(T)\cdot|s-t|^{\varepsilon}\ (s,t \in [0,T)) \text{ for all } T > 0.$$

This is an immediate consequence of Proposition 3.1.

REMARK 3.3. a) In the case when $\alpha = -1$ a much stronger result can be obtained (see Da Prato-Iannelli [5]). In particular, if A is densely defined and η is the characteristic function of $(0,\infty)$, then A is the generator of a holomorphic semigroup.

b) Assume that $(S(t))_{t\geq 0}$ is the n-times integrated solution family of $P(A,\eta)$. Then

$$\|(\lambda-\hat{\eta}(\lambda)A)^{-1}\| = \|\int_0^\infty \lambda^n e^{-\lambda t} S(t)\, dt\| \leq M|\lambda|^n/(\mathrm{Re}\lambda - w) \text{ for } \mathrm{Re}\,\lambda > w$$

seems to be the best possible estimate on a right half plane, and so Theorem 3.2 would merely yield an n+2 -times integrated solution family. So Theorem 3.2 gives a convenient criterion; however, in general one looses an integration order of 2.

SECTION 4. RELATION TO DISTRIBUTION SEMIGROUPS

In 1960 J.L. Lions developed the theory of distribution semigroups as an approach to abstract Cauchy problems $u' = Au$, $u(0) = x$ which are well posed in a weaker sense than the classical one that leads to C_0-semigroups ([11]). As a particular class he defined the "distribution semigroups of exponential growth" and for their generators he gave a very simple characterization.

THEOREM 4.1 ([11, Theorem 6.1]). Let $A:D(A) \to X$ be a densely defined linear operator. The following are equivalent.

(i) A is the generator of a distribution semigroup of exponential growth.

(ii) For some positive constants and c and some $n \in \mathbb{N}$ $\{\lambda:\mathrm{Re}\,\lambda > \omega\} \subset \rho(A)$ and $\|(\lambda-A)^{-1}\| \leq c|\lambda|^n$ for $\mathrm{Re}\,\lambda > \omega$.

By Theorem 3.2 we immediately obtain the following.

THEOREM 4.2. Let $A:D(A) \to X$ be linear and densely defined. The following are equivalent.

(i) A generates a distribution semigroup of exponential growth.

(ii) There exists some $n \in \mathbb{N}$ such that A generates an n-times integrated semigroup.

A more detailed analysis is possible by the theory of "smooth distribution semigroups" developed by M. Balabane and H. Emami-Rad ([3], [4]).

DEFINITION 4.3 ([3], [4]). Let $\mathcal{D}(\mathbb{R}_+)$ denote the test functions with support in $(0,\infty)$ and let $k \in \mathbb{N}$. By $p_k(\phi) := \Sigma_{i=1}^{k} \| t^i \phi^{(i)} \|_{L^1}$ a norm is defined on $\mathcal{D}(\mathbb{R}_+)$. Let T_k denote the completion of $\mathcal{D}(\mathbb{R}_+)$ with respect to p_k.

A smooth distribution semigroup of order k is a continuous linear mapping G from T_k to $L(X)$ such that

(i) $G(\phi * \Psi) = G(\phi)G(\Psi)$ for $\phi, \Psi \in T_k$.

(ii) There exists a dense subspace D in X such that for $x \in D$ there is a continuous function $g_x : \mathbb{R}_+ \to X$ with $g_x(0) = x$ and $G(\phi)x = \int_0^\infty \phi(t) g_x(t) dt$ for all $\phi \in T_k$.

In the following we study how smoothness of order k translates into terms of integrated semigroups.

THEOREM 4.4. Let A be linear and densely defined. The following are equivalent.

(i) A generates a smooth distribution semigroup of order k.

(ii) A generates a k-times integrated semigroup $(S(t))_{t \geqq 0}$ with the additional property $\|S(t)\| \leq ct^k$ for $t \geq 0$ (and a suitable constant c).

To prove this we need a few lemmas.

LEMMA 4.5. Let $T \in \mathcal{D}(\mathbb{R}_+)'$ and suppose $T(\phi^{(k)}) = 0$ for all $\phi \in \mathcal{D}(\mathbb{R}_+)$ and some fixed $k \in \mathbb{N}$.

Then T is a polynomial $\Sigma_{i=0}^{k-1} a_i t^i$ for suitable scalar constants $a_0, \ldots, a_{k-1}$.

For the proof we refer to books about the theory of distributions.

LEMMA 4.6. (i) The norms $p_k(\phi) := \Sigma_{i=1}^{k} \|t^i \phi^{(i)}\|_{L^1}$ and $\phi \to \|t^k \phi^{(k)}\|_{L^1}$ are equivalent norms on $\mathcal{D}(\mathbb{R}_+)$.

(ii) The set $\{\phi^{(k)} : \phi \in \mathcal{D}(\mathbb{R}_+)\}$ is dense in $L^1(\mathbb{R}_+, t^k dt)$.

PROOF. (i): For all $\phi \in \mathcal{D}(\mathbb{R}_+)$ the function $t \to |\phi(t)|$ is absolutely continuous and $|\phi(t)|' = (\text{sign } \phi(t))\phi(t)'$.

This implies for $i < k$ and $\phi \in \mathcal{D}(\mathbb{R}_+)$

$$\int_0^\infty t^i |\phi^{(i)}(t)| dt = -\frac{1}{i+1}\int_0^\infty t^{i+1}(\text{sign } \phi^{(i)}(t))\phi^{(i+1)}(t)dt$$

$$\leq \frac{1}{i+1}\int_0^\infty t^{i+1}|\phi^{(i+1)}(t)|dt.$$

Therefore all terms in the sum $p_k(\phi)$ can be estimated by $c\,\|t^k\phi^{(k)}\|_{L^1}$ for some constant c.

(ii): Let $h \in L^\infty(\mathbb{R}_+)$ be such that $\int_0^\infty h(t)\phi^{(k)}(t)t^k\, dt = 0$ for $\phi \in \mathcal{D}(\mathbb{R}_+)$. By Lemma 4.5 we have $h(t)t^k = \Sigma_{i=0}^{k-1}\, a_i t^i$ a.e. in $(0,\infty)$ or $h(t) = \Sigma_{i=0}^{k-1}\, a_i\, t^{i-k}$. The right hand side is bounded only if all a_i's vanish. Therefore $h = 0$.

PROOF OF THEOREM 4.4. (i) $\Rightarrow$ (ii): Let G be the smooth distribution semigroup of order k generated by A. For all x in some dense subspace D we have a continuous function $g_x : \mathbb{R}_+ \to X$ such that

$$G(\phi)x = \int_0^\infty \phi(t)g_x(t)\, dt \text{ for } \phi \in \mathcal{D}(\mathbb{R}_+).$$

Integration by parts yields

$$G(\phi)x = \int_0^\infty (-1)^k\phi^{(k)}(t)H_x(t)t^k\, dt$$

where $H_x(t) := (1/t^k)\int_0^t \frac{1}{(k-1)!}(t-s)^{k-1}\, g_x(s)ds$. Since the function g_x is unique $x \to H_x(t)$ defines a linear operator from D to X. We show that this operator is bounded. By Lemma 4.6(i) the boundedness of G means $\|G(\phi)\| \leq c\|t^k\phi^{(k)}\|_{L^1}$ for all $\phi \in \mathcal{D}(\mathbb{R}_+)$. Fix some $x \in D$ and some $x' \in X'$. Then

$$\left|\int_0^\infty (-1)^k\phi^{(k)}(t)\ \langle H_x(t), x'\rangle\ t^k\, dt\right| = |\langle G(\phi)x, x'\rangle|$$

$$\leq c\,\|t^k\phi^{(k)}\|_{L^1}\|x\|\,\|x'\| = c\,\|\phi^{(k)}\|_{L^1(t^k dt)}\,\|x\|\,\|x'\|\,.$$

Since by Lemma 4.6(ii) the functions $\phi^{(k)}$ are dense in $L^1(t^k dt)$ the functional $\phi^{(k)} \to \int_0 (-1)^k \phi^{(k)}(t) \langle H_x(t),x'\rangle\, t^k\, dt$ extends to the whole of $L^1(t^k dt)$. Therefore $\langle H_x(\cdot),x'\rangle \in L^\infty$ and $\|\langle H_x(.),x'\rangle\|_\infty \leq c\,\|x\|\,\|x'\|$. Hence $\|H_x(t)\| \leq c\,\|x\|$ for $t > 0$. This shows that $x \to H_x(t)$ is bounded. Let $S(t)$ denote the bounded extension of $x \to t^k H_x(t) = \int_0^t \frac{1}{(k-1)!}\, t^k\, g_x(s)ds$ to the whole space X. Then $\|S(t)\| \leq ct^k$. Since $t \to S(t)$ is strongly continuous on D we obtain a strongly continuous family $(S(t))_{t\geq 0}$ of bounded operators $S(t)$ that fulfills $G(\phi) = \int_0^\infty (-1)^k \phi^{(k)}(t)\, S(t)dt$ for $\phi \in T_k$. Plugging in $\phi(t) = e^{-\lambda t}$ (which is a function in T_k for $\lambda > 0$) we get $(\lambda - A)^{-1} = G(e^{-\lambda t}) = \lambda^k \int_0^\infty e^{-\lambda t}\, S(t)dt$ for $\lambda > 0$. By definition this implies that $(S(t))_{t\geq 0}$ is a k-times integrated semigroup with generator A.

(ii) => (i): Define $G(\phi) := \int_0^\infty (-1)^k \phi^{(k)}(t)\, S(t)dt$. It is not difficult to prove that G is a smooth distribution semigroup of order k (For the functional equation use [2, Th. 3.1]). To see that Def. 4.3 (ii) holds note that for $x \in D(A^k)$ the function $t \to S(t)x$ is k-times continuously differentiable with $S^{(k)}(0)x = x$ by Theorem 1.2 and Theorem 1.8 for $f = 0$ and $\eta = 1_{(0,\infty)}$. Integration by parts yields

$$G(\phi) = \int_0^\infty \phi(t)\, S^{(k)}(t)x\, dt \text{ for } \phi \in \mathcal{D}(\mathbb{R}_+).$$

It remains to show that A coincides with the generator B of G. By definition ([3, Definition 4]) the set $\{G(\phi)x : x \in X,\ \phi \in \mathcal{D}(\mathbb{R}_+)\}$ is a core of B and $BG(\phi)x = G(-\phi')x$. By Lemma 1.5, $\int_0^t S(s)x\, ds \in D(A)$ and

$$A\int_0^t S(s)x\, ds = S(t)x - (t^k/k!)x \text{ for all } x \in X. \text{ Hence}$$

$$G(\phi)x = \int_0^\infty (-1)^{k+1}\phi^{(k+1)}(t) \int_0^t S(s)x\, dsdt \in D(A) \text{ and } AG(\phi)x = G(-\phi')x$$

for all $x \in X$, $\phi \in \mathcal{D}(\mathbb{R}_+)$. This implies $A = B$. □

SECTION 5. INTEGRATED COSINE FUNCTIONS

We assume $\eta(t) = t$ $(t \geq 0)$ throughout this section so that $P(A,\eta)$ is equivalent to

$$P_2(A)\begin{cases} u''(t) = Au(t), \ t \geqq 0 \\ u(0) = x \\ u'(0) = 0 \end{cases}$$

in the sense that u is a classical solution of $P(A,\eta)$ if and only if $u \in C^2([0,\infty),X) \cap C([0,\infty), D(A))$ and u satisfies $P_2(A)$.

Now a solution family of $P_2(A)$ is the same as a cosine function (in fact, one can show, A is the genrator of a cosine function (in the classical sense) if and only if there exist $\omega \geqq 0$ and $M \geqq 0$ and a strongly continuous function $C:[0,\infty) \to L(X)$ satisfying $\|C(t)\| \leq Me^{\omega t}$ $(t \geqq 0)$ such that $(\lambda^2-A)^{-1}$ exists for $\lambda > \omega$ and $\lambda(\lambda^2-A)^{-1} = \int_0^\infty e^{-\lambda t}C(t)dt$. Of course, in that case $(C(t))_{t\geqq 0}$ is the cosine function generated by A.). So in the case considered here we use the term n-times integrated cosine function (n = 0, 1,2,...) for n-times integrated solution families of $P_2(A)$. We also use the term <u>sine function</u> for once-integrated cosine function. For convenience we recollect the precise meaning from Definition 1.1: An operator A is called <u>the generator of an n-times integrated cosine function</u> if $(\omega,\infty) \subset \rho(A)$ (the resolvent set of A) for some $\omega \geqq 0$ and the function $\lambda \to (\lambda^2-A)^{-1}/\lambda^{n-1}$ $(\lambda > \omega)$ is a Laplace transform (i.e. there exists a strongly continuous family $(S(t))_{t\geqq 0}$ of bounded operators in $L(X)$ such that $\|S(t)\| \leq Me^{\omega t}$ $(t \geqq 0)$ for some $M \geqq 0$ and $(\lambda^2-A)^{-1}/\lambda^{n-1} = \int_0^\infty e^{-\lambda t}S(t)dt$). Of course, in that case, $(S(t))_{t\geqq 0}$ is the n-times integrated cosine function generated by A.

<u>THEOREM 5.1</u>. Let $n \in \mathbb{N} \cup \{0\}$. Suppose that A and -A are generators of an n-times integrated semigroup. Then A^2 is the generator of an n-times integrated cosine function.

<u>Proof</u>. There exists $\omega \geqq 0$ such that $(\omega,\infty) \subset \rho(A) \cap \rho(-A)$ and $(\lambda-A)^{-1}/\lambda^n$ and $(\lambda+A)^{-1}/\lambda^n$ are Laplace transforms (in the sense made precise above). So $(\lambda^2-A^2)^{-1} = (\lambda-A)^{-1}(\lambda+A)^{-1}$ exists for $\lambda > \omega$ and by the resolvent equation we have

$$(\lambda^2-A^2)^{-1} = -(\lambda-A)^{-1}(-\lambda-A)^{-1} = \frac{1}{2\lambda}\left((\lambda+A)^{-1} + (\lambda-A)^{-1}\right).$$

So $(\lambda^2-A^2)^{-1}/\lambda^{n-1}$ is a Laplace transform. □

In the following theorem we consider semigroups other than C_0-semigroups.

We say, that a linear operator A is the generator of a semigroup, if there exists a strongly continuous function $T:(0,\infty) \to L(X)$ and $\omega \geq 0$ such that $(\omega,\infty) \subset \rho(A)$, $\int_0^\infty e^{-\lambda t} \|T(t)x\| \, dt < \infty$ for every $x \in X$, $\lambda > \omega$ and $(\lambda-A)^{-1}x = \int_0^\infty e^{-\lambda t}T(t)x \, dt$ $(x \in X, \lambda > \omega)$. Clearly such T is unique and one shows as in [2, Theorem 1.3] that T has the semigroup property $T(s + t) = T(s)T(t)$ $(s,t > 0)$. Thus we call $(T(t))_{t>0}$ the semigroup generated by A. (Of course, then $\int_0^t T(s)ds = S(t)$ defines an integrated semigroup $(S(t))_{t\geq 0}$ whose generator is A).

<u>THEOREM 5.2</u>. Let A be the generator of a sine function $(S(t))_{t\geq 0}$. Then A generates a semigroup $(T(t))_{t>0}$ which is given by

$$T(t)x = 1/(2\sqrt{\pi}\, t^{3/2}) \int_0^\infty se^{-s^2/4t} S(s)x \, ds \quad (t > 0, x \in X) \tag{5.1}$$

where the integral exists in the sense of Bochner).

In particular $(T(t))_{t>0}$ has an analytic extension to $[\mathrm{Re}\, t > 0]$. If in addition D(A) is dense and $\overline{\lim}_{t\to 0} \|S(t)\| / t < \infty$ then A generates a C_0-semigroup.

<u>Proof</u>. Let $\Psi(s,t) := 1/(2\sqrt{\pi}\, t^{3/2})\, se^{-s^2/4t}$ $(s \geq 0, t > 0)$. Then $\int_0^\infty e^{-\lambda t} \Psi(s,t)dt = e^{-s\sqrt{\lambda}}$ $(s > 0, \lambda > 0)$ (see [7, p. 50-51]). There exists $\omega \geq 0$, $M \geq 0$ such that $\|S(t)\| \leq Me^{\omega t}$ $(t \geq 0)$. The argument given in [5, p. 139] shows that $\int_0^\infty se^{-s^2/4t} \|S(s)x\| \, ds < \infty$ for all $t > 0$, $x \in X$. Moreover by Fubini's theorem

$$\int_0^\infty e^{-\lambda t} \|T(t)x\| \, dt \leq \int_0^\infty e^{-\lambda t} \int_0^\infty 1/(2\sqrt{\pi}\, t^{3/2}) se^{-s^2/4t} \|S(s)x\| \, ds \, dt$$

$$= \int_0^\infty e^{-\lambda t} \|S(s)x\| \; 1/(2\sqrt{\pi}\, t^{3/2})\, se^{-s^2/4t} \, dtds$$

$$= \int_0^\infty \|S(s)x\| \, e^{-s\sqrt{\lambda}} \, ds < \infty \quad (\lambda > \omega^2).$$

Hence $\int_0^\infty e^{-\lambda t}\, T(t)x\, dt$ exists for $\lambda > \omega^2$ and (by Fubini's theorem again)

$$\int_0^\infty e^{-\lambda t} T(t)x\, dt = \int_0^\infty e^{-s\sqrt{\lambda}} S(s)x\, ds = (\lambda - A)^{-1}x.$$

To see the additional assertion note that $t \to T(t)$ is strongly continuous at $t = 0$ on $D(A)$ (Observe that for large enough λ we have $T(t)(\lambda-A)^{-1} = \int_0^\infty e^{-\lambda s}\, T(s+t)ds = e^{\lambda t} \int_t^\infty e^{-\lambda s}\, T(s)ds$).
Further $\overline{\lim}_{t\to 0} \|S(t)\| / t < \infty$ implies $\|S(t)\| \leqq cte^{\omega t}$ $(t \geqq 0)$. Since

$$\begin{aligned}
\|T(t)\| &= \|1/(2\sqrt{\pi}\, t^{3/2}) \int_0^\infty se^{-s^2/4t}\, S(s)x\, ds\| \\
&= \|1/(2\sqrt{\pi}\, t^{3/2}) \int_0^\infty 4tve^{-v^2}\, S(2v\sqrt{t})\, dv\| \\
&\leqq 4/(\sqrt{\pi}t) \int_0^\infty c\sqrt{t}ve^{-v^2}\, e^{2\omega v\sqrt{t}}\, dv \\
&\leqq 4/(\sqrt{\pi}) \int_0 cve^{-v^2}\, e^{2\omega v}\, dv \text{ for } 0 < t \leqq 1
\end{aligned}$$

the strong continuity of $(T(t))_{t>0}$ (at $t = 0$) extends to the whole space. □

The preceding theorem shows that generators of sine functions are also of interest in the context of Cauchy problems of first order. In addition to the regularity property described above they have nice properties with respect to perturbations (which is not true for all generators of integrated semigroups, see the example in [9, Section 5]). The following perturbation theorem will illustrate this. First we prepare some notation. We denote by $(S(t))_{t\geqq 0}$ the sine function generated by the (unperturbed) operator A and by $(S^B(t))_{t\geqq 0}$ the sine function generated by $A + B$.

Since we shall also make a statement about the spaces where the sine functions are differentiable we use the abbreviations

$$C^n := \{x \in X : t \to S(t)x \text{ is a } C^n\text{-function}\} \quad \text{and}$$

$$C^n_B := \{x \in X : t \to S^B(t)x \text{ is a } C^n\text{-function}\}.$$

<u>THEOREM 5.3.</u> Let A generate a sine function and $B \in L(\overline{D(A)}, X)$. Then

$A + B$ generates a sine function, too. Furthermore $C_B^n = C^n$ for $n = 1,2$.

Proof. Let $(S(t))_{t\geq 0}$ be the sine function generated by A. For some positive constants M and ω we have $\|S(t)\| \leq Me^{\omega t}$ for $t \geq 0$. Consider now the integral equation on $\mathbb{R}_+$

$$S^B(t) = S(t) + \int_0^t S^B(s)BS(t-s)\,ds. \tag{5.2}$$

We show first that the term $BS(t-s)$ in (5.2) is well defined. By Lemma 1.5 we have $\int_0^t \int_0^s S(r)x\,drds \in D(A)$ for all $x \in X$ and $t \geq 0$. Differentiating this twice we obtain $S(t)x \in \overline{D(A)}$.

Multiplying (5.2) by $e^{-(\omega+\alpha)t}$ we obtain the equivalent integral equation

$$U(t) = e^{-(\omega+\alpha)t}S(t) + \int_0^t U(s)Be^{-(\omega+\alpha)(t-s)}S(t-s)ds. \tag{5.3}$$

Consider now the Banach space

$$E := \{V:\mathbb{R}_+ \to L(X) : V \text{ is strongly continuous and bounded}$$

(equipped with the norm $|V| := \sup_{t\geq 0} \|V(t)\|$) and fix some $\alpha > M\|B\|$. It is not difficult to verify that $(JV)(t) := \int_0^t V(s)Be^{-(\omega+\alpha)(t-s)}S(t-s)ds$ defines a bounded linear operator J on E with $\|J\|_{L(E)} \leq \frac{M\|B\|}{\alpha} < 1$. Write now the integral equation (5.3) in the form

$$(I - J)U = (e^{-(\omega+\alpha)t}S(t))_{t\geq 0}. \tag{5.4}$$

Obviously the operator function on the right hand side is in E and the operator $I - J$ is invertible. This shows that we have a unique strongly continuous and bounded solution $(U(t))_{t\geq 0}$ of (5.3). Therefore $(S^B(t))_{t\geq 0} := (e^{(\omega+\alpha)t}U(t))_{t\geq 0}$ solves (5.2) and fulfills $\|S^B(t)\| \leq M'e^{(\omega+\alpha)t}$ for $t \geq 0$.

Let $\lambda > \omega + \alpha$. With the abbreviations $Q(\lambda) := \int_0^\infty e^{-\lambda t}S^B(t)dt$ and $R(\lambda^2) := (\lambda^2 - A)^{-1}$ we have

$$
\begin{aligned}
Q(\lambda) - R(\lambda^2) &= \int_0^\infty e^{-\lambda t} (S^B(t) - S(t))\, dt \\
&= \int_0^\infty e^{-\lambda t} \int_0^t S^B(s)BS(t-s)\, dsdt \\
&= \int_0^\infty \int_0^t e^{-\lambda s} S^B(s)B\, e^{-\lambda(t-s)} S(t-s) dsdt \\
&= \int_0^\infty \int_s^\infty e^{-\lambda s} S^B(s)B\, e^{-\lambda(t-s)} S(t-s) dtds \\
&= \int_0^\infty e^{-\lambda s} S^B(s)B \int_0^\infty e^{-\lambda u} S(u) duds \\
&= Q(\lambda)BR(\lambda^2).
\end{aligned}
$$

This shows $Q(\lambda)(I - BR(\lambda^2)) = R(\lambda^2)$. The operator $I - BR(\lambda^2)$ is invertible since $\|BR(\lambda^2)\| \leq \|B\|\, \|\int_0^\infty e^{-\lambda t} S_t dt\| \leq \frac{\|B\|M}{\lambda - \omega} < 1$. A simple calculation shows now that $Q(\lambda) = R(\lambda^2)(I - BR(\lambda^2))^{-1}$ is the inverse of λ^2-A-B. By definition this implies that $(S^B(t))_{t \geqq 0}$ is a sine function with generator A + B.

It remains to show the additional assertion. The inclusions $C^n \subset C^n_B$ for $n = 1,2$ can be verified by simply differentiating (5.2). Since $-B \in L(\overline{D(A + B)}, X)$ we also have $S(t) = S^B(t) - \int_0^t S(s)BS^B(t-s)ds$ which shows that the converse inclusions hold, too. □

SECTION 6. APPLICATIONS

I. Differential Operators on $\mathbb{R}^n$ as Generators of Integrated Semigroups

We consider elliptic differential operators with constant coefficients $A = \Sigma_{|\alpha| \leq m} a_\alpha D^\alpha$ on $L^p(\mathbb{R}^n)$, $1 \leqq p \leqq \infty$, $C_b(\mathbb{R}^n)$, $UC_b(\mathbb{R}^n)$ (= the space of bounded, uniformly continuous functions) and $C_o(\mathbb{R}^n)$.

Such an operator A is called elliptic if the polynomial $\Sigma_{|\alpha| \leqq m} a_\alpha i^{|\alpha|} x^\alpha$ is ellitpic (i.e. $\Sigma_{|\alpha| = m} a_\alpha x_1^{\alpha_1} \dots x_n^{\alpha_n} = 0 \Rightarrow x_1 = \dots = x_n = 0$).

It is known that A (with some suitable domain) generates an analytic C_o-semigroup on $L^p(\mathbb{R}^n)$, $1 \leqq p < \infty$, if m is even and $\operatorname{Re} \Sigma_{|\alpha| = m} a_\alpha i^{|\alpha|} x^\alpha < 0$ for $x \neq 0$ (see e.g. [8, Chapter 2, Theorem 12.7]).

On the other hand if $\mathrm{Re}\,\Sigma_{|\alpha|\leq m}\, a_\alpha\, i^{|\alpha|}x^\alpha = 0$ for $x \in \mathbf{R}^n$ and $m \geq 2$ then it follows from [10, Theorem 1.14] that A does not generate a semigroup on $L^p(\mathbf{R}^n)$ unless $p = 2$.

However M. Balabane and H. Emami-Rad showed that in this case A and -A (with suitable domain) generate smooth distribution semigroups of order k where $k > \frac{n}{2}$ on $L^p(\mathbf{R}^n)$, $1 < p < \infty$. By Theorem 4.4 this is equivalent to saying that A and -A generate a k-times integrated semigroup $(S(t))_{t\geq 0}$ on $L^p(\mathbf{R}^n)$, $1 < p < \infty$, with $\|S(t)\| \leq ct^k$ for $t \geq 0$ and with $k > \frac{n}{2}$.

The following theorem deals with a more general situation. We will give a direct proof which remains completely within the framework of integrated semigroups. Our proof (like the proof of the Balabane-Emami-Rad result) is based on a lemma on Fourier multipliers due to Mikhlin. We are indebted to Mikhael Balabane who demonstrated to us how one can - by sophisticated enough techniques - eliminate problems, so to say, with a twinkle of the eye.

THEOREM 6.1. Let A be elliptic on $L^p(\mathbf{R}^n)$, $1 < p < \infty$, and suppose

$$\sup_{x\in\mathbf{R}^n} \mathrm{Re}\,\Sigma_{|\alpha|\leq m}\, a_\alpha i^{|\alpha|}x^\alpha < \infty\,. \tag{6.1}$$

Then $A := \overline{A_{|S}}$ (where S denotes the space of rapidly decreasing functions) generates a k-times integrated semigroup for $k > \frac{n}{2}$.

We state the lemma mentioned above in the following form.

LEMMA 6.2 ([16, p. 96]). Let $k > \frac{n}{2}$ and define

$$E := \{u:\mathbf{R}^n \to \mathbb{C} : u \text{ is a } C^k\text{-function and } |x|^{|\alpha|}|D^\alpha u(x)| \leq c$$

$$\text{for } x \in \mathbf{R}^n, |\alpha| \leq k \text{ and some constant } c \geq 0\}.$$

If E is equipped with the norm $\|u\|_E := \Sigma_{|\alpha|\leq k}\, \|\,|x|^{|\alpha|}D^\alpha u\|_\infty$ then $u \to T_u$, where $T_u f := F^{-1}(uFf)$, defines a bounded operator from E to M_p for $1 < p < \infty$.

Here F means the Fourier transform and $M_p \subset L(L^p(\mathbf{R}^n))$ is the set of bounded L^p-multipliers. For the theory of multipliers we refer to [10] and [16].

PROOF OF THEOREM 6.1. Let us abbreviate $p(x) := \Sigma_{|\alpha|\leq m} a_\alpha i^{|\alpha|} x^\alpha$. We show that the functions

$$u_t := \frac{1}{(k-1)!} \int_0^t (t-s)^{k-1} e^{sp}\, ds \tag{6.2}$$

give, by $S(t) := T_{u_t}$, an exponentially bounded k-times integrated semigroup $(S(t))_{t\geq 0}$.

Let w be a positive constant such that $\operatorname{Re} p(x) \leq w$ for $x \in \mathbb{R}^n$. We show now that $u_t \in E$ for $t \geq 0$ and $\|u_t\|_E \leq Me^{\omega t}$ for some M and $\omega \geq 0$. Ellipticity of p implies that for some constants c, $L \geq 0$

$$|p(x)| \geq c|x|^m \quad \text{for } |x| > L. \tag{6.3}$$

We divide now $\mathbb{R}^n$ in two parts, the closed ball with radius L centered at the origin and the rest.

For $|x| \leq L$ we obviously have

$$|x|^{|\alpha|}|D^\alpha u_t(x)| \leq M'e^{\omega' t} \text{ for } |\alpha| \leq k \text{ and } t \geq 0 \text{ (} M' \text{ and } \omega' \text{ suitable).}$$

Let, therefore $|x| > L$. Carrying out the integration in (6.2) one obtains

$$u_t = e^{tp}/p^k - \Sigma_{j=1}^k \frac{1}{(k-j)!} t^{k-j}/p^j. \tag{6.4}$$

By induction one shows the following.

$D^\alpha(e^{tp}/p^k) = e^{tp}\, q/p^r$ where $r \in \mathbb{N}$ and q is some polynomial of $\operatorname{ord}(q) \leq rm - (k - |\alpha|)m - |\alpha|$. (6.5)

For any $j \in \mathbb{N}$ we have $D^\alpha(1/p^j) = q/p^r$ where $r \in \mathbb{N}$ and q is some polynomial of $\operatorname{ord}(q) \leq rm - jm - |\alpha|$. (6.6)

Using this in combination with (6.3) and $|q(x)| \leq c|x|^{\operatorname{ord}(q)}$ one easily shows that $|x|^{|\alpha|}|D^\alpha u_t(x)| \leq M''e^{\omega'' t}$ for $|\alpha| \leq k$, $|x| \geq L$ and $t \geq 0$ (M'', ω'' suitable).

Lemma 6.2 implies now that the multipliers T_u are bounded and fulfill

$$\|T_{u_t}\| \leq Ce^{\omega t} \text{ for } t \geq 0 \text{ (and constants } C,\ \omega \geq w). \tag{6.7}$$

Next we prove that for $\lambda > \omega$ the operator $\lambda - A$ is invertible with inverse $T_{r(\lambda)}$, where $r(\lambda) := \frac{1}{\lambda - p}$. The fact that $T_{r(\lambda)}$ is a bounded multiplier can be seen by means of (6.6), just as before. On S we further have $(\lambda-A)T_{r(\lambda)} = T_{r(\lambda)}(\lambda-A) = I$. Since $T_{r(\lambda)}$ is bounded and A is the closure of $A|_S$ this implies $T_{r(\lambda)} = (\lambda-A)^{-1}$.

To see strong continuity of the family $(S(t))_{t\geq 0}$ observe first that we have

$$u_t = p \int_0^t u_s \, ds + t^k/k!. \tag{6.8}$$

We fix now some $\mu > \omega$ and obtain from (6.8)

$$u_t r(\mu) = (\mu r(\mu) - 1) \int_0^t u_s \, ds + (t^k/k!)r(\mu). \tag{6.9}$$

Recalling that $\|u_t\|_E \leq Me^{\omega t}$ for $t \geq 0$ we see that the right hand side of (6.9) is continuous with respect to $\| \ \|_E$. By Lemma 6.2 we obtain that $T_{u_t r(\mu)} = T_{u_t} T_{r(\mu)} = T_{u_t}(\mu-A)^{-1}$ is continuous with respect to the M_p-norm. This implies in particular that $t \to T_{u_t} = S(t)$ is strongly continuous on the dense subspace $D(A)$. By the exponential boundedness of the family $(S(t))_{t\geq 0}$ we obtain strong continuity on the whole space.

The only thing that remains to show is $(\lambda-A)^{-1} = \lambda^k \int_0^\infty e^{-\lambda t} S(t)dt$ for $\lambda > \omega$. Obviously $r(\lambda) = \lambda^k \int_0^\infty e^{-\lambda t} u_t \, dt$ for $\lambda > \omega$. Fixing again some $\mu > \omega$ we obtain $r(\lambda)r(\mu) = \lambda^k \int_0^\infty e^{-\lambda t} u_t r(\mu)dt$. As we just showed $t \to u_t r(\mu)$ is $\| \ \|_E$-continuous. Therefore by Lemma 6.2 we have

$$T_{r(\lambda)} T_{r(\mu)} = \int_0^\infty e^{-\lambda t} T_{u_t r(\mu)} \, dt = \int_0^\infty e^{-\lambda t} T_{u_t} \, dt \; T_{r(\mu)}$$

or $(\lambda-A)^{-1} = \int_0^\infty e^{-\lambda t} S_t dt$ on $D(A)$. Boundedness of both sides implies the same on the whole space. □

The situation looks different on L^1 and L^∞, since Lemma 6.2 does not apply here anymore. But the following technique can be used. One shows that (for some large enough $k \in \mathbb{N}$) the functions

$$F^{-1}u_t = F^{-1} \frac{1}{(k-1)!} \int_0^t (t - s)^{k-1} e^{sp} \, ds$$

are in L^1 and that the family of convolution operators $(S(t))_{t \geq 0}$, given by $S(t)f := (\frac{1}{2\pi})^{n/2} F^{-1}u_t * f$ is a k-times integrated semigroup with generator A. One way to show that the functions $F^{-1}u_t$ are in L^1 is to show that the functions m_t are in $H^{k,2}$ (= the k-th Sobolev space in L^2) for $k > \frac{n}{2}$ and then to use the fact that F^{-1} maps $H^{k,2}$ boundedly into L^1. This procedure demands a severe restriction on the polynomial p, namely $\mathrm{ord}(p) > \frac{n}{2}$. The result is then

<u>THEOREM 6.3 ([10, Part II, Theorem 1.9])</u>. Let $p(x) := \Sigma_{|\alpha| \leq m} a_\alpha \, i^{|\alpha|} x^\alpha$ be an elliptic polynomial of $\mathrm{ord}(p) > \frac{n}{2}$ such that $\sup_{x \in \mathbb{R}^n} \mathrm{Re}\, p(x) < \infty$. Then for $k > \frac{n}{2} + 1$ the operator $A = \Sigma_{|\alpha| \leq m} a_\alpha D^\alpha$ (with distributional domain) generates a norm continuous k-times integrated semigroup on the spaces $L^1(\mathbb{R}^n)$, $L^\infty(\mathbb{R}^n)$, $C_b(\mathbb{R}^n)$, $UC_b(\mathbb{R}^n)$ and $C_o(\mathbb{R}^n)$.

A better result is possible for $n = 1$.

<u>THEOREM 6.4 ([10, Part II, Theorem 1.1])</u>. Suppose that $\sup_{x \in \mathbb{R}} \mathrm{Re}\, \Sigma_{j=0}^m a_j (ix)^j < \infty$. Then $A = \Sigma_{j=0}^m a_j (\frac{d}{dx})^j$ (with distributional domain) generates a once integrated semigroup on the spaces $L^1(\mathbb{R})$, $L^\infty(\mathbb{R})$, $C_b(\mathbb{R})$, $UC_b(\mathbb{R})$ and $C_o(\mathbb{R})$.

II. <u>Differential Operators on $\mathbb{R}^n$ as Generators of Integrated Cosine Functions</u>

As in 6.I we consider elliptic differential operators $A = \Sigma_{|\alpha| \leq m} a_\alpha D^\alpha$ on $L^p(\mathbb{R}^n)$, $1 \leq p \leq \infty$, $C_b(\mathbb{R}^n)$, $UC_b(\mathbb{R}^n)$ and $C_o(\mathbb{R}^n)$. We study the question under what conditions they generate k-times integrated cosine functions. Since the techniques are essentially the same as in I. We do not go so much into detail, but only show the various points where the proofs have to be changed to yield results for this setting here. Again we treat the reflexive L^p-spaces separately and we obtain

THEOREM 6.5. Let A be elliptic on $L^p(\mathbb{R}^n)$, $1 < p < \infty$, and suppose that $\Sigma_{|\alpha|\leq m} a_\alpha i^{|\alpha|} x^\alpha$ is real valued and bounded above.
Then $A := \overline{A|_S}$ generates a k-times integrated cosine function for $k > \frac{n}{2}$.

Proof. Define $p(x) := \Sigma_{|\alpha|\leq m} a_\alpha i^{|\alpha|} x^\alpha$. The proof consists in showing that the functions

$$u_t := \frac{1}{(k-1)!} \int_0^t (t-s)^{k-1} \cosh(s\sqrt{p})ds \tag{6.10}$$

give, by $S(t) := T_{u_t}$, an exponentially bounded family of L^p-multipliers $(S(t))_{t\geq 0}$. The rest is more or less a copy of the proof of Theorem 6.1. As in the proof of Theorem 6.1 we divide $\mathbb{R}^n$ in the ball $\{x \in \mathbb{R}^n : |x| \leq L\}$ and the rest. Since p is a polynomial with $p(x) \leq c^2$ $(x \in \mathbb{R}^n)$ for some $c \geq 0$, we have

$$\mathrm{Re}\,\sqrt{p(x)} \leq c \text{ for } x \in \mathbb{R}^n. \tag{6.11}$$

By (6.11) we have $|x|^{|\alpha|}|D^\alpha u_t(x)| \leq Me^{\omega t}$ for $|x| \leq L$, $|\alpha| \leq k$ and $t \geq 0$.
For $|x| > L$ we write (6.10) in the form

$$u_t = \begin{cases} \sinh(t\sqrt{p})/\sqrt{p} & \text{for } k = 1, \\ \sinh(t\sqrt{p})/p^{k/2} - \Sigma_{i=1}^{(k-1)/2} \frac{1}{(k-2i)!} t^{k-2i}/p^i, & k \geq 3, \text{ odd} \\ \cosh(t\sqrt{p})/p^{k/2} - \Sigma_{i=1}^{(k-1)/2} \frac{1}{(k-2i)!} t^{k-2i}/p^i, & k \text{ even.} \end{cases} \tag{6.12}$$

We ignore now the difference between the functions cosh and sinh since we are only interested in estimating the right hand side of (6.12) and we denote both functions by the same symbol U.
By induction one shows then

$D^\alpha(U(t\sqrt{p})/p^{k/2}) = U(t\sqrt{p})q/p^{r/2}$ where $r \in \mathbb{N}$ and q is some polynomial of order $\mathrm{ord}(q) \leq \frac{rm}{2} - (k - |\alpha|)\frac{m}{2} - |\alpha|$.

Together with (6.11) and (6.6) this implies

$|x|^{|\alpha|}|D^\alpha u_t(x)| \leqq M'e^{\omega' t}$ for $|\alpha| \leqq k$, $|x| \geqq L$ and $t \geqq 0$ (M' and ω' suitable). Lemma 6.2 implies now that the multipliers T_{u_t} are bounded and fulfill $\|T_{u_t}\| \leqq Ce^{\omega'' t}$ for $t \geqq 0$ (and suitable constants C and ω''). If one replaces now (6.8) by $u_t = p\int_0^t (t - s)u_s\, ds + t^k/k!$ the rest consists in copying the proof of Theorem 6.1 with minor changes. □

We turn now to the cases L^1 and L^∞ applying the same technique as in 6.I. Again the restriction $\mathrm{ord}(p) > \frac{n}{2}$ comes up and we obtain

THEOREM 6.6 Let $p(x) := \Sigma_{|\alpha|\leq m}\, a_\alpha\, i^{|\alpha|} x^\alpha$ be a real valued polynomial of order $> \frac{n}{2}$ which is elliptic and bounded above.

Then for $k > \frac{n}{2} + 2$ the operator $A = \Sigma_{|\alpha|\leq m}\, a_\alpha D^\alpha$ (with distributional domain) generates a norm continuous k-times integrated cosine function on the spaces $L^1(\mathbb{R}^n)$, $L^\infty(\mathbb{R}^n)$, $C_b(\mathbb{R}^n)$, $UC_b(\mathbb{R}^n)$ and $C_o(\mathbb{R}^n)$.

For the case $N = 1$ we can again improve our result.

THEOREM 6.7. Suppose that the polynomial $\Sigma_{j=0}^m\, a_j(ix)^j$ is real valued and bounded above.
Then $A = \Sigma_{j=0}^m\, a_j(\frac{d}{dx})^j$ (with distributional domain) generates a norm continuous sine function on the spaces $L^1(\mathbb{R})$, $L^\infty(\mathbb{R})$, $C_b(\mathbb{R})$, $UC_b(\mathbb{R})$ and $C_o(\mathbb{R})$.

References

[1] W. Arendt; Resolvent Positive Operators. Proc. London Math. Soc. (3) 54 (1987) 321-349.

[2] W. Arendt; Vector Laplace Transforms and Cauchy Problems. Israel J. of Math. 59 (1987), 327-352.

[3] M. Balabane, H. Emami-Rad; Smooth Distribution Semigroup and Schrödinger Equation in $L^p(\mathbb{R}^n)$. J. Math. Anal. Appl. 70 (1) (1979), 61-71.

[4] M. Balabane, H. Emami-Rad; Pseudo Differential Parabolic Systems in $L^p(\mathbb{R}^n)$. in: Contributions to Nonlinear Partial Differential Equations, Pitman (1983).

[5] G. Da Prato, M. Iannelli; Linear Integro-Differential Equations in Banach Spaces. Rend. Sem. Mat. Univ. Padova 62 (1980), 207-219.

[6] G. Da Prato, E. Sinestrari; Differential Operators with non dense Domain. Ann. Scuola Norm. Sup. Pisa (1986).

[7] G. Doetsch; Handbuch der Laplace Transformation. Birkhäuser (1971).

[8] J. Goldstein; Semigroups of Linear Operators and Applications. Oxford Press (1985).

[9] M. Hieber, H. Kellermann; Integrated Semigroups. J. Funct. Analysis, to appear.

[10] L. Hörmander; Estimates for translation invariant operators in L^p spaces. Acta Math. 104 (1960), 93-139.

[11] J.L. Lions; Semi-groupes distributions. Portugalae Math. 19 (1960), 141-164.

[12] R. Nagel (Ed.); One-parameter Semigroups of Positive Operators. Lecture Notes Math. 1184 Springer Berlin (1986).

[13] F. Neubrander; Integrated Semigroups and their Applications to the Abstract Cauchy Problem. Pacific J. Math. to appear.

[14] J. Prüss; Lineare Volterra-Gleichungen in Banachräumen. Habilitationsschrift, Paderborn (1984).

[15] J. Prüss; Positivity and Regularity of Hyperbolic Volterra Equations in Banach Spaces. Math. Ann. 279 (1987) 317-344.

[16] E. Stein; Singular Integrals and Differentiability Properties of Functions. Princeton University Press (1970).

Wolfgang Arendt
Equipe de Mathématiques
Université de Franche Comté
25030 Besançon Cedex
France

Hermann Kellermann
Mathematisches Institut
Auf der Morgenstelle 10
7400 Tübingen
West Germany.

M. CHIPOT & G. VERGARA-CAFFARELLI

Viscoelasticity without initial condition

I. INTRODUCTION

If Ω is a smooth, bounded open set of $\underline{R}^n$ we would like to consider the following model problem:

Find $u : \Omega \times (-\infty,0] \to \underline{R}$ such that:

$$\rho u_{tt} - G(0)\Delta u - \int_{-\infty}^{t} \dot{G}(t-\tau)\Delta u(x,\tau)d\tau = f \text{ in } \Omega \times (-\infty,0] \qquad (P)$$

$$u(x,t) = 0 \text{ on } \partial\Omega,\ t \in (-\infty,0].$$

For the time being, we assume that u is of class C^2, G is positive and of class C^1 on $[0, +\infty)$, $\dot{G}$ denotes the derivative of G, and that the integral makes sense. ρ is a positive constant.

In the case $n = 1$, $\Omega = (0,\ell)$, $\ell > 0$, u represents the displacement of 1-dimensional viscoelastic material of denstiy ρ, whose reference configuration is Ω, and which is subject to the action of a force f. (See for instance [5], [11], [13]).

In the case where the past history of the material is known say, up to time t_0, $t_0 < 0$, then we can absorb the term

$$\int_{-\infty}^{t_0} \dot{G}(t-\tau)\Delta u(x,\tau)d\tau$$

into f and rewrite the problem as:

$$\rho u_{tt} - G(0)\Delta u - \int_{t_0}^{t} \dot{G}(t-\tau)\Delta u(x,\tau)d\tau = f \quad \text{in } \Omega \times (t_0,0] \qquad (V)$$

$$u(x,t) = 0 \text{ on } \partial\Omega,\ t \in (t_0,0].$$

This problem together with the initial conditions:

$$u(x,t_0) = u_0(x), \quad u_t(x,t_0) = v_0(x) \tag{1}$$

has been considered by a large number of authors (see for instance [6], [7] and [8], [13], [14], [15], [16], [17] for his nonlinear version).

In the problem (P) the behaviour of u(x,t) for t tending to $-\infty$ plays an essential role.

For instance, if we consider (P) in a space of functions with bounded support, i.e. if we assume that the external force f(x,t) is identically equal to zero for $t \leq t_0$ and if we look for solutions which are also identically equal to zero for $t \leq t_0$, then we obtain the Volterra integro-differential equation (V) in which the past history of the material is known up to time t_0, $t_0 < 0$.

On the other hand, if we are interested in investigating how the fading memory assumptions on G are connected with the solvability of the dynamical problem of linear viscoelasticity we need to consider the problem (P).

In [9], [10], [11], [12] Fichera considered the quasi-static version of (P), i.e. the case where $\rho = 0$, arguing with more general operators than the Laplacian. In [12] he pointed out that the problem regarding the fading memory of a viscoelastic material has meaning only if referred to a fixed functional space. More precisely he stated:

"Come conseguenza dell'analisi svolta, si può dire che il concetto di evanescenza della memoria di un corpo elastico assume senso solo se riferita ad un prefissato spazio funzionale, dotdato di una struttura topologica, nel quale si conviene di studiare il problema.

Ma quali sono i criteri fisici che conducono a scegliere tale spazio funzionale e ad introdurvi una struttura topologica?"

So, it is clear, when considering viscoelastic materials without initial condition, that the most important thing is to find a physically meaningful space of functions in which the problem (P) is well posed. In this direction G. Capriz and G. Vergara-Caffarelli [1] proved an existence result of a weak solution to (P) in a peculiar Sobolev space. At the same time G. Capriz and E. Virga [2] and later G. Vergara-Caffarelli and E. Virga [18] obtained, in the case n = 1, some conditions for uniqueness of a solution to (P).

Our goal, in this note is to introduce a suitable class of functions in which (P) is well posed, and with a different approach and in an unified way, to prove existence, uniqueness and some smoothness of the solution.

For simplicity we have considered only the case of the Laplacian and the reader is referred to [4] for more general results including in particular the case of two different families of operators, i.e. in front of G(0) and under the integral.

II. SOME OBSTRUCTIONS

Let us first explain why some assumptions are in order if we want to have uniqueness of a solution to (P). First, from an heuristic point of view, one should note that having to deal with an integral starting at $-\infty$ makes a lot of difference. For instance Gronwall Inequality fails if one bound of the integral is $-\infty$. i.e.

$$0 \leqq \alpha(t) \leqq \int_{-\infty}^{t} \alpha(s)ds$$

does not imply $\alpha = 0$.

Now, for B,α positive let us, for example, consider the following kernel:

$$G(s) = B(\alpha - se^{-\nu s})$$

(This kernel was used by Fichera [12] in the quasi-static case (see also [3])). We assume $\nu > 0$, $\alpha\nu > 1/e$ in such a way that $G(s) > 0$ for $s \in [0,\infty)$, and $\dot{G} \in L^1(0,+\infty)$).

Assume that f is equal to 0. Then, clearly 0 is a solution to (P). Now, look for a solution of the type:

$$u(x,t) = e^{\sigma t}v(x)$$

where v is an eigenfunction of the Laplace operator i.e. satisfies:

$$-\Delta v = \lambda v \quad (\lambda > 0)$$

and σ is a complex number.

Then u is a solution to (P) if and only if:

$$\rho\sigma^2 e^{\sigma t} v + B\alpha e^{\sigma t}\lambda v + \int_{-\infty}^{t} \dot{G}(t-\tau)e^{-\sigma(t-\tau)} e^{\sigma t}\lambda v \, d\tau = 0.$$

So in particular if:

$$\rho\sigma^2 + B\alpha\lambda + \lambda\int_{-\infty}^{t} \dot{G}(t-\tau)e^{-\sigma(t-\tau)}d\tau = 0$$

Replacing G by its value and performing some integrations we get that u is a solution to (P) if

$$\rho\sigma^2(\nu + \sigma)^2 + B\lambda\alpha(\nu + \sigma)^2 - B\lambda\sigma = 0$$

which, after dividing by $B\lambda$ gives

$$\rho\sigma^2(\nu + \sigma)^2/B\lambda + \alpha(\nu + \sigma)^2 - \sigma = 0.$$

Now if we let $B\lambda/\rho$ go to $+\infty$, two of the roots of this polynomial are going to converge toward the roots of the equation:

$$\alpha(\nu + \sigma)^2 - \sigma = 0 \iff \alpha\sigma^2 + (2\alpha\nu - 1)\sigma + \alpha\nu^2 = 0.$$

But if we select α, ν in such a way that $1/e < \alpha\nu < 1/2$, then clearly such equation has its complex roots with a positive real part. So, for $B\lambda/\rho$ large enough it is possible to find a solution to (P) of the type

$$u(x,t) = e^{\sigma t}v(x)$$

with $\text{Re}(\sigma) > 0$. This solution is decaying exponentially at $-\infty$ and is a <u>non-zero</u> solution to (P). (To have a real solution of (P) it is enough to consider the real part of u). So, in order to get uniqueness for a solution to (P) in a space of functions which decay exponentially at $-\infty$ we need to impose some condition on G. Note that the decay of the solution somehow replaces the initial condition (1).

Of course, the decay here has to depend on the kernel G (see the above example). More precisely we are going to look for a u which decays like $e^{\nu t/2}$ where ν and G are related by the following fundamental assumption:

$$\frac{G(0)}{2} > 2\int_0^{+\infty}|\dot{G}(s)|e^{-\frac{\nu}{2}s}\,ds + \frac{|\dot{G}(0)|}{\nu} + \frac{1}{\nu}\int_0^{+\infty}|\ddot{G}(s)|e^{-\frac{\nu}{2}s}\,ds \qquad (G)$$

We suppose here and in what follows that G is of class C^2, although all our results hold for $\dot{G} \in W^{1,1}((0,+\infty))$.

REMARK. Note that if G is a C^2 function such that:

$$\dot{G},\ \ddot{G} \in L^1((0,+\infty))$$

then the above inequality holds for ν large enough (by Lebesgue Theorem).

In order to control the decay of u we introduce the following spaces:

DEFINITION. If V is a Banach space normed by $|\ \ |_V$, we denote by $L^{\infty,\nu}(V)$ the set of bounded, measurable functions u defined on $(-\infty,0)$ with values in V such that for some constant C one has:

$$|v(t)|_V \leqq C\,e^{\frac{\nu}{2}t} \qquad \text{a.e. } t \in (-\infty,0).$$

Moreover, we define on $L^{\infty,\nu}(V)$ the norm $|\ \ |_{\infty,\nu}$ by:

$$|v|_{\infty,\nu} = \operatorname*{ess\text{-}sup}_{t\in(-\infty,0)} \{|v(t)|_V\, e^{-\frac{\nu}{2}t}\}.$$

It is easy to check that $L^{\infty,\nu}(V)$ is a Banach space for the above norm.

III. EXISTENCE AND UNIQUENESS

In this section we would like to show that the problem (P) is well posed in the framework of the space $L^{\infty,\nu}((L^2(\Omega))$. More precisely, we would like to show that it has a unique weak solution. Assuming the data are smooth, we will see that this solution is also a strong one. Let us begin by the following definition:

DEFINITION. For f in $L^{\infty,\nu}((L^2(\Omega))$, we will say that u is a weak solution to (P) if

$$u \in L^{\infty,\nu}(L^2(\Omega))$$

$$\int_{-\infty}^{0}\int_{\Omega}\{\rho u\varphi_{tt} - G(0)u\Delta\varphi - \int_{-\infty}^{t}\dot{G}(t-\tau)u(x,\tau)d\tau\Delta\varphi\}\ dxdt = \int_{-\infty}^{0}\int_{\Omega} f\varphi dxdt \quad (W)$$

for all $\varphi \in C_c^{\infty}((-\infty,0),H^2(\Omega) \cap H_0^1(\Omega))$.

($C_c^{\infty}((-\infty,0),H^2(\Omega) \cap H_0^1(\Omega))$ denotes the set of C^{∞} functions with compact support in $(-\infty,0)$ and with values in $H^2(\Omega) \cap H_0^1(\Omega)$).

REMARK. Note that we do not impose any boundary condition on u but, as we will see, the set of test functions that we are using will force u to be equal to 0 on the boundary of Ω. Note also that the integral

$$\int_{-\infty}^{t}\dot{G}(t-\tau)v(x,\tau)d\tau$$

makes sense, and in fact is in $L^{\infty,\nu}(L^2(\Omega))$, as soon as $v(.,t)$ is in $L^{\infty,\nu}(L^2(\Omega))$. If $|\ |_2$ denotes the usual norm on $L^2(\Omega)$, this is indeed an immediate consequence of the following inequality:

$$\left|\int_{-\infty}^{t}\dot{G}(t-\tau)v(.,\tau)d\tau\right|_2 \leq \int_{-\infty}^{t}|\dot{G}(t-\tau)|\ |v(.,\tau)|_2\ d\tau \leq$$

$$|v|_{\infty,\nu}\int_{-\infty}^{t}\dot{G}(t-\tau)|e^{\frac{\nu}{2}\tau}\ d\tau \leq |v|_{\infty,\nu}\ e^{\frac{\nu}{2}t}\int_{-\infty}^{t}|\dot{G}(t-\tau)|e^{\frac{\nu}{2}(t-\tau)}\ d\tau \leq$$

$$|v|_{\infty,\nu}\ e^{\frac{\nu}{2}t}\int_{0}^{+\infty}|\dot{G}(s)|e^{-\frac{\nu}{2}s}\ ds$$

(This last integral is finite, by (G). $|\ |_{\infty,\nu}$ denotes the norm in $L^{\infty,\nu}(L^2(\Omega))$).

THEOREM 1: Assume that $f \in L^{\infty,\nu}(L^2(\Omega))$ and that G is a C^2 function satisfying (G), then there exists a unique weak solution u to (P). Moreover

$u \in L^{\infty,\nu}(H_0^1(\Omega)) \cap W^{1,\infty}((-\infty,0), L^2(\Omega))$.

REMARK. We denote by $W^{k,p}((-\infty,0),V)$ the usual Sobolev spaces of functions defined on $(-\infty,0)$, with values in V, whose derivatives exist up to the order k almost everywhere and are in $L^p((-\infty,0),V)$.

Before going into the proof of this theorem let us introduce the following definition:

DEFINITION. For f in $L^{\infty,\nu}(L^2(\Omega))$, we will say that u is a strong solution to (P) if

$$u \in L^{\infty}((-\infty,0),H^2(\Omega)) \cap W^{1,\infty}((-\infty,0),H_0^1(\Omega)) \cap W^{2,\infty}((-\infty,0),L^2(\Omega))$$

$$\Delta u \in L^{\infty,\nu}(L^2(\Omega))$$

$$\rho u_{tt} - G(0)\Delta u - \int_{-\infty}^{t} \dot{G}(t-\tau)\, \Delta u(x,\tau)d\tau = f \quad \text{a.e. in } \Omega \times (-\infty,0).$$

For a strong solution we have the following useful energy estimate:

LEMMA 1. Assume that $f \in L^{\infty,\nu}(L^2(\Omega))$, u is a strong solution to (P), and that (G) holds. Set

$$E(t) = \frac{1}{2} \{\rho|u_t(t)|_2^2 + G(0) \; \||\nabla u(t)|\|_2^2\}.$$

If for some constant K one has:

$$E(t) \leq Ke^{\nu t} \quad \text{a.e.} \tag{2}$$

then there exists a constant C such that:

$$E(t) \leq C(|f|_{\infty,\nu})^2 e^{\nu t} \quad \text{a.e.} \tag{3}$$

($|\nabla u|$ denotes the Euclidean norm of the gradient of u, for simplicity we have denoted by $u_t(t)$ the function $u_t(\cdot,t)$ and will do so for all the functions appearing below).

Proof. Multiplying the equation satisfied by u by u_t, and integrating over Ω we get:

$$\frac{d}{dt}E(t) = -\int_{-\infty}^{t} \dot{G}(t-\tau) \int_{\Omega} \nabla u(\tau).\nabla u_t(t)dx\, d\tau + \int_{\Omega} f(t)u_t(t)dx \leq$$

$$-\int_{-\infty}^{t} \dot{G}(t-\tau) \frac{d}{dt} \int_{\Omega} \nabla u(\tau).\nabla u(t)\, dx\, d\tau + |f(t)|_2 |u_t(t)|_2$$

(By Cauchy-Schwarz Inequality). This leads to:

$$\frac{d}{dt}E(t) \leq -\frac{d}{dt}\int_{-\infty}^{t} \dot{G}(t-\tau) \int_{\Omega} \nabla u(\tau).\nabla u(t)\, dx\, d\tau + \dot{G}(0)\, \|\nabla u(t)\|_2^2$$

$$+\int_{-\infty}^{t} \ddot{G}(t-\tau) \int_{\Omega} \nabla u(\tau).\nabla u(t)dx\, d\tau + |f(t)|_2 |u_t(t)|_2$$

Integrating between $-\infty$ and t and since, by (2), $E(-\infty) = 0$, we deduce:

$$E(t) \leq \int_{-\infty}^{t} |\dot{G}(t-\tau)|\ \|\nabla u(\tau)\|_2\ \|\nabla u(t)\|_2 d\tau + |\dot{G}(0)| \int_{-\infty}^{t} \|\nabla u(t)\|_2^2 dt$$

$$+ \int_{-\infty}^{t} \int_{-\infty}^{\xi} |\ddot{G}(\xi-\tau)|\ \|\nabla u(\tau)\|_2 \|\nabla u(\xi)\|_2\, d\tau d\xi + \int_{-\infty}^{t} |f(\xi)|_2 |u_t(\xi)|_2 d\xi$$

Assume that:

$$K = \operatorname*{ess-sup}_{t\in(-\infty,0)} e^{-\nu t}E(t).$$

Then for almost every t we have (see the definition of E(t)):

$$|u_t(t)|_2 \leq \sqrt{\frac{2}{\rho}}\sqrt{K}\, e^{\frac{\nu}{2}t}$$

$$\|\nabla u(t)\|_2 \leq \sqrt{\frac{2}{G(0)}}\sqrt{K}\ \ e^{\frac{\nu}{2}t}$$

Plugging this into the above inequality, we get, after some easy computations:

$$E(t) \leqq \frac{2}{G(0)} \{\int_0^{+\infty} |\dot{G}(s)| e^{-\frac{\nu}{2}s} ds + \frac{|\dot{G}(0)|}{\nu} + \frac{1}{\nu}\int_0^{+\infty} |\ddot{G}(s)| e^{-\frac{\nu}{2}s} ds\} K e^{\nu t}$$

$$+ \frac{1}{\nu} |f|_{\infty,\nu} \sqrt{\frac{2}{\rho}} . \sqrt{K} . e^{\nu t}$$

$$\leqq c\ K e^{\nu t} + \frac{1}{\nu} |f|_{\infty,\nu} \sqrt{\frac{2}{\rho}} . \sqrt{K} . e^{\nu t}$$

with an obvious definition for c. But, now, by (G), $c < 1$. So, multiplying both sides of this inequality by $e^{-\nu t}$ and taking the supremum in t we arrive at:

$$K \leqq c\ K + \frac{1}{\nu} \sqrt{\frac{2}{\rho}} \sqrt{K}\ |f|_{\infty,\nu} \iff K \leqq \frac{2}{(1-c)^2 \nu^2 \rho} |f|^2_{\infty,\nu}$$

and the result follows.

Let us now turn to the proof of Theorem 1.

PROOF OF THEOREM 1. Let us denote by $\{v_k\}$ an orthonormal decomposition of $L^2(\Omega)$ where the v_k, $k = 1,2,\dots$, are the eigenfunctions of $-\Delta$ with Dirichlet boundary conditions, i.e. one has:

$$- \Delta v_k = \lambda_k v_k,\quad v_k \in H_0^1(\Omega),\quad \lambda_k > 0,\quad |v_k|_2 = 1.$$

Then, since u is supposed to be in $L^{\infty,\nu}(L^2(\Omega))$, one can write:

$$u(x,t) = \sum_{k=1}^{+\infty} \alpha_k(t) v_k(x) \tag{4}$$

(the convergence of the series taking place in $L^2(\Omega)$).

If Ω is smooth enough, then clearly, $v_k \in H^2(\Omega) \cap H_0^1(\Omega)$. Thus for Ψ of class C^∞ with compact support in $(-\infty,0)$ the function

$$\varphi = v_k(x)\Psi(t)$$

is a suitable test function for (W). Plugging it in, we easily deduce that if u is a solution to (P) one has:

$$\int_{-\infty}^{0} \{\rho\alpha_k\Psi''+G(0)\lambda_k\alpha_k\psi + \int_{-\infty}^{t} \dot{G}(t-\tau)\lambda_k\alpha_k(\tau)d\tau\ \psi\}\ dt = \int_{-\infty}^{0} \beta_k\ \psi dt$$

where the $\beta_k(t)$ are functions such that:

$$f(x,t) = \sum_{k=1}^{+\infty} \beta_k(t)v_k(x)$$

(Note that the α_k, β_k belong to $L^{\infty,\nu}$ ($\underline{R}$). Indeed this is an easy consequence of the inequality

$$|\alpha_k| = |\int_{\Omega} u(x,t)v_k(x)dx| \leqq |u|_2\ |v_k|_2 \leqq |u|_{\infty,\nu}\ e^{\frac{\nu}{2}t}$$

where $|\ \ |_{\infty,\nu}$ denotes the norm of u in $L^{\infty,\nu}(\underline{R})$. The argument is exactly the same for β_k.)

So α_k is a solution of a second order integrodifferential equation in the sense of distributions. One can show using in particular (G), that this equation has a unique solution α_k in $L^{\infty,\nu}(\underline{R})$. This solution is also the solution in the usual sense almost everywhere (see [4] for details). At this point we already know that if (W) has a solution, this solution is <u>unique</u>.

Now, let us remark that from the equation satisfied by α_k, we easily deduce

$$\alpha_k,\ \dot{\alpha}_k,\ \ddot{\alpha}_k \in L^{\infty,\nu}(\underline{R}). \tag{5}$$

and if we set

$$u_N = \sum_{k=1}^{N} \alpha_k\ v_k,\quad f_N = \sum_{k=1}^{N} \beta_k v_k.$$

then, u_N is $W^{2,\infty}((-\Omega,0),\ H^2(\Omega) \cap H_o^{\ 1}(\Omega))$ and satisfies a.e.:

$$\rho(u_N)_{tt} - G(0)\Delta u_N - \int_{-\infty}^{t} \dot{G}(t-\tau)\Delta u_N(x,\tau)d\tau = f_N\ .$$

We have also (see (5):

$$|(u_N)_t|_2,\quad \|\nabla u_N\|_2 \in L^{\infty,\nu}(\underline{R}).$$

So, we can use the energy estimate (3) to obtain:

$$\rho|(u_N)_t|_2^2 + G(0)\,\|\nabla u_N\|_2^2 \leq C|f_N|^2_{\infty,\nu}\; e^{\nu t} \leq C\;|f|^2_{\infty,\nu}\; e^{\nu t} \quad \text{a.e.} \qquad (6)$$

which gives

$$\sum_{k=1}^{N} \lambda_k\alpha_k^2(t) \leq C\;|f|^2_{\infty,\nu}\; e^{\nu t} \quad \text{a.e.} \qquad (7)$$

Thus, the series in the left side of this inequality converges. Since the λ_k are non decreasing with k, this implies that the series of the squares of α_k converges and so we can define u by (4). But, from (7), and again since the λ_k are non decreasing with k, we get also:

$$\lambda_N \sum_{k\geq N} \alpha_k^2(t) \leq C\;|f|^2_{\infty,\nu}\, e^{\nu t} \quad \text{a.e.}$$

Hence,

$$|u_N-u|_2^2 \leq \sum_{k\geq N} \alpha_k^2(t) \leq C\;|f|^2_{\infty,\nu}\;\frac{e^{\nu t}}{\lambda_N} \quad \text{a.e.} \qquad (8)$$

and $u_N(t) \to u(t)$ in $L^{\infty,\nu}(L^2(\Omega))$, since $\lambda_N \to +\infty$ when $N \to +\infty$. This, together with $u_N \in L^{\infty,\nu}(L^2(\Omega))$ implies $u \in L^{\infty,\nu}(L^2(\Omega))$. Now, u_N satisfies

$$\int_{-\infty}^{0}\int_\Omega \rho u_N\varphi_{tt} - G(0)u_N\,\Delta\varphi - \int_{-\infty}^{t} \dot{G}(t-\tau)u_N(x,\tau)\;d\tau\Delta\varphi = \int_{-\infty}^{0}\int_\Omega f_N\;\varphi$$

for every $\varphi \in C_c^\infty((-\infty,0),\, H^2(\Omega) \cap H_0^{\;1}(\Omega))$. So, to conclude the proof, it is enough to pass to the limit in the above equation. This clearly can be done if

$$u_N,\; f_N \to u,\;\; f \text{ in } L^2(\Omega \times (-\infty,0))$$

$$\int_{-\infty}^{t} \dot{G}(t-\tau)u_N(x,\tau)d\tau \;\to \int_{-\infty}^{t} \dot{G}(t-\tau)u(x,\tau)d\tau \text{ in } L^2(\Omega \times \;(-\infty,0)).$$

But, from (8), we have:

$$\int_{-\infty}^{0}\int_{\Omega}|u_N - u|^2 \leq C \frac{|f|^2_{\infty,\nu}}{\lambda_N}\int_{-\infty}^{0} e^{\nu t} \to 0.$$

Thus, u_N converges toward u in $L^2(\Omega \times (-\infty,0))$. We have also

$$|f_N - f|_2^2 \leq \sum_{k\geq N} \beta_k^2(t) \to 0 \qquad \text{for a.e. } t$$

$$|f_N - f|_2^2 \leq |f|_2^2 \leq C\, e^{\nu t}.$$

Thus, by Lebesgue convergence theorem, f_N converges toward f in $L^2(\Omega \times (-\infty,0))$. Finally, we have:

$$\left|\int_{-\infty}^{t} \dot{G}(t-\tau)(u_N-u)(\tau)d\tau\right|_2 \leq \int_{-\infty}^{t} |\dot{G}(t-\tau)|\,|u_N-u(\tau)|_2 d\tau$$

$$\leq \frac{C}{\sqrt{\lambda_N}} |f|_{\infty,\nu} \int_{-\infty}^{t} |\dot{G}(t-\tau)| e^{\frac{\nu}{2}\tau}\, d\tau \leq \frac{C'}{\sqrt{\lambda_N}} e^{\frac{\nu}{2}t}$$

(see (G)). Squaring and integrating over $(-\infty,0)$ this allows us to conclude as above. The fact that $u \in L^{\infty,\nu}(H_0^1(\Omega)) \cap W^{1,\infty}((-\infty,0), L^2(\Omega))$ is an easy consequence of (6). This completes the proof of the theorem.

REMARK. One can show also that $u \in C^0((-\infty,0], H_0^1(\Omega)) \cap C^1((-\infty,0], L^2(\Omega))$.

Let us prove now the following about strong solution:

THEOREM 2. Assume that $f_t \in L^{\infty,\nu}(L^2(\Omega))$ then u is a strong solution to (P).

Proof. Consider the equation satisfied by u_N, multiply it by $\Delta(u_N)_t$ and integrate over Ω. If E'(t) denotes the quantity

$$E'(t) = \frac{\rho}{2} |\nabla(u_N)_t|_2^2 + \frac{G(0)}{2}|\Delta u_N|_2^2$$

and if we drop the index N we get:

$$\frac{d}{dt} E'(t) = -\int_{-\infty}^{t} \dot{G}(t-\tau) \int_{\Omega} \Delta u(\tau).\Delta u_t(t)dx\, d\tau - \int_{\Omega} f(t).\Delta u_t(t)dx$$

which reads also:

$$\frac{d}{dt}E'(t) = -\frac{d}{dt}\int_{-\infty}^{t}\dot{G}(t-\tau)\int_{\Omega}\Delta u(\tau).\Delta u(t)dx\,d\tau + \dot{G}(0)|\Delta u(t)|_2^2$$

$$+\int_{-\infty}^{t}\ddot{G}(t-\tau)\int_{\Omega}\Delta u(\tau).\Delta u(t)dx\,d\tau - \frac{d}{dt}\int_{\Omega}f.\Delta u\,dx + \int_{\Omega}f_t.\Delta u\,dx$$

Integrating between $-\infty$ and t and since, by (5), $E'(-\infty) = 0$, we deduce:

$$E'(t) \leqq \int_{-\infty}^{t}|\dot{G}(t-\tau)||\Delta u(\tau)|_2|\Delta u(t)|_2\,d\tau + |\dot{G}(0)|\int_{-\infty}^{t}|\Delta u(t)|_2^2dt$$

$$+\int_{-\infty}^{t}\int_{-\infty}^{\xi}|\ddot{G}(\xi-\tau)||\Delta u(\tau)|_2|\Delta u(\xi)|_2d\tau d\xi + \int_{\Omega}|f(t)||\Delta u(t)|dx$$

$$+\int_{-\infty}^{t}\int_{\Omega}|f_t(\xi)||\Delta u(\xi)|\,dx\,d\xi.$$

Then, arguing as in Lemma 1, we deduce:

$$E'(t) \leqq C(|f_N|_{\infty,\nu} + \frac{|(f_N)_t|_{\infty,\nu}}{\nu})^2\,e^{\nu t} \leqq C(|f|_{\infty,\nu} + \frac{|f_t|_{\infty,\nu}}{\nu})^2\,e^{\nu t}$$

This implies easily, taking the limit in N, that

$$\Delta u \in L^{\infty,\nu}(L^2(\Omega)), \quad \nabla u_t \in (L^{\infty,\nu}(L^2(\Omega)))^n. \tag{9}$$

So, we obtain from (W), that u satisfies

$$\rho u_{tt} - G(0)\Delta u - \int_{-\infty}^{t}\dot{G}(t-\tau)\Delta u(x,\tau)\,d\tau = f$$

in the sense of distributions, but also a.e. thanks to the fact that $\Delta u \in L^{\infty,\nu}(L^2(\Omega))$. (The fact that $u \in W^{2,\infty}((-\infty,0), L^2(\Omega))$ results from the equation, the fact that $u \in W^{1,\infty}((-\infty,0), H_0^{\,1}(\Omega))$ from (9)).

REMARK. With the same proof one could have assumed $f \in L^{\infty,\nu}(H_0^{\,1}(\Omega))$ in Theorem 2 and obtained the same result. Now, by standard arguments, it can be shown that if f and its derivatives are smooth and have the right decay so does u.

ADDED IN PROOF. J. Prüss pointed out that part of our results could be recovered, under some other assumptions on G using a semi-group approach. We refer the reader to his contribution in this issue.

References

[1] G. Capriz, G. Vergara-Caffarelli: Alcune osservazioni sul problema dinamico in viscoelasticità lineare. Atti VIII Congresso AIMETA Torino 1, (1986), p. 11-13.

[2] G. Capriz, E. Virga: Un teorema di unicità in viscoelasticità lineare (to appear in Rend. Sem. Mat. Univ. Padova 1986).

[3] G. Capriz, E. Virga: Esempi di non-unicità in viscoelasticità lineare (to appear in Atti Accad. Sci. Torino 1986).

[4] M. Chipot, G. Vergara-Caffarelli (to appear).

[5] B.D. Coleman, M.E. Gurtin: Waves in materials with memory II. On the growth and decay of one dimensional acceleration waves. Arch. Rat. Mech. Anal., 19, (1965), p. 239-265.

[6] C.M. Dafermos: An abstract Volterra equation with application to linear viscoelasticity. J. Diff. Eq., 7, (1970), p. 554-569.

[7] C.M. Dafermos: Asymptotic Stability in viscoelasticity. Arch. Rat. Mech. Anal., 37, (1970), p. 297-308.

[8] C.M. Dafermos, J.A. Nohel: Energy Methods for Nonlinear Hyperbolic Volterra Integrodifferential equations. Comm. in P.D.E., 4 (3), (1979), p. 219-278.

[9] G. Fichera: Integro-Differential Problems of Hereditary Elasticity. Atti Sem. Mat. Modena 26, (1977), p. 363-370.

[10] G. Fichera: Avere Una Memoria Tenace Crea Gravi Problemi Arch. Rat. Mech. Anal., 70, (1979), p. 101-112.

[11] G. Fichera: Analytic problems of materials with memory. Proc. Int. Congres IUTAM (1980), p. 223-230.

[12] G. Fichera: Sul principlo della memoria evanescente. Rend. Sem. Mat. Univ. Padova, 68, (1982), p. 245-259.

[13] W.J. Hrusa, J.A. Nohel: Global Existence and Asymptotics in One Dimensional Nonlinear Viscoelasticity. Trends and Applications of Pure Mathematics to Mechanics Lecture Notes in Physics #195 (P.G. Ciarlet - M. Roseau Eds.).

[14] W. Hrusa, J. Nohel, M. Renardy: Mathematical Problems in Viscoelasticity - Pitman Monographs #35, Longman eds. (1987).

[15] R.C. MacCamy: A model for one-dimensional, nonlinear Viscoelasticity. Quarterly of Appl. Math. 35, (1977), p. 21-33.

[16] J.A. Nohel: This issue

[17] M. Renardy: This issue.

[18] G. Vergara-Caffarelli, E. Virga: Sull' unicità della soluzione del problema dinamico della viscoelasticità lineare (to appear in Atti Accad. Naz. Lincei 1987.)

M. Chipot
Université de Metz
Département de Mathématique
et Informatique
Ile du Saulcy,
57045 Metz Cedex
France.

G. Vergara-Caffarelli
Università di Trento
Dipartimento di Matematica
Povo,
I-38050 Trento
Italy.

PH. CLÉMENT, O. DIEKMANN, M. GYLLENBERG,
H.J.A.M. HEIJMANS & H.R. THIEME

Perturbation theory for dual semigroups. III. Nonlinear Lipschitz continuous perturbations in the sun-reflexive

We consider nonlinear Lipschitz perturbations of the infinitesimal generator of a linear C_0-semigroup on a non-reflexive Banach space. It is allowed that the perturbation maps the space into a bigger space which arises in a natural way when considering dual semigroups. Using a generalized variation - of - constants formula we show that the perturbed operator generates a strongly continuous nonlinear semigroup. We study regularity properties of this semigroup and prove the principle of linearized stability.

1. INTRODUCTION

The aim of this paper is to demonstrate how pertrubation theory for dual semigroups can be used to write certain nonlinear evolution equations as abstract semilinear equations. Becoming familiar with our duality framework (the game of suns and stars) requires, as always, a certain investment of time and energy of the reader. The refund includes, among other things, 1) the possibility to use standard techniques to prove local stability and bifurcation results, and 2) the unification of theories for various equations, like delay equations [13, 10] and age dependent population equations [19, 2].

In this paper we concentrate on the principle of linearized stability while postponing the derivation of results concerning invariant manifolds to future papers. As a prelude to semilinear problems we present in Section 2 some results about inhomogeneous linear equations which are, in our opinion, interesting by themselves. We hope to return to this topic too at some later time.

In this paper we try to convey the flavour of our recent work which was largely motivated by physiologically structured population models [17]. It should be noted that Desch, Schappacher and coworkers [7,9] as well as Greiner [11, 12] and Kellermann [16] developed alternative approaches for the study of these kind of problems.

Consider a semigroup $T_0(t)$ of bounded linear operators acting on some Banach space X. Assume that $T_0(t)$ is strongly continuous, i.e. the orbit

$t \to T_0(t)x$ is continuous for each initial value $x \in X$. If X is non-reflexive, the dual semigroup $T_0^*(t)$ need not have the same property. All one can say in general is that the orbits are continuous in the weak * topology. At the level of the generators this phenomenon is reflected in the fact that A_0^*, the adjoint of the infinitesimal generator A_0 of $T_0(t)$, need not have dense domain and that A_0^* is the weak * generator of $T_0^*(t)$ (see Butzer & Berens [1]).

The basic idea now is to take advantage of this phenomenon and to construct a second dual space for X based on the behaviour of the semigroup for $t \downarrow 0$ (or, equivalently, the behaviour of the resolvent of the generator for $\lambda \to \infty$).

To make this more precise we introduce

$$X^{\odot} = \{x^* \in X^* : \lim_{t\downarrow 0} \|T_0^*(t)x^* - x^*\| = 0\}.$$

The space $X^{\odot}$ is the maximal invariant subspace of X^* on which $T_0^*(t)$ is strongly continuous. Moreover, it is known that $X^{\odot} = \overline{\mathcal{D}(A_0^*)}$ (see [1, 14]). The restriction $T_0^{\odot}(t)$ of $T_0^*(t)$ to $X^{\odot}$ is strongly continuous and is generated by $A_0^{\odot}$, the part of A_0^* in $X^{\odot}$ (that is, $\mathcal{D}(A_0^{\odot}) = \{x^{\odot} \in X^{\odot}: x^{\odot} \in \mathcal{D}(A_0^*)$ and $A_0^*x^{\odot} \in X^{\odot}\}$). So on $X^{\odot}$ we now have the same situation as we had on X at the start. Consequently we can play the same game once more. In self-explanatory notation we obtain $X^{\odot}$, $T_0^{\odot}(t)$, $A_0^{\odot *}$ and $X^{\odot\odot}$, $T_0^{\odot\odot}(t)$ and $A_0^{\odot\odot}$. The pairing between elements of X and $X^{\odot}$ defines an embedding j of X into $X^{\odot *}$. Note that $j(x) \subset X^{\odot\odot}$.

<u>DEFINITION</u>. X is $\odot$-reflexive (pronounce: sun-reflexive) with respect to $T_0(t)$ if and only if $j(X) = X^{\odot\odot}$.

In this paper we shall restrict our attention to the $\odot$-reflexive case. Moreover, we shall in the following identify X with its embedding j(X) and omit the symbol j.

Integrals of the form $\int_0^t T_0^{\odot *}(t-\tau)h(\tau)d\tau$, where h is a given weakly * continuous $X^{\odot *}$-valued function, are now defined as weak * Riemann integrals.

This means that the integral is by definition the continuous linear functional on $X^{\odot}$ which assigns to an arbitrary $x^{\odot} \in X^{\odot}$ the value

$$\int_0^t \langle h(\tau), T_0^{\odot}(t-\tau)x^{\odot}\rangle d\tau.$$

Section 2 provides further information on these integrals.

In our previous paper [2] (also see [4]) we showed that any bounded linear operator

$$B : X \to X^{\odot *}$$

defines via solutions of the variation - of - constants equation

$$T(t)x = T_0(t)x + \int_0^t T_0^{\odot *}(t-\tau)BT(\tau)xd\tau$$

a "perturbed" strongly continuous semigroup $T(t)$ on X such that

(i) the spaces $X^{\odot}$ and $X^{\odot\odot}$ do not change, so in particular they are independent of B,

(ii) on the "big" spaces X^* and $X^{\odot *}$ the domains of the weak * generators remain unchanged, i.e.

$$\mathcal{D}(A^*) = \mathcal{D}(A_0^*), \text{ and } \mathcal{D}(A^{\odot *}) = \mathcal{D}(A_0^{\odot *}),$$

(iii) $A^{\odot *} = A_0^{\odot *} + B$ and $A^* = A_0^* + B^*$.

Since A is the part of $A^{\odot *}$ in X the domain $\mathcal{D}(A)$ in the "small" space X may depend on B, even to the extent that all information about B is contained in $\mathcal{D}(A)$, the action of A_0 and A being the same (this actually happens for functional differential equations and age-dependent population equations). Thus we see quite clearly how the duality frame-wrok is exploited to handle the equation $dx/dt = Ax$ by perturbation methods. In the following we shall consider nonlinear perturbations

$$F : X \to X^{\odot *}$$

and use basically the same approach (see [3] for linear time-dependent perturbations B(t)).

In conclusion of this introduction we mention a useful fact which follows from dual semigroup theory. The so-called Favard class of $T_0(t)$ is defined by

$$\mathrm{Fav}(T_0) = \{x \in X : \limsup_{t \downarrow 0} \frac{1}{t} \|T(t)x - x\| < \infty\}.$$

It is known ([1, 4]) that Fav $(T_0) = \mathcal{D}(A_0^{\odot *})$. Moreover, Fav (T_0) is precisely the set of initial data which yield Lipschitz continuous orbits. It is this latter non-local characterization which immediately guarantees the invariance and which, therefore, is very useful in nonlinear situations.

Section 2 deals with linear inhomogeneous equations. In Section 3 we derive results about the existence, uniqueness and differentiability of solutions of semilinear equations and Section 4 is devoted to the principle of linearized stability. In Section 5, finally, we show how to obtain solutions corresponding to initial data in the 'big' space $X^{\odot *}$ if the nonlinear perturbation is somewhat special (in particular it must have finite dimensional range). This is achieved by solving a nonlinear Volterra convolution equation in L_∞.

2. THE LINEAR INHOMOGENEOUS INITIAL VALUE PROBLEM

Let $T_0(t)$ be a strongly continuous semigroup on the (non-reflexive) Banach space X, and assume that X is sun-reflexive with respect to A_0, the generator of $T_0(t)$. There exist constants $\omega \in \mathbb{R}$ and $M \geq 1$ such that the following estimate holds:

$$\|T_0(t)\| \leq Me^{\omega t}, \quad t \geq 0.$$

In this section we consider the inhomogeneous initial value problem

$$\begin{cases} \frac{du}{dt}(t) = A_0^{\odot *} u(t) + f(t), & 0 < t < T, \\ u(0) = x \in X, \end{cases} \tag{2.1}$$

where $f: [0,T] \to X^{\odot *}$ is a given weakly * continuous function. We define the so-called mild solution to (2.1) by

$$u(t) = T_0(t)x + \int_0^t T_0^{\odot *}(t-s)f(s)ds, \quad 0 \leq t \leq T, \tag{2.2}$$

where the integral is a weak * Riemann integral. We also define

$$v(t) = \int_0^t T_0^{\odot *}(t-s)f(s)ds, \quad 0 \leq t \leq T. \tag{2.3}$$

We will successively consider what can be said about v if f is weakly * continuous, norm continuous, Lipschitz continuous, and continuously differentiable.

PROPOSITION 2.1. If f is weakly * continuous then v is weakly * continuous with values in $X^{\odot *}$. If f is norm continuous then v is norm continuous as well and takes values in X. In both cases the following estimate holds:

$$\|v(t)\| \leq \frac{M}{\omega}(e^{\omega t}-1) \cdot \sup_{0 \leq s \leq t} \|f(s)\| .$$

A proof of this result can be found in [2].

PROPOSITION 2.2. Let f be (locally) Lipschitz continuous. Then v is weakly * continuously differentiable (in particular, v is Lipschitz continuous), v takes values in $\mathcal{D}(A_0^{\odot *})$, and $w^* - \frac{dv}{dt}(t) = A_0^{\odot *}v(t) + f(t)$. Here $w^* - \frac{d}{dt}$ denotes the weak * derivative.

Proof. To show that v takes values in $\mathcal{D}(A_0^{\odot *})$ we use that $\mathcal{D}(A_0^{\odot *}) = \text{Fav}(T_0)$. Let $t \geq 0$ and $h > 0$. Then

$$\frac{1}{h}\|T_0(h)v(t)-v(t)\| \leq \frac{1}{h}\left\|\int_t^{t+h} T_0^{\odot *}(s)f(t+h-s)ds\right\|$$

$$+\frac{1}{h}\left\|\int_h^t T_0^{\odot *}(s)\ f(t+h-s)-f(t-s)\}ds\right\| + \frac{1}{h}\left\|\int_0^h T^{\odot *}{}_0(s)f(t-s)ds\right\| ,$$

which stays bounded as $h \downarrow 0$. It also follows from this expression that $\|A_0^{\odot *}v(t)\|$ is uniformly bounded on compact t-intervals. Now we show that $t \to A_0^{\odot *}v(t)$ is weakly * continuous. First take $x^{\odot} \in \mathcal{D}(A_0^{\odot})$. Then

$$\langle A_0^{\odot *}v(t), x^{\odot}\rangle = \langle v(t), A_0^{\odot}x^{\odot}\rangle$$

is continuous. The continuity of $\langle A_0^{\odot *} v(\cdot), x^{\odot}\rangle$ for arbitrary $x^{\odot} \in X^{\odot}$ now easily follows from the observation that $\mathcal{D}(A_0^{\odot})$ lies dense in $X^{\odot}$ and that $\|A_0^{\odot *} v(t)\|$ is bounded on compact t-intervals. Finally we show that $v(t) = \int_0^t \{A_0^{\odot *} v(s) + f(s)\}ds$, which completes the proof. Let $x^{\odot} \in \mathcal{D}(A_0^{\odot})$.

$$\langle v(t), x^{\odot}\rangle - \int_0^t \langle f(s), x^{\odot}\rangle ds = \int_0^t \langle f(s), T^{\odot}(t-s)x^{\odot} - x^{\odot}\rangle ds =$$

$$= \int_0^t \langle f(s), \int_0^{t-s} T_0^{\odot}(\sigma) A_0^{\odot} x^{\odot} d\sigma\rangle \; ds = \int_0^t \langle f(s), \int_s^t T_0^{\odot}(\sigma - s) A_0^{\odot} x d\sigma\rangle \; ds =$$

$$= \int_0^t \int_s^t \langle T_0^{\odot *}(\sigma - s) f(s), A_0^{\odot} x^{\odot}\rangle d\sigma ds = \int_0^t \int_0^{\sigma} \langle T_0^{\odot *}(\sigma - s) f(s), A_0^{\odot} x^{\odot}\rangle ds d\sigma =$$

$$= \int_0^t \langle v(\sigma), A_0^{\odot} x^{\odot}\rangle d\sigma = \int_0^t \langle A_0^{\odot *} v(\sigma), x^{\odot}\rangle d\sigma.$$

Since $\mathcal{D}(A_0^{\odot})$ is dense in $X^{\odot}$ the same identity holds for arbitrary $x^{\odot} \in X^{\odot}$. □

PROPOSITION 2.3. Assume that f is continuously differentiable on [0,T) and that $f(0) \in X$. Then v is continuously differentiable on [0,T) and takes values in $\mathcal{D}(A_0^{\odot *})$. Furthermore

$$v'(t) = A_0^{\odot *} v(t) + f(t) = T_0(t) f(0) + \int_0^t T_0^{\odot *}(t-s) f'(s) ds \in X.$$

Proof. From Proposition 2.2 we know that $v(t) \in \mathcal{D}(A_0^{\odot *})$, $0 \leq t \leq T$. We compute $A_0^{\odot *} v(t)$ explicitly.

$$\frac{1}{h}(T_0(h) - I) v(t) = \frac{1}{h} \int_{-h}^{0} T_0^{\odot *}(t-s) f(s+h) ds$$

$$- \frac{1}{h} \int_{t-h}^{t} T_0^{\odot *}(t-s) f(s+h) ds + \frac{1}{h} \int_0^t T_0^{\odot *}(t-s) \{f(s+h) - f(s)\} ds$$

$$\rightarrow T_0^{\odot *}(t) f(0) - f(t) + \int_0^t T_0^{\odot *}(t-s) f'(s) ds,$$

in the weak * topology, as $h \downarrow 0$, and by definition this limit equals $A_0^{\odot *} v(t)$. Next we show that v is continuously differentiable.

$$\frac{1}{h}(v(t+h)-v(t)) = \frac{1}{h}\int_{-h}^{0} T_0^{\odot *}(t-s)f(s+h)ds + \frac{1}{h}\int_0^1 T_0^{\odot *}(t-s)\{f(s+h)-f(s)\}ds =$$

$$T_0(t)\frac{1}{h}\int_0^h T_0(s)f(0)ds + T_0(t)\frac{1}{h}\int_0^h T_0^{\odot *}(s)\{f(h-s)-f(0)\}ds +$$

$$\int_0^t T_0^{\odot *}(t-s)f'(s)ds + \int_0^t T_0^{\odot *}(t-s)\ \{\frac{f(s+h)-f(s)}{h} - f'(s)\}ds$$

$$\to T_0(t)f(0) + \int_0^t T_0^{\odot *}(t-s)f'(s)ds$$

in norm as $h \downarrow 0$. So v is right-differentiable and the right-derivative is continuous. Therefore v(t) is continuously differentiable and

$$v'(t) = T_0(t)f(0) + \int_0^t T_0^{\odot *}(t-s)f'(s)ds = A_0^{\odot *}v(t) + f(t).$$

This proves the proposition. □

COROLLARY 2.4. Assume that f is continuously differentiable. Let $x \in \mathcal{D}(A_0^{\odot *})$ and $A_0^{\odot *}x + f(0) \in X$. Then u(t) given by (2.2) is continuously differentiable and takes values in $\mathcal{D}(A_0^{\odot *})$. Furthermore $A_0^{\odot *}u(t) + f(t) \in X$, $0 \leq t \leq T$, and u satisfies (2.1). In other words, (2.1) admits a classical solution.

Proof. Define $\tilde{f}(t) = A_0^{\odot}x + f(t)$, and let

$$\tilde{v}(t) = \int_0^t T_0^{\odot *}(t-s)\tilde{f}(t)dt = \int_0^t T_0^{\odot *}(t-s)\{A_0^{\odot *}x + v(t)\}dt$$

$$= T_0(t)x - x + v(t) = u(t) - x,$$

where v, u are given by (2.3) and (2.2) respectively. Now $\tilde{f}$ satisfies the assumption of Proposition 2.3, hence $\tilde{v}$ is C^1 and $\tilde{v}'(t) = A_0^{\odot *}\tilde{v}(t) + \tilde{f}(t)$, or equivalently, u is C^1 and $u'(t) = A_0^{\odot *}u(t) + f(t)$. □

We refer to Da Prato and Sinestrari [6] for a more general result. Also see Kellermann [15].

When studying the equation $du/dt = A_0^{\odot *}u + Bu + f$ with B a given bounded linear operator from X into $X^{\odot *}$, we may either consider Bu + f as a perturbation and solve a variation-of-constants integral equation involving $T_0^{\odot *}(t)$ or first define the semigroup T(t) on X generated by the part of

$A_0^{\odot*} + B$ in X and write down the solution explicitly as a variation-of-constants formula involving T(t). Of course, both procedures should define the same solution. That they actually do so is the content of the following proposition.

PROPOSITION 2.5. Let $B:X \to X^{\odot*}$ be a bounded linear operator and let T(t) be the C_0-semigroup on X generated by A, the part of $A_0^{\odot*} + B$ in X. Let $x \in X$ and let $f:[0,T] \to X^{\odot*}$ be an arbitrary continuous function. Let u(t) be the norm-continuous solution of the integral equation

$$u(t) = T_0(t)x + \int_0^t T_0^{\odot*}(t-s)\{Bu(s) + f(s)\}ds, \; 0 \leqq t \leqq T, \tag{2.4}$$

then

$$u(t) = T(t)x + \int_0^t T^{\odot*}(t-s)f(s)ds.$$

To prove this proposition we need a lemma which is of interest by itself. Let A,B be as in Proposition 2.5 and define the integrated semigroups (see [15]) $W_0(t)$, $W(t):X^{\odot*} \to X$ by

$$W_0(t)x^{\odot*} = \int_0^t T_0^{\odot*}(s)x^{\odot*}ds$$

$$W(t)x^{\odot*} = \int_0^t T^{\odot*}(s)x^{\odot*}ds.$$

LEMMA 2.6. For every $x^{\odot*} \in X^{\odot*}$ and $t \geqq 0$,

$$W(t)x^{\odot*} = W_0(t)x^{\odot*} + \int_0^t T_0^{\odot*}(t-s)BW(s)x^{\odot*}ds.$$

Proof. From the variation-of-constants formula

$$T(t)x = T_0(t)x + \int_0^t T_0^{\odot*}(s)BT(t-s)xds$$

it follows that for every $x \in X$, $x^\odot \in X^\odot$,

$$\langle\int_0^t T(\tau)xd\tau,x^\odot\rangle = \langle\int_0^t T_0(\tau)xd\tau,x^\odot\rangle + \int_0^t \int_0^\tau \langle T(\tau-s)x,B^*T_0^\odot(s)x^\odot\rangle dsd\tau$$

where $B^*:X^{\odot} \to X^*$ is the restriction of the dual operator of B to $X^{\odot}$. By changing the order of integration in the last integral we find that

$$<W(t)x,x^{\odot}> = <W_0(t)x,x^{\odot}> + \int_0^t <W(t-s)x,B^*T_0^{\odot}(s)x^{\odot}>ds.$$

Let $x^{\odot *} \in X^{\odot *}$ and let $\{x_n\}$ be a sequence in X converging to $x^{\odot *}$ in the weak * sense. It is easily seen that $W_0(t)^* = \int_0^t T_0^*(\tau)d\tau$ and $W(t)^* = \int_0^t T^*(\tau)d\tau$ and that they map X* into $X^{\odot}$ and consequently that $W_0(t)x_n \to W_0(t)x^{\odot *}$ and $W(t)x_n \to W(t)x^{\odot *}$ with respect to the weak topology. By the dominated convergence theorem it follows that

$$\int_0^t <W(t-s)x_n,B^*T_0^{\odot}(s)x^{\odot}>ds \to \int_0^t <W(t-s)x^{\odot *},B^*T_0^{\odot}(s)x^{\odot}>ds.$$

From this, the result follows immediately. □

PROOF OF PROPOSITION 2.5. It is easy to show that (2.4) has a unique continuous solution. Therefore we have to check that substitution of $u(t) = T(t)x + \int_0^t T^{\odot *}(t-s)f(s)ds$ into equation (2.4) gives an identity. This amounts to verifying that

$$\int_0^t T^{\odot *}(t-s)f(s)ds = \int_0^t T_0^{\odot *}(t-s)f(s)ds + \int_0^t T_0^{\odot *}(s)B\{\int_0^{t-s} T^{\odot *}(t-s-\sigma)f(\sigma)d\sigma\}ds. \quad (*)$$

In order to achieve this we integrate both sides of the equation from 0 to t. Clearly,

$$\int_0^t \{\int_0^\tau T^{\odot *}(\tau-s)f(s)ds\}d\tau = \int_0^t W(t-s)f(s)ds,$$

and we find a similar identity for the integral of the first term at the right-hand-side. We evaluate the integral of the last term by pairing it with an element $x^{\odot} \in X^{\odot}$.

$$<\int_0^t \{\int_0^\tau T_0^{\odot *}(s)B\{ \int_0^{\tau-s} T^{\odot *}(\tau-s-\sigma)f(\sigma)d\sigma\}ds\}d\tau,x^{\odot}> =$$

$$\int_0^t \int_0^\tau \langle \int_0^{\tau-s} T^{\odot *}(\tau-s-\sigma)f(\sigma)d\sigma, B^*T^*T_0^{\odot}(s)x^{\odot}\rangle ds d\tau =$$

$$\int_0^t \int_s^t \langle \int_0^{\tau-s} T^{\odot *}(\tau-s-\sigma)f(\sigma)d\sigma, B^*T_0^{\odot}(s)x^{\odot}\rangle d\tau ds =$$

$$\int_0^t \langle \int_s^t \int_0^{\tau-s} T^{\odot *}(\tau-s-\sigma)f(\sigma)d\sigma d\tau, B^*T_0^{\odot}(s)x^{\odot}\rangle ds =$$

$$\int_0^t \langle \int_0^{t-s} \int_0^{t-s-\sigma} T^{\odot *}(\tau)f(\sigma)d\tau d\sigma, B^*T_0^{\odot}(s)x^{\odot}\rangle ds =$$

$$\int_0^t \langle \int_0^{t-s} W(t-s-\sigma)f(\sigma)d\sigma, B^*T_0^{\odot}(s)x^{\odot}\rangle ds =$$

$$\langle \int_0^t T_0^{\odot *}(s)B\{\int_0^{t-s} W(t-s-\sigma)f(\sigma)d\sigma\}ds, x^{\odot}\rangle =$$

$$\langle \int_0^t \{\int_0^{t-\sigma} T_0^{\odot}(t-\sigma-\tau)BW(\tau)\}d\tau f(\sigma)d\sigma, x^{\odot}\rangle =$$

$$\langle \int_0^t W(t-\sigma)f(\sigma)d\sigma - \int_0^t W_0(t-\sigma)f(\sigma)d\sigma, x^{\odot}\rangle$$

because of Lemma 2.6. Hence, the integral from 0 to t of (*) yields an identity. Moreover (*) becomes an identity for $t = 0$, and from this the result follows. □

3. THE SEMILINEAR EQUATION

Let $F:X \to X^{\odot *}$ be a (nonlinear) function which is globally Lipschitz continuous with Lipschitz constant L, i.e.

$$\|F(x) - F(y)\| \leq L\,\|x-y\|, \quad x,y \in X.$$

We assume global Lipschitz continuity for ease of formulation. Appropriate variants of our results can be formulated when F is only locally Lipschitz continuous but our aim here is to explain the essence of our approach in as simple a setting as possible.

In this section we examine the initial value problem

$$\frac{du}{dt}(t) = A_0^{\odot *}u(t) + F(u(t)),\ t > 0$$

$$u(0) = x. \tag{3.1}$$

Note that even when orbits $t \to u(t;x)$ in X are considered, the differential equation should be interpreted as an identity of $X^{\odot *}$-valued functions. We can reformulate (3.1) as an integral equation:

$$u(t) = T_0(t)x + \int_0^t T_0^{\odot *}(t-s)F(u(s))ds. \tag{3.2}$$

We shall first study this integral equation and subsequently discuss the relation with the differential equation (3.1).

If $u(\cdot)$ is a norm continuous X-valued function then $F(u(\cdot))$ is a norm continuous $X^{\odot *}$-valued function, and from Proposition 2.1 we conclude that the integral at the right-hand-side of (3.2) defines a norm continuous X-valued function. Moreover, if $u_1(\cdot)$, $u_2(\cdot)$ are two norm continuous functions, then

$$\left\|\int_0^t T_0^{\odot *}(t-s)F(u_1(s))ds - \int_0^t T_0^{\odot *}(t-s)F(u_2(s))ds\right\| \leq$$

$$\int_0^t Me^{\omega(t-s)}L\ \|u_1(s)-u_2(s)\|\,ds \leq ML\,\frac{e^{\omega t}-1}{\omega}\sup_{0\leq s\leq t}\|u_1(s)-u_2(s)\|.$$

Therefore, we can invoke standard contraction and continuation arguments to prove existence and uniqueness of solutions to (3.2).

<u>THEOREM 3.1</u>. For every $x \in X$ there exists a unique continuous solution $u(\cdot,x)$ to the integral equation (3.2). This solution has the semigroup property

$$u(t+s;x) = u(t;u(s;x)),\quad s,t \geq 0.$$

Furthermore, the following estimates hold:

$$\|u(t;x)\| \leq (M\|x\| + \frac{M}{\omega}\|F(0)\|)e^{(\omega+ML)t}, \quad t \geq 0, \tag{3.3}$$

$$\|u(t;x)-u(t;y)\| \leq M\|x-y\| e^{(\omega+ML)t}, \quad t \geq 0, \tag{3.4}$$

for every $x,y \in X$.

The estimates (3.3) and (3.4) follow easily from the Gronwall lemma. We define the non-linear strongly continuous semigroup $S(t):X \to X$ by

$$S(t)x := u(t;x), \quad t \geq 0, \ x \in X.$$

The remainder of this section is devoted to the question in which sense $S(t)x$ satisfies the differential equation (3.1). To this end we need some definitions. The weak * generator $A_S^{\odot *}$ (note the abuse of notation) is defined as follows: $x \in \mathcal{D}(A_S^{\odot *})$ if $h^{-1}(S(h)x-x)$ converges with respect to the weak * topology of $X^{\odot *}$ to some element $y^{\odot *} \in X^{\odot *}$ as $h \downarrow 0$ and in this case $A_S^{\odot *}x = y^{\odot *}$. The strong generator is defined as usual. Finally, we define the Favard class Fav(S) as in the linear case:

$$\mathrm{Fav}(S) = \{x \in X: \limsup_{h \downarrow 0} \frac{1}{h}\|S(h)x-x\| < \infty\}$$

We can now prove the following important result.

<u>THEOREM 3.2</u>. $\mathcal{D}(A_S^{\odot *}) = \mathrm{Fav}(S) = \mathrm{Fav}(T_0) = \mathcal{D}(A_0^{\odot *})$, and $A_S^{\odot *}x = A_0^{\odot *}x + F(x)$ for $x \in \mathcal{D}(A_0^{\odot *})$.

<u>Proof</u>. For any $x \in X$, the expression $h^{-1}\int_0^h T_0^{\odot *}(h-s)F(S(s)x)ds$ converges towards $F(x)$ with respect to the weak * topology as $h \downarrow 0$. This implies that $\mathcal{D}(A_S^{\odot *}) = \mathcal{D}(A_0^{\odot *})$ and that $A_S^{\odot *}x = A_0^{\odot *}x + F(x)$ for $x \in \mathcal{D}(A_0^{\odot})$. But we may also conclude that $\|h^{-1}\int_0^h T_0^{\odot *}(h-s)F(S(s)x)ds\|$ remains bounded as $h \downarrow 0$, which yields that Fav (S) = Fav (T_0). From the linear theory we know that Fav (T_0) = $\mathcal{D}(A_0^{\odot *})$ and the result is proved. □

<u>PROPOSITION 3.3</u>. $x \in$ Fav (S) if and only if the orbit $t \to S(t)x$ is locally Lipschitz continuous.

The proof of this result, which is relatively easy, can be found in a paper by Crandall [5], who calls Fav (S) the generalized domain of S. This result implies in particular that $\mathcal{D}(A_S^{\odot *})$ is invariant under S(t): namely Fav (S) is invariant under S(t). Proposition 2.2 implies that $S(\cdot)x$ is the solution to the differential equation (3.1) interpreted in the weak * sense if $x \in \mathcal{D}(A_S^{\odot *})$:

THEOREM 3.4. If $x \in \mathcal{D}(A_S^{\odot *})$, then $u = S(\cdot)x$ is weakly * continuously differentiable and

$$w^* - \frac{du}{dt}(t) = A_0^{\odot *} u(t) + F(u(t)).$$

THEOREM 3.5. A_S is the part of $A_S^{\odot *}$ in X, i.e. $\mathcal{D}(A_S) = \{x \in \mathcal{D}(A_S^{\odot *}) : A_S^{\odot *} x \in X\}$ and $A_S x = A_S^{\odot *} x$.

Proof. It is clear that $\mathcal{D}(A_S) \subseteq \{x \in \mathcal{D}(A_S^{\odot *}) : A_S^{\odot *} x \in X\}$. Conversely, assume that $x \in \mathcal{D}(A_S^{\odot *})$ and $A_S^{\odot *} x \in X$. Then

$$\frac{1}{h}(S(h)x - x) = \frac{1}{h}(T_0(h)x - x) + \frac{1}{h}\int_0^h T_0^{\odot *}(h-s)F(S(s)x)ds =$$

$$\frac{1}{h}\int_0^h T_0^{\odot *}(h-s)\{A_0^{\odot *}x + F(x)\}ds + \frac{1}{h}\int_0^h T_0^{\odot *}(h-s)\{F(S(s)x) - F(x)\}ds.$$

The first integral can be replaced by $h^{-1}\int_0^h T_0(h-s)\{A_0^{\odot *}x + F(x)\}ds$ and converges in norm to $A_0^{\odot}x + F(x)$, whereas the second integral converges in norm to 0. This concludes the proof. □

To conclude this section we state a result which gives sufficient conditions in order that $t \to S(t)x$ is a classical solution (see [18]) of the differential equation (3.1).

THEOREM 3.6. Assume that F is continuously Frechet differentiable. If $x \in \mathcal{D}(A_S)$ then $u(t) := S(t)x$ is continuously differentiable and

$$u'(t) = A_0^{\odot *} u(t) + F(u(t)).$$

Proof. Let $x \in \mathcal{D}(A_S)$. Below we shall prove that $u(t) = S(t)x$ is continuously differentiable on $\mathbb{R}_+$. Then f given by $f(t) := F(u(t))$ is continuously differentiable as well, and, moreover, $A_0^{\odot *}x + f(0) = A_0^{\odot *}x + F(x) \in X$. From Corollary 2.4 we may conclude that $\tilde{u}(t)$ given by $\tilde{u}(t) = T_0(t)x + \int_0^t T_0^{\odot *}(t-s)$ $f(s)ds = T_0(t)x + \int_0^t T_0^{\odot *}(t-s)F(u(s))ds$ is C^1, takes values in $\mathcal{D}(A_0^{\odot *})$ and $\tilde{u}'(t) = A_0^{\odot *}\tilde{u}(t) + f(t)$. But $\tilde{u}(t) = u(t)$ and the result is proved.

So it remains to show that $u(\cdot) \in C^1$. Define $B(t) := F'(u(t))$, $t \geq 0$. Then $B(t) : X \to X^{\odot *}$ is a bounded linear operator for every $t \geq 0$ and $t \to B(t)$ is norm continuous. We also define

$$\eta(h,s) = F(u(s + h)) - F(u(s)) - B(s)(u(s + h) - u(s)).$$

Then $h^{-1} \|\eta(h,s)\| \to 0$, $h \downarrow 0$, uniformly on bounded s-intervals. In [3] we have proved that the family $A_0^{\odot *} + B(t)$ "generates" a strongly continuous evolutionary system $U(t,s)$ on X. The function $w(t) := U(t,0)A_S x$ is a solution of the integral equation

$$w(t) = T_0(t)A_S x + \int_0^t T_0^{\odot *}(t-s)B(s)w(s)ds.$$

We also define $w_h(t) := h^{-1}(u(t+h)-u(t))-w(t)$. Then

$$\begin{aligned}
w_h(t) &= \frac{1}{h} (T_0(t+h)x-T_0(t)x) + \frac{1}{h} T_0(t) \int_0^h T_0^{\odot *}(h-s)F(u(s))ds \\
&\quad + \frac{1}{h} \int_0^t T_0^{\odot *}(t-s)\{F(u(s+h))-F(u(s))\}ds - T_0(t)A_S x \\
&\quad - \int_0^t T_0^{\odot *}(t-s)B(s)w(s)ds \\
&= T_0(t)\{ \frac{1}{h} (T_0(h)x-x) + \frac{1}{h} \int_0^h T_0^{\odot *}(h-s)F(u(s))ds-A_S x\} \\
&\quad + \int_0^t T_0^{\odot *}(t-s)\{B(s)(\frac{u(s+h)-u(s)}{h}) + \frac{1}{h} \eta(h,s)\}ds \\
&\quad - \int_0^t T_0^{\odot *}(t-s)B(s)w(s)ds \\
&= T_0(t)\{\frac{1}{h}(S(h)x-x) - A_S x\} + \frac{1}{h} \int_0^t T_0^{\odot *}(t-s)\eta(h,s)ds
\end{aligned}$$

$$+ \int_0^t T_0^{\odot *}(t-s)B(s)w_h(s)ds.$$

Fix $t > 0$. It is easy to check that

$$\|T_0(t')\{\frac{1}{h}(S(h)x-x) - A_S x\}\| + \frac{1}{h}\|\int_0^{t'} T_0^{\odot *}(t'-s)\eta(h,s)ds\| \leq \varepsilon_t(h)$$

for $t' \leq t$. Here $\varepsilon_t(h)$ is a function of h (for fixed t) which goes to 0 as $h \downarrow 0$. Thus we get

$$\|w_h(t')\| \leq \varepsilon_t(h) + C_t \int_0^{t'} \|w_h(s)\| ds,$$

for some constant $C_t > 0$. Application of Gronwall's lemma yields that

$$\|w_h(t')\| \leq \varepsilon_t(h)e^{t'C_t}, \; t' \leq t,$$

from which we deduce that $w_h(t') \to 0$, $h \downarrow 0$, for every $t' \geq 0$. This implies that u(t) is right-differentiable and it's right derivative is $w(t) = U(t,0)A_S x$ which is a continuous function. Thus we have proved that $u \in C^1$. □

4. LINEARIZED STABILITY

Let $\bar{x} \in X$ be an equilibrium of the nonlinear semigroup S(t), i.e.

$$S(t)\bar{x} = \bar{x}, \; t \geq 0,$$

or equivalently,

$$\bar{x} \in \mathcal{D}(A_S) \text{ and } A_S\bar{x} = 0.$$

In this section we prove the principle of linearized stability which says that local (in-)stability of the equilibrium $\bar{x}$ is completely determined by the spectral properties of the linearization of the nonlinear operator "$A_0^{\odot *} + F(\cdot)$" around $u = \bar{x}$.

In what follows we assume that F is continuously Fréchet differentiable at $\bar{x}$, and we denote the Fréchet derivative $DF(\bar{x})$ by B. Thus B is a bounded linear operator from X into $X^{\odot *}$. In [2] we showed that the part of $A_0^{\odot} + B$

in X, which we denote by A, generates a linear C_0-semigroup $T(t)$, $t \geq 0$, on X. It satisfies the variation-of-constants formula

$$T(t)x = T_0(t)x + \int_0^t T_0^{\odot *}(t-s)BT(s)xds, \ t \geq 0, \ x \in X. \tag{4.1}$$

The following theorem gives the relation between $S(t)$ and $T(t)$.

THEOREM 4.1. The nonlinear operator $S(t)$ is Fréchet differentiable at $\bar{x}$ and

$$(DS(t))(\bar{x}) = T(t)$$

for every $t \geq 0$.

Proof. Assume that $\bar{x} = 0$. This can always be achieved by setting $\tilde{u} = u - \bar{x}$, $\tilde{F}(\tilde{u}) = F(\bar{x} + \tilde{u}) - F(\bar{x})$. Let $x \in X$ and $u(t) = S(t)x$. Then $u(t)$ is a solution of the integral equation

$$u(t) = T_0(t)x + \int_0^t T_0^{\odot *}(t-s)Bu(s)ds + \int_0^t T_0^{\odot *}(t-s)\{F(u(s))-Bu(s)\}ds$$

and from Proposition 2.5 we get that

$$u(t) = T(t)x + \int_0^t T^{\odot *}(t-s)\{F(u(s)) - Bu(s)\}ds,$$

hence

$$S(t)x - T(t)x = \int_0^t T^{\odot *}(t-s)\{F(S(s)x) - BS(s)x\}ds.$$

With this observation the proof becomes straightforward. □

It is not difficult to give a proof of this theorem without taking recourse to Proposition 2.5, but instead using Gronwall's lemma.

Let $\omega_0(A)$ denote the type of the semigroup $T(t)$: recall that A is the generator of $T(t)$. We are now ready to prove the following version of one part of the principle of linearized stability which is due to Desch and Schappacher [8].

THEOREM 4.2. Let $\omega_0(A) < 0$. For every ε with $0 \leqq \varepsilon < -\omega_0(A)$ there exists a $\delta > 0$ such that $\|x\| \leqq \delta$ implies that

$$e^{\varepsilon t}(S(t)x-\bar{x}) \to 0, t \to \infty.$$

Proof. Assume $\bar{x} = 0$. Let $\varepsilon \in [0,-\omega_0(A))$ and choose $\gamma \in (\varepsilon,-\omega_0(A))$. There is a constant $M_\gamma \geqq 1$ such that

$$\|T(t)\| \leqq M_\gamma e^{-\gamma t}, \quad t \geqq 0.$$

Choose t_0 so large that $M_\gamma e^{-(\gamma-\varepsilon)t_0} \leqq 1/4$, and choose $\delta \in (0,1]$ so small that

$$\|S(t_0)x - T(t_0)x\| \leqq \frac{1}{4} e^{-\varepsilon t_0} \|x\| \leqq \delta.$$

Then, if $\|x\| \leqq \delta$,

$$e^{\varepsilon t_0} \|S(t_0)x\| \leqq e^{\varepsilon t_0} \|S(t_0)x-T(t_0)x\| + e^{\varepsilon t_0} \|T(t_0)x\|$$

$$\leqq \frac{1}{4} \|x\| + \frac{1}{4} \|x\| = \frac{1}{2} \|x\|.$$

Thus, for $k \geqq 1$,

$$e^{k\varepsilon t_0} \|S(kt_0)x\| \leqq (\tfrac{1}{2})^k \|x\|, \quad \|x\| \leqq \delta.$$

Every $t \geqq 0$ can be written as $t = kt_0 + \tau$ for some integer k and some $\tau \in [0,t_0)$. Then

$$e^{\varepsilon t} \|S(t)x\| = e^{\varepsilon\tau} e^{k\varepsilon t_0} \|S(kt_0)S(\tau)x\| \leqq$$

$$e^{\varepsilon\tau}(\tfrac{1}{2})^k \|S(\tau)x\| \leqq e^{\varepsilon t_0}(\tfrac{1}{2})^{t/t_0-1} \cdot\delta, \quad t \geqq 0$$

if $\|x\| \leqq M^{-1}e^{-(\omega+ML)t_0}\delta$. □

Desch and Schappacher also proved an instability result in [8], and it is easy to show that this result carries over to our case immediately, so we omit the proof.

THEOREM 4.3. Suppose that $\omega_0(A) > 0$, and assume that $X = X_1 \oplus X_2$, where X_1 is finite-dimensional and where both X_1 and X_2 are invariant under $T(t)$. Let $T_i(t)$ denote the restriction of $T(t)$ to X_i and let A_i be the corresponding infinitesimal generator. If

$$\omega_0(A_2) < \min\{\mathrm{Re}\lambda : \lambda \in \sigma(A_1)\}$$

then there exists a constant $\varepsilon > 0$ and a sequence $\{x_n\}_{n \geq 1} \subset X$ converging to $\bar{x}$ and a sequence $\{t_n\}_{n\geq 1} \subset \mathbb{R}$, converging to ∞ such that $\|S(t_n)x_n - \bar{x}\| \geq \varepsilon$.

5. PERTURBATIONS WITH FINITE DIMENSIONAL RANGE AND EXTENSIONS TO THE "BIG" SPACE

In the linear case we can, for an arbitrary C_0-semigroup $T(t)$, define the extension $T^{\odot *}(t)$ to the "big" space $X^{\odot *}$ by taking second semigroup adjoints. This technique breaks down in the nonlinear case. In this section we introduce a more direct alternative technique which does work in the non-linear case, but which is bound to special perturbations with finite dimensional range. Still other techniques are conceivable but we will not go into these here.

For ease of formulation we will restrict ourselves to the one-dimensional case. More precisely we take $F: X \to X^{\odot *}$ to be of the special form

$$F(x) = G(\langle x, r^* \rangle) r^{\odot *} \tag{5.1}$$

for some $r^* \in X^*$, $r^{\odot *} \in X^{\odot *}$ and $G:\mathbb{R} \to \mathbb{R}$ a globally Lipschitz continuous function. As in [2, Section 5] we note that

$$Q(t) = \langle \int_0^t T_0^{\odot *}(\tau) r^{\odot *} d\tau, r^* \rangle \tag{5.2}$$

is locally Lipschitz continuous and hence can be written as

$$Q(t) = \int_0^t K(\tau) d\tau \tag{5.3}$$

for some $K \in L_\infty^{\mathrm{loc}}$. For given $x \in X$ we define

$$h(t) = \langle T_0(t)x, r^* \rangle, \quad y(t) = \langle S(t)x, r^* \rangle. \tag{5.4}$$

Lemma 5.1 of [2] then implies that

$$y = h + K*G(y), \tag{5.5}$$

where * denotes the convolution product. In other words, y is the solution of a nonlinear renewal (i.e. Volterra convolution) equation. Conversely the formula

$$S(t)x = T_0(t)x + \int_0^t T_0^{\odot *}(t-\tau)r^{\odot *}G(y(\tau))d\tau \tag{5.6}$$

enables us to reconstruct the semigroup $S(t)x$ from the solution y of (5.5), the unperturbed semigroup $T_0^{\odot}(t)$ and $r^{\odot *}$. This observation suggests that we may define

$$S^{\odot *}(t)x^{\odot *} = T_0^{\odot *}(t)x^{\odot *} + \int_0^t T_0^{\odot *}(t-\tau)r^{\odot *}G(y(\tau))d\tau \tag{5.7}$$

provided (i) we can give a proper definition of h for initial data in $X^{\odot *}$, (ii) we are able to solve (5.5) for such h, (iii) we can give a meaning to the integral in (5.7).

The definition of h proceeds exactly as the definition of K: for given $x^{\odot *} \in X^{\odot *}$ there exists $h \in L_\infty^{loc}$ such that

$$\left\langle \int_0^t T_0^{\odot *}(\tau)x^{\odot *}d\tau, r^* \right\rangle = \int_0^t h(\tau)d\tau, \tag{5.8}$$

for all $t \geqq 0$.

<u>THEOREM 5.1</u>. For given $x^{\odot *} \in X^{\odot *}$ define $S^{\odot *}(t)x^{\odot *}$ by (5.7) where y is the unique L_∞^{loc} solution of (5.5) corresponding to $h \in L_\infty^{loc}$ defined by (5.8). Then $S^{\odot *}(t)$ is a semigroup whose restriction to X is exactly S(t). Moreover, $t \to S^{\odot *}(t)x^{\odot *}$ is weakly * continuous.

<u>Proof</u>. We start by proving the last assertion. The Lebesgue integral

$$\left\langle \int_0^t T_0^{\odot *}(t-\tau)r^{\odot *}G(y(\tau))d\tau, x^{\odot} \right\rangle = \int_0^t \langle r^{\odot *}, T_0^{\odot}(t-\tau)x^{\odot}\rangle G(y(\tau))d\tau$$

is the conv olution of a continuous function and an L_∞-function and is therefore a continuous function of t. Since $t \to T_0^{\odot*}(t)x^{\odot*}$ is weakly * continuous formula (5.7) then implies that $t \to S^{\odot*}(t)x^{\odot*}$ has the same property. Since

$$S^{\odot*}(t+s)x^{\odot*} = T_0^{\odot*}(t+s)x^{\odot*} + \int_0^{t+s} T_0^{\odot*}(t+s-\tau)r^{\odot*}G(y(\tau))d\tau$$

$$= T_0^{\odot*}(t+s)x^{\odot*} + \int_0^{s} T_0^{\odot*}(t+s-\tau)r^{\odot*}G(y(\tau))d\tau + \int_0^{t} T_0^{\odot*}(t-\tau)r^{\odot*}G(y(\tau+s))d\tau$$

$$= T_0^{\odot*}(t)S^{\odot*}(s)x^{\odot*} + \int_0^{t} T_0^{\odot*}(t-\tau)r^{\odot*}G(y(\tau+s))d\tau,$$

the identity $S^{\odot*}(t+s)x^{\odot*} = S^{\odot*}(t)S^{\odot*}(s)x^{\odot*}$ follows provided $y(s+\cdot)$ is the solution of the renewal equation (5.5) with forcing function h_s such that

$$\left\langle\int_0^t T_0^{\odot*}(\tau)S^{\odot*}(s)x^{\odot*}d\tau,\ r^*\right\rangle = \int_0^t h_s(\tau)d\tau.$$

On the other hand the fact that y satisfies (5.5) with the original forcing function h implies that

$$y(t + s) = h(t + s) + \int_0^{t+s} K(t + s - \tau)G(y(\tau))d\tau$$

$$= h(t + s) + \int_0^{s} K(t + s-\tau)G(y(\tau))d\tau + \int_0^{t} K(t-\tau)G(y(\tau+s))d\tau,$$

or, in other words, that $y(s+\cdot)$ is the solution of the Volterra equation (5.5) with forcing function $h(t+s) + \int_0^s K(t+s-\tau)G(y(\tau))d\tau$. So it only remains to be shown that

$$\int_0^t h(\tau+s)d\tau + \int_0^t \int_0^s K(\tau+s-\sigma)G(y(\sigma))d\sigma d\tau = \left\langle\int_0^t T_0^{\odot*}(\tau)S^{\odot*}(s)x^{\odot*}d\tau, r^*\right\rangle.$$

Now observe that

$$\left\langle\int_0^t T_0^{\odot*}(\tau)T_0^{\odot*}(s)x^{\odot*}d\tau, r^*\right\rangle = \left\langle\int_s^{t+s} T_0^{\odot*}(\sigma)x^{\odot*}d\sigma, r^*\right\rangle =$$

$$= \int_s^{t+s} h(\sigma)d\sigma = \int_0^t h(\tau+s)d\tau,$$

and that

$$<\int_0^t T_0^{\odot *}(\tau)\int_0^s T_0^{\odot *}(s-\sigma)r^{\odot *}G(y(\sigma))d\sigma d\tau, r^*> = \int_0^s <\int_0^t T_0^{\odot *}(\tau+s-\sigma)r^{\odot *}d\tau, r^*>G(y(\sigma))d\sigma$$

$$= \int_0^s \int_{s-\sigma}^{t+s-\sigma} K(\tau)d\tau G(y(\sigma))d\sigma = \int_0^t \int_0^s K(\tau+s-\sigma)G(y(\sigma))d\sigma.$$

When combined with (5.7) these two observations yield the required identity.

Finally, that S(t) is the restriction of $S^{\odot *}(t)$ to X is precisely the content of formula (5.6). □

REMARKS i) Many delay equations as well as many age dependent population equations are described by perturbations of the form (5.1).
ii) In both of these applications $\int_0^t T_0^{\odot *}(t-\tau)r^{\odot *}h(\tau)d\tau \in X$ for any L_∞-function h and, moreover, $T_0^{\odot *}(t)x^{\odot *} \to 0$ for $t \to \infty$. As a consequence all of the interesting dynamics occurs in X.

References

[1] P.L. Butzer and H. Berens, Semigroups of Operators and Approximation, Springer, Berlin, 1967.

[2] Ph. Clement, O. Diekmann, M. Gyllenberg, H.J.A.M. Heijmans and H.R. Thieme, Perturbation theory for dual semigroups, I. The sun-reflexive case, Math. Ann. 277 (1987) 709-725.

[3] Ph. Clement, O. Diekmann, M. Gyllenberg, H.J.A.M. Heijmans and H.R. Thieme, Perturbation theory for dual semigroups, II. Time-dependent perturbations in the sun-reflexive case, Proc. Roy. Soc. Edinb. to appear.

[4] Ph. Clement, H.J.A.M. Heijmans et al., One-parameter Semigroups, CWI Monograph 5, North-Holland, Amsterdam, 1987.

[5] M.G. Crandall, A generalized domain for semigroup generators, Proc. AMS 37 (1973) 434-439.

[6] G. Da Prato and E. Sinestrari, Differential operators with non-dense domain and evolution equations. Preprint.

[7] W. Desch, I. Lasiecka and W. Schappacher, Feedback boundary control problems for linear semigroups, Isr. J. Math. 51 (1985) 177-207.

[8] W. Desch and W. Schappacher, Linearized stability for nonlinear semigroups, in: Differential Equations in Banach Spaces (A. Favini & E. Obrecht, eds.) Springer LNiM 1223 (1986) 61-73.

[9] W. Desch, W. Schappacher and Kang Pei Zhang, Semilinear evolution equations, Preprint.

[10] O. Diekmann, Perturbed dual semigroups and delay equations, to appear in the proceedings of a meeting on Infinite Dimensional Dynamical Systems, Lisbon, 1986.

[11] G. Greiner, Perturbing the boundary conditions of a generator, Houston J. Math. to appear.

[12] G. Greiner, Semilinear boundary conditions for evolution equations, Semesterbericht Funktionalanalysis, Tübingen, Wintersemester 1986/87, 127-153.

[13] J.K. Hale, Theory of Functional Differential Equations, Springer, New York, 1977.

[14] E. Hille and R.S. Phillips, Functional Analysis and Semi-groups, AMS, Providence, 1957.

[15] H. Kellermann, Integrated Semigroups, Ph.D. Thesis, Tübingen, 1986.

[16] H. Kellermann, Linear evolution equations with time-dependent domain, Semesterbericht Funktionalanalysis, Tübingen, Wintersemester 1985/86, 15-44.

[17] J.A.J. Metz and O. Diekmann (eds), Dynamics of Physiologically Structured Populations, Springer Lecture Notes in Biomathematics 68, Springer, Berlin, 1986.

[18] A. Pazy, Semigroups of Linear Operators and Applications to Partial Differential Equations, Springer, New York, 1983.

[19] G.F. Webb, Theory of Nonlinear Age-dependent Population Dynamics, Marcel Dekker, New York, 1985.

Ph. Clément
Delft University of Technology
Department of Mathematics and
Informatics
Julianalaan 132, Postbus 356
2600 AJ Delft
The Netherlands.

O. Diekmann
Centre for Mathematics and
Computer Science
PO Box 4079
1009 AB Amsterdam
The Netherlands

and

University of Leiden
Institute of Theoretical Biology
Groenhovenstraat 5
2311 BT Leiden
The Netherlands.

M. Gyllenberg
Helsinki University of Technology
Department of Mathematics and
Systems Analysis
SF - 02150 Espoo
Finland

H.J.A.M. Heijmans
Centre for Mathematics and
Computer Science
P.O. Box 4079
1009 AB Amsterdam
The Netherlands

H.R. Thieme
Universität Heidelberg
Sonderforschungsbereich 123
Im Neuenheimer Feld 294
D-6900 Heidelberg
FRG.

C. CORDUNEANU

Some global problems for Volterra functional differential equations

In this paper we shall deal with linear functional differential equations of the form

$$\dot{x}(t) = (Lx)(t) + f(t), \tag{1}$$

where L stands for a linear continuous Volterra operator on the space $L^2(J,R^n)$, and $f \in L^2(J,R^n)$, with $J = [0,T]$, $T > 0$, or $J = R_+ = [0,\infty)$. The case when f is substituted by a nonlinear term, say $f = (fx)(t)$, $t \in J$, will be also considered:

$$\dot{x}(t) = (Lx)(t) + (fx)(t). \tag{2}$$

The problems to be investigated in regard to equations (1) or (2) are typical boundary value problems in case $J = [0,T]$, such as finding solutions to (1) or (2) verifying the condition

$$Ax(0) + Bx(T) = c, \tag{3}$$

or admissibility-type problems [1], [6] in case $J = R_+$.

Before we get into details concerning the problems sketched above, we shall briefly discuss the integral representation of the solution of the linear equation (1), satisfying the usual initial condition $x(0) = x^0 \in R^n$.

It is shown in [2] that, under above specified conditions on L and f, there exists a unique absolutely continuous solution of (1) verifying $x(0) = x^0$, and this solution is given by the formula

$$x(t) = X(t,0)x^0 + \int_0^t X(t,s)f(s)ds. \tag{4}$$

The matrix-valued kernel $X(t,s)$ is defined for $0 \leq s \leq t$, $t \in J$, and enjoys other properties: $X(t,t) = I$, $X(t,s)$ is absolutely continuous in t for fixed s etc. Further properties will be discussed below.

The validity of (4), including the existence of the kernel $X(t,s)$, results from the existence of resolvent kernels attached to Volterra classical operators

$$(Kx)(t) = \int_0^t k(t,s)\, x(s)ds,$$

in the space $L^2(J,R^n)$, or in the space $L^2_{loc}(J,R^n)$ - when $J = R_+$.

Let us emphasize the fact that in (1) or (2), L stands for an abstract Volterra operator. This means that $x(s) = y(s)$ a.e. on $[0,t]$ implies $(Lx)(s) = (Ly)(s)$ a.e. on $[0,t]$, for any $t \in J$.

We can consider now the problem (1), (3), which is a typical boundary value problem within the class of functional-differential equations, without functional initial data.

Let us substitute $x(t)$ given by (4) in (3), in order to determine the initial value x^0 that produces the solution verifying (3). One obtains for x^0 the equation

$$[A+BX(T,0)]x^0 = c - B \int_0^T X(T,s)f(s)ds. \tag{5}$$

If we want unique solution to problem (1), (3), then necessarily the condition

$$\det[A + BX(T,0)] \neq 0 \tag{6}$$

must hold true. This leads to

$$x^0 = [A + BX(T,0)^{-1}[c-B \int_0^T X(T,s)f(s)ds], \tag{7}$$

and if this value of x^0 is substituted in (4), one obtains

$$x(t) = x^0(t) + \int_0^T \tilde{X}(t,s)f(s)ds, \tag{8}$$

with

$$x^0(t) = X(t,0)[A + BX(T,0)]^{-1}c, \tag{9}$$

and

$$\tilde{X}(t,s) = \begin{cases} X(t,s) - X(t,0)[A+BX(T,0)]^{-1}BX(T,s), & 0 \leqq s \leqq t \leqq T, \\ -X(t,0)[A+BX(T,0)]^{-1}BX(T,s), & 0 \leqq t < s \leqq T. \end{cases} \tag{10}$$

Let us notice that the matrix $\tilde{X}(t,s)$ has a first kind discontinuity along the diagonal $t = s$, the jump being the unit matrix I.

Therefore, we can state the following result:

LEMMA 1. Assume that the operator $L:L^2([0,T],R^n) \to L^2([0,T],R^n)$ is of Volterra type, linear and continuous, and $f \in L^2([0,T],R^n)$.

Then the boundary value problem (1), (3) has unique solution, given by (8), provided condition (6) holds true.

A more general boundary value condition is

$$\int_0^T [dA(s)]x(s) = b \in R^n, \tag{3'}$$

where $A(s)$ is a matrix with bounded variation on $[0,T]$, which we can assume to be left continuous. One obtains as condition of existence for the solution of (1), (3'), for any $b \in R^n$

$$\det\left(\int_0^T [dA(s)]X(s,0)\right) \neq 0, \tag{6'}$$

which should substitute condition (6).

We will consider now the nonlinear system (2), under the same boundary value condition (3). A fixed point method seems to be most adequate in investigating the solvability of (2), (3), due to the fact that we already possess an existence result for the linear problem (1), (3).

It is natural to assume that the operator L verifies the conditions specified in Lemma 1, as well as condition (6). Of course, this condition is also related to the operator L, because $X(t,s)$ is completely determined by L.

In regard to the (nonlinear) operator f, we can make various kinds of assumptions. We could, for instance, assume that f is an operator acting on $L^2([0,T],R^n)$, satisfying other (growth) conditions. But x needs to be absolutely continuous, and consequently we will not have any advantage in assuming f given on a "larger" space than necessary. Therefore, we can assume that f is defined only on the space of continuous functions $C([0,T],R^n)$, and takes values into the same space. Of course, it would be possible to accept operators f that are defined only on the space of absolutely continuous functions, while their range is in $L^2([0,T],R^n)$. But assuming $f: C([0,T],R^n \rightarrow C([0,T],R^n)$ will help us handling the proof of existence in a more direct way.

THEOREM 2. Let us consider the problem (2), (3), and assume that the operator L verifies the conditions of Lemma 1. Moreover, let condition (6) be satisfied.

If $f: C([0,T],R^n) \rightarrow C([0,T],R^n)$ is a continuous operator such that

$$\sup\{|fx|_C;\ |x|_C \leq \rho\} = \Phi(\rho) < \infty \tag{11}$$

for any $\rho > 0$, and the inequality

$$|x^0(t)|_C + \gamma\sqrt{T}\,\phi(\rho) \leq \rho, \tag{12}$$

with $x^0(t)$ given by (9), and γ defined by

$$\sup\{\int_0^T |\tilde{X}(t,s)|^2\,ds;\ t \in [0,T]\} = \gamma^2 \quad , \tag{13}$$

has at least one positive solution ρ^*, then there exists at least one solution x(t) of (2), (3) in Carathéodory sense, such that $|x(t)|_C \leq \rho^*$.

REMARK. A condition that will assure the existence of a positive solution to (12), for any $x^0(t)$ continuous on [0,T], is

$$\lim_{\rho\to\infty} \frac{\phi(\rho)}{\rho} = 0. \tag{14}$$

PROOF OF THEOREM 2. The plan of the proof is very simple. One considers the operator

$$(Ux)(t) = x^0(t) + \int_0^T \tilde{X}(t,s)(fx)(s)ds \tag{15}$$

on the space $C([0,T],R^n)$, for which a fixed point is a solution of (2), (3). The problem (2), (3) is obviously equivalent to the integral equation

$$x(t) = x^0(t) + \int_0^T \tilde{X}(t,s)(fx)(s)ds, \tag{16}$$

with $x^0(t)$ and $\tilde{X}(t,s)$ defined by the formulas (9) and (10). This easily follows from the validity of the formula (8) for the solution of (2), (3).

In view of application of Schauder-Tikhonov fixed point theorem to the operator U defined by formula (15), let us establish first the continuity of this operator on the space $C([0,T],R^n)$. If we take (13) into account, then (15) leads immediately to the inequality

$$|Ux-Uy|_C \leqq \gamma\sqrt{T}\ |fx-fy|_C. \tag{17}$$

From (17) one sees that U is continuous, because so is f.

Of course, it remains to show that γ in (13) does exist. Since $X(t,s)$ is given [2] by the formula.

$$X(t,s) = I + \int_s^t \tilde{k}(t,u)du,$$

with $\tilde{k}(t,u)$ the resolvent (matrix) kernel corresponding to the kernel $k(t,s)$ in the representation

$$\int_0^t (Lx)(s)ds = \int_0^t k(t,s)x(s)ds$$

valid on $L^2([0,T],R^n)$, there results that $k(t,s)$ is a Hilbert-Schmidt square integrables kernel on $0 \leqq s \leqq t \leqq T$. The same property is enjoyed by the resolvent kernel $\tilde{k}(t,s)$, which leads to the conclusion that the map $t \to X(t,\cdot)$ of $[0,T]$ in $L^2([0,T],R^n)$ is continuous. By means of (10) one derivesthe conclusion that $t \to X(t,\cdot)$ is also continuous from $[0,T]$, into $L^2\ [0,T],\ R^n)$. This implies the existence of γ in (13).

Next property we have to prove for the operator U is the comapctness (also in the sense of the topology of the space $C([0,T],R^n)$). Let $v = Ux$ for $x \in C([0,T],R^n)$, with $|x|_C \leqq \rho$. On behalf of (15), (13) and (11) one obtains

$$|v|_C \leqq |x^0(t)|_C + \gamma \sqrt{T}\phi(\rho). \tag{18}$$

The inequality (18) implies the boundedness of the operator U on $C([0,R],R^n)$, which in turn implies the boundedness of the set $\{v; v = Ux, |x|_C \leqq \rho\}$ in $L^2([0,T],R^n)$. Consequently, the set $\{Lv; v = Ux, |x|_C \leqq \rho\}$ is also bounded in $L^2([0,T],R^n)$. But

$$v'(t) = (Lv)(t) + (fx)(t),$$

which means that the set $\{v'(t);\ v = Ux,\ |x|_C \leqq \rho\}$ is bounded in $L^2([0,T],R^n)$. This property immediately implies the equicontinuity of the set $\{v;\ v = Ux,\ |x|_C \leqq \rho\}$ in $[0,T]$.

Indeed, one has to rely on the inequality

$$|v(t) - v(s)|^2 \leqq |t-s| \int_0^T |v'(u)|^2 du,$$

which follows from $v(t) - v(s) = \int_s^t v'(u)du$.

The only property we have to secure for the operator U on $C([0,T],R^n)$ is the existence of a closed convex set, which is taken by U into itself. Since the estimate (18) actually means

$$|Ux|_C \leqq |x^0(t)|_C + \gamma \sqrt{T}\ \phi(\rho) \text{ for } |x|_C \leqq \rho ,$$

there results that the ball of radius $\rho > 0$, centred at the origin of the space $C([0,T],R^n)$, is taken into itself by the operator U if ρ satisfies (12). Or, by assumption, there is at least one positive solution ρ^* to (12), which means that $U\Sigma_{\rho^*} \subset \Sigma_{\rho^*}$, where Σ_ρ denotes the ball of radius ρ, centred at the origin of the space $C([0,T],R^n)$.

Therefore, the conditions of Theorem 2 assure the existence of a solution to the problem (2), (3). This ends the proof of Theorem 2.

Let us consider now the following boundary value problem, depending on a parameter. This kind of problem constitutes a generalization of the classical Sturm-Liouville problem on eigenvalues and eigenfunctions. Namely, let us deal with the system

$$x'(t) = (Lx)(t) + \lambda M(t)x(t), \tag{19}$$

under the homogeneous boundary value condition

$$Ax(0) + Bx(T) = \theta, \tag{20}$$

where $\theta \in R^n$ is the zero vector. The notations have the same meaning as above, while $M(t)$ is a matrix $n \times n$, whose entries are assumed, for simplicity, to be continuous fron $[0,T]$ into R or into the complex field.

A natural question is whether (19), (20) admits nonzero solutions, at least for some values of λ.

Of course, a more general problem would consist in substituting to $M(t)x(t)$ an operator $(Mx)(t)$, and to the boundary value condition (20) a condition of the form (3'). We do not pursue now the greatest generality in this regard.

If we make use again of the formula (4), and we assume the existence of a solution to (19), (20), then we find out easily that such a solution must verify the integral equation

$$x(t) = \lambda \int_0^T \tilde{X}(t,s)M(s)x(s)ds, \tag{21}$$

which is a Fredholm type equation in the classical sense, with piece-wise continuous (matrix) kernel.

The answer to the generalized Sturm-Liouville problem formulated above is now obvious: excepting a set of values of λ at most countable, the Sturm-Liouville problem (19), (20) has only the trivial solution $x \equiv \theta$. Since the kernel in (21) is not symmetric, we cannot infer the existence of at least one λ such that (21) has non-zero solutions. Of course, the same is true for the problem (19), (20).

In the remaining part of this paper we shall consider the case when $J = R_+$. As mentioned above, the hypotheses on L and f occurring in equation (1) can be formulated as follows:

a) L is a Volterra type operator from $L^2_{loc}(R_+,R^n)$ into itself, linear and continuous.

b) $f \in L^2_{loc}(R_+,R^n)$.

As shown in [2], under assumptions a) and b), there exists a unique (locally absolutely continuous) solution of (1), defined on R_+, and such that $x(0) = x^0 \in R^n$. Moreover, this solution is given by the formula (4), where $X(t,s)$ enjoys locally the same properties as in the case $J = [0,T]$.

The kind of problem we shall consider now pertains to the so-called admissibility type [1], [6], and it is concerned with establishing connections in between the behaviour of solutions and that of the forcing term f. There are many results available in this regard for the case of ordinary differential systems [6], as well as in the case of integral operators and equations [1]. In particular, it is shown in [1] that some admissibility problems for delay equations can be reduced to convenient admissibility problems for integral operators/equations.

Briefly, let (B,D) a pair of function spaces, such that $B,D \subset L^2_{loc}(R_+,R^n)$. We will assume that both B,D are endowed with topologies stronger than the topology of the space $L^2_{loc}(R_+,R^n)$.

We shall say that the pair (B,D) is admissible with respect to the system (1), if for every $f \in B$ there exists a solution $x \in D$ for (1).

The concept of admissibility has originated with Perron (1930). In the 1940's Bellman has brought interesting contributions, showing the role of functional analytic methods. The concept has been further developed in the 1950's and 1960's by J.L. Massera and J.J. Schaffer [6], who have reached a very high level of generality. Further developments of the admissibility concept, mainly in the case of integral equations, are due to C. Corduneanu [1]. The abstract case, with general operators, has been dealt with in J.M. Cushing [3], [4].

Let us now assume that (B,D) is an admissible pair with respect to the system (1). Since the solution of the corresponding homogeneous system ($f \equiv \theta \in D$) can be represented as $x(t) = X(t,0)x^0$, it is obvious that $X(t,0)x^0 \in D$ for $t \in R_+$ is satisfied for $x^0 = \xi^0 \in X_{0D}$, where X_{0D} stands for a subspace of R^n (hence $X_{0D} = R^k$, $0 \leq k \leq n$). We denote by P the projection operator of R^n onto X_{0D}. Of course, $R^n = X_{0D} \oplus X_{1D}$, with X_{1D} any complement of X_{0D} in R^n.

The following result can be established for an admissible pair, with respect to (1).

LEMMA 3. Assume (B,D) is admissible with respect to (1). Then there exists unique solution $x(t) \in D$ of (1), such that

$$Px(0) = \xi^0 \in X_{0D}. \tag{22}$$

The following estinate holds ture:

$$|x|_D \leqq C \|\xi^0\| + M|f|_B, \tag{23}$$

with C and M positive constants depending only on L, B, and D.

Proof. Since existence of D-solutions is a consequence of the admissibility hypothesis of the pair (B,D), it remains to prove the uniqueness of the solution verifying the partial initial condition (22). This uniqueness is a consequence of the fact that (1) has exactly one solution such that $x(0) \in X_{1D}$. This solution added to any solution of the form $X(t,0)\xi^0$, $\xi^0 \in X_{0D}$, will provide all D-solutions of (1), for fixed $f \in B$. The unique solution of (1), satisfying (22) is obtained adding to the solution $X(t,0)\xi^0$ of the homogeneous equation the unique solution of (1) with $x(0) \in X_{1D}$.

Then, an application of the closed graph theorem to the map $(\xi_0,f) \to x$ provides the continuity of this map and the estimate (23). We omit the details, which in a slightly modified form appear in [2]-[6].

By means of the Lemma 3, we can establish a nonlinear result corresponding to the system (2). The existence (and unqiueness) of a D-solution to (2) will be proven, assuming that $f:D \to B$, and verifies other conditions.

THEOREM 4. Assume that the pair (B,D) is admissible with respect to the system (1). Let $f : D \to B$ be a map verifying on D the Lipschitz condition

$$|fx-fy|_B \leqq \lambda|x-y|_D. \tag{24}$$

If $\lambda < M^{-1}$, then the system (2) has a unique D-solution which satisfies the initial condition (22).

Proof. We shall consider on D the operator $u \to x$, where x is the D-solution of the equation

$$\dot{x}(t) = (Lx)(t) + (fu)(t), \qquad (25)$$

satisfying the condition (22). The Lemma 3 guarantees the existence and uniqueness of x.

If we denote by T the operator $u \to x = Tu$, then we can write for arbitrary $u,v \in D$ the inequality (see Lemma 3)

$$|Tu-Tv|_D \leqq M|fu-fr|_B,$$

and according to (24) one obtains

$$|Tu-Tv|_D \leqq \lambda M|u-v|_D.$$

Since $\lambda M < 1$ by hypotheses, there results that T is a contraction on D.

<u>REMARK</u>. The spaces B and D can be chosen in a variety of ways. One possible choice is $B = D = L^\infty(R_+,R^n)$, which leads to the problem of bounded solutions for bounded forcing terms f. This is the case to be considered for the first time by Perron (1930), for the class of ordinary differential equations.

Unfortunately, in our (general) case, it does not seem possible to relate the admissibility with the properties of $X(t,s)$ because the semigroup property $X(t,s)X(s,u) = X(t,u)$ does not hold anymore.

If we specialize the admissibility concept, to strengthen it to <u>full admissibility</u>, i.e. to require all solutions to be in D for each fixed $f \in B$, then in the case $B = D = L^\infty(R_+,R^n)$ one obtains as necessary and sufficient conditions the following:

$$\|X(t,0)\| \leqq A, \int_0^t \|X(t,s)\| \, ds \leqq B, \ t \in R_+,$$

with A,B positive constants. See [1] for details.

<u>References</u>

[1] Corduneanu, C: Integral Equations and Stability of Feedback Systems. Academic Press, New York, 1973.

[2] Corduneanu, C: Integral Equations and Applications. Cambridge University Press (in Preparation).

[3] Cushing, J.M: Admissible operators and solutions of perturbed operator equations. Funk. Ekvacioj, 19 (1976), 79-84.

[4] Cushing, J.M: Strongly admissible operators and Banach spaces solutions of nonlinear equations. Ibidem, 20 (1977), 237-245.

[5] Hartman, Ph. and Onuchic, Nelson: On the asymptotic integration of ordinary differential equations. Pacific J. Math., 13 (1963), 1193-1207.

[6] Massera, J.L. and Schäffer, J.J: Linear Differential Equations and Function Spaces. Academic Press, New York, 1966.

C. Corduneanu
Department of Mathematics
University of Texas
Arlington
Texas 76019
USA.

W. DESCH, R. GRIMMER & W. SCHAPPACHER

Propagation of singularities by solutions of second order integrodifferential equations

1. INTRODUCTION

In this paper we examine propagation of singularities by solutions of second order linear hyperbolic partial differential integral equations. Our method involves posing the problem as an abstract second order linear integrodifferential equation

$$u''(t) = Au(t) + \int_0^t B(t-s)u(s)ds, \ t \geq 0, \qquad u(0) = u_0, \ u'(0) = u_1 \tag{1.1}$$

in a Banach space $(X, \|\cdot\|)$. It shall be assumed throughout that A generates a cosine family C(t) on X and that B(t) is a linear operator on X with domain at least D(A).

Earlier work by two of the authors, [2, 3], dealt with similar questions for first order integrodifferential equations. This work was applied to a variety of one-dimensional problems arising in the study of viscoelastic materials and to certain special two-dimensional problems. However, general two- and three-dimensional problems of the type occurring in viscoelasticity could not be treated with the techniques developed earlier. The purpose of this paper is to present a technique based on the theory of second order differential equations in Banach space which will apply to various problems involving two or three space dimensions.

Additional work on related topics appears in [1,6,7,8,11,12,13].

2. PRELIMINARIES

Throughout this paper $(X, \|\cdot\|)$ will denote a Banach space. In addition we shall assume the operators A and B on X satisfy:

(H1) A is the infinitesimal generator of a cosine family C(t) on X. Thus, the domain of A endowed with the graph norm $\|x\| + \|Ax\|$ is again a Banach space which we denote by $(Y, \|\cdot\|_Y)$.

(H2) For each $t \geq 0$, $B(t)$ is a bounded linear operator from $(Y, \|\cdot\|_y)$ to $(X, \|\cdot\|)$.

(H3) For each $x \in Y$, the map $(Bx)(t) = B(t)x$ is in $W^{3,1}_{loc}([0,\infty);X)$.

We note that it follows from (H2) that $B(t)$ can be written $B(t) = F(t)A + K(t)$ where $F(t)$ and $K(t)$ are bounded operators on X. One way of obtaining such a factorization of $B(t)$ is to write $B(t) = -B(t)(\lambda-A)^{-1}A + \lambda B(t)(\lambda-A)^{-1}$ where λ is in $\rho(A)$. One easily sees that $B(t)(\lambda-A)^{-1}$ is a bounded linear operator on X and, in addition, for each $x \in X$, $(Fx)(t)$ and $(Kx)(t)$ are in $W^{3,1}_{loc}([0,\infty);X)$. This is not usually a very useful factorization, however a more useful factorization often presents itself. For this reason we shall restrict ourselv es to the case $B(t) = F(t)A + K(t)$ in this paper.

One can show well-posedness of the initial value problem for (1.1) with assumptions (H1)-(H3). In fact, $(Bx)(t) \in W^{2,1}_{loc}([0,\infty);X)$ will suffice, [9], [16]. Also see [14] where existence is shown for $u_0 \in D(A)$ and $u_1 \in E = \{x \in X: C(t)x$ is once continuously differentiable on $R\}$. For our purposes here it will be convenient to follow the approach of Tsuruta [16]. In particular, Tsuruta shows that under somewhat weaker hypotheses than (H1)-(H3) there exist two one-parameter families $U_C(t)$, $U_S(t)$ of bounded operators on X which satisfy (1.1) in the strong sense. In particular, for the equation

$$u''(t) = Au(t) + \int_0^t B(t-s)u(s)ds + f(t), \quad t \geq 0,$$

$$u(0) = u_0,\ u'(0) = u_1,$$

one has the variation of parameters formula

$$u(t) = U_C(t)u_0 + U_S(t)u_1 + \int_0^t U_S(t-\tau)f(\tau)d\tau. \tag{2.1}$$

This formula yields a strong solution in case $u_0 \in D(A)$, $u_1 \in E_1 = \{x:U_C(t)x$ is continuously differentiable$\}$ and $f \in W^{1,1}_{loc}(R^+;X)$.

The main result is in the spirit of our earlier work and will compare solutions of (1.1) with those of an associated differential equation of the form

$$v''(t) = [A+F_1]v(t) + F_2v'(t)$$
$$v(0) = u_0, \; v'(0) = u_1, \tag{2.2}$$

where F_1 and F_2 are bounded operators. The presence of the first order term, $F_2v'(t)$, necessitates some preliminary discussion, however.

Associated with (2.2) are solution operators $\mathcal{C}(t)$ and $\mathcal{S}(t) \in \mathcal{B}(X)$ where $v(t) = \mathcal{C}(t)u_0$ is the solution of (2.2) with $v(0) = u_0$, $v'(0) = 0$ while $v(t) = \mathcal{S}(t)u_1$ is the solution of (2.2) with $v(0) = 0$, $v'(0) = u_1$. Thus, the arbitrary solution $v(t)$ of (2.2) is given by $v(t) = \mathcal{C}(t)u_0 + \mathcal{S}(t)u_1$. We shall be particularly interested in the variation of parameters formula and when this formula yields a (classical) solution of (2.2) rather than a weak solution.

<u>LEMMA 2.1</u>. Suppose A generates a cosine family on X and that F_1, F_2 are bounded operators on X. Then there exist strongly continuous families of bounded operators $\mathcal{C}(t)$ and $\mathcal{S}(t)$ on X so that if $u_0 \in D(A)$ and $u_1 \in E = \{x \in X: \mathcal{C}(t)x \text{ is continuously differentiable}\}$, then

$$v(t) = \mathcal{C}(t)u_0 + \mathcal{S}(t)u_1 \tag{2.3}$$

is a solution of (2.2). In addition, if $0 \leq t \leq T$, there exists K(T) so that

$$\|v(t)\| \leq K(T)(\|u_0\| + \|u_1\|), \; 0 \leq t \leq T. \tag{2.4}$$

Finally, if $f \in W^{1,1}_{loc}(R^+,X)$, $u_0 \in D(A)$ and $u_1 \in E$, then

$$v(t) = \mathcal{C}(t)u_0 + \mathcal{S}(t)u_1 + \int_0^t \mathcal{S}(t-\tau)f(\tau)d\tau \tag{2.5}$$

is a solution of

$$v''(t) = [A+F_1]v(t) + F_2v'(t) + f(t)$$
$$v(0) = u_0, \; v'(0) = u_1. \tag{2.6}$$

In particular, $v(t)$ is twice differentiable, $v(t) \in D(A)$, and $Av(t)$ is continuous, $t \geq 0$. Also, if $f(t)$ is continuous and $v(t)$ is a solution of (2.6), then $v(t)$ is given by (2.5).

Proof. It is known [5, p. 46] that E becomes a Banach space with norm

$$\|x\|_0 = \max \{\|x\|, \sup_{t\geq 0} e^{-wt}(1+t)^{-1} \|C'(t)x\|\}$$

where w is the growth rate of $C(t)$. Also,

$$\mathcal{A} = \begin{bmatrix} 0 & I \\ A & 0 \end{bmatrix}$$

generates a semigroup on $E \times X$ with $D(\mathcal{A}) = D(A) \times E$. As F_1 and F_2 are bounded operators on X, F_1 is a bounded operator on E into X and so

$$\mathcal{A}_1 = \begin{bmatrix} 0 & I \\ A+F_1 & F_2 \end{bmatrix}$$

with $D(\mathcal{A}_1) = D(\mathcal{A})$ generates a C_0 semigroup $\mathcal{T}(t) = [T_{ij}(t)]$ i,j = 1,2 with $T_{11}(t) = \mathcal{C}(t)$ and $T_{12}(t) = \mathcal{S}(t)$. If now Eqs. (2.2) or (2.6) are written in system form for the vector $(v(t),v'(t))^t$ one obtains (2.3) and (2.5) as solutions of (2.2) and (2.6), respectively, (cf. [4]). To extend $\mathcal{C}(t)$ to X and obtain (2.4) we note that $A + F_1$ generates cosine and sine families, $C_1(t)$ and $S_1(t)$, ([10] or [15]), and for $u_0 \in D(A)$, $u_1 \in E$,

$$v(t) = C_1(t)u_0 + S_1(t)u_1 + \int_0^t S_1(t-\tau)F_2v'(\tau)d\tau$$

or

$$v(t) = C_1(t)u_0 + S_1(t)u_1 - S_1(t)F_2u_0 + \int_0^t C_1(t-\tau)F_2v(\tau)d\tau. \qquad (2.7)$$

Thus, for $u_0 \in D(A)$,

$$C(t)u_0 = C_1(t)u_0 - S_1(t)F_2u_0 + \int_0^t C_1(t-\tau)F_2C(\tau)u_0 d\tau. \qquad (2.8)$$

As D(A) is dense and $C_1(t)$, $S_1(t)$ and F_2 are bounded on [0,T], the expression given by (2.8) extends to $u_0 \in X$, defining $C(t)$ on X. Also, the expression (2.7) yields

$$S(t)u_1 = S_1(t)u_1 + \int_0^t C_1(t-\tau)F_2S(\tau)u_1 d\tau$$

which together with (2.8) justifies (2.4). The last statement regarding the variation of parameters formula is standard from semigroup theory.

3. MAIN RESULT

As mentioned earlier, our result compares the solutions of (1.1) with those of a second order differential equation. We shall assume that B(t) is written in the form B(t) = F(t)A + K(t). In this case (1.1) can be written

$$\begin{aligned} u''(t) &= Au(t) + \int_0^t F(t-s)Au(s)ds + \int_0^t K(t-s)u(s)ds \\ u(0) &= u_0,\ u'(0) = u_1. \end{aligned} \qquad (3.1)$$

The differential equation of interest in this case is

$$\begin{aligned} v''(t) &= [A+F'(0)-F^2(0)]v(t) + F(0)v'(t) \\ v(0) &= u_0,\ v'(0) = u_1. \end{aligned} \qquad (3.2)$$

THEOREM 3.1. Suppose B(t) = F(t)A + K(t) satisfies (H1)-(H3). Let $U_C(t)$, $U_S(t)$ be the solution operators of (3.1) and $C(t)$, $S(t)$ the solution operators of (3.2). Then $(U_C(t)-C(t))$ and $(U_S(t)-S(t))$ map X into D(A). In addition, for $u_0 \in X$, $(U_C(t)-C(t))u_0$, $(U_S(t)-S(t))u_0 \in C^2(R^+;X) \cap C(R^+;Y)$.

Proof. For convenience we use the convolution notation, $K*H(t) = \int_0^t K(t-u)H(u)du$. Let R(t) be the solution of $R + F + R*F = 0$, so that $(\delta+R)*(\delta+F) = \delta$ where $\delta*H = H$. In convolution notation (3.1) is written

$$u'' = Au + F*Au + K*u$$

$$= (\delta+F)*Au + K*u,$$

$$u(0) = u_0,\ u'(0) = u_1.$$

Thus, if $u_0 \in D(A)$ and $u_1 \in E_1$,

$$u''(t) + R*u''(t) = Au(t) + (K+R*K)*u(t)$$

and integration by parts yields

$$u''(t) = Au(t) - R'(0)u(t) - R(0)u'(t)$$

$$+ ((K+R*K)-R'')*u(t) + R(t)u'(0) + R'(t)u(0).$$

From the relation $R + F + R*F = 0$, we see that $R(0) = -F(0)$ while $R'(0) = -F'(0) + F^2(0)$. Thus, $u(t)$ satisfies

$$u''(t) = [A + F'(0)-F^2(0)]u(t) + F(0)u'(t) + H*u(t) + J(t)$$

where $H = K + R*K - R''$ and $J(t) = R'(t)u_0 + R(t)u_1$. Now $F(t)u_0 \in W^{3,1}_{loc}(\mathbb{R}^+;X)$ implies $R(t)u_0 \in W^{3,1}_{loc}(\mathbb{R}^+,X)$ and, hence, $u(t)$ satisfies

$$u(t) = C(t)u_0 + S(t)u_1 + S*H*u(t) + S*J(t).$$

Also, $u(t) = U_C(t)u_0 + U_S(t)u_1$. This yields

$$\begin{aligned} U_C(t)u_0 + U_S(t)u_1 &= C(t)u_0 + S(t)u_1 \\ &\quad + S*H*[U_C(t)u_0+U_S(t)u_1] \qquad (3.3) \\ &\quad + S*[R'(t)u_0+R(t)u_1]. \end{aligned}$$

However, as $D(A)$ and E_1 are dense in X, [16], we see that (3.3) can be extended to all $u_0,u_1 \in X$. As $H(t)x$, $R(t)x$ and $R'(t)x$ are in $W^{1,1}_{loc}(\mathbb{R}^+;X)$ for each $x \in X$ we see that the result follows from Lemma 2.1 by alternately taking $u_1 = 0$ and $u_0 = 0$.

REMARK 3.2. An examination of the proof of Theorem 3.1 reveals that it is sufficient to assume $F(t)x \in W^{3,1}_{loc}(\mathbb{R}^+;X)$ and $K(t)x \in W^{1,1}_{loc}(\mathbb{R}^+;X)$ for each $x \in X$.

REMARK 3.3. It is possible to obtain a result in the spirit of Theorem 3.1 for cases when a convenient factorization of $B(t)$ is not available. We shall present this result in a later paper. This result will apply directly to the momentum equation for a homogeneous isotropic viscoelastic material mentioned in the next section.

4. APPLICATIONS

The problem of anti-plane strain leads naturally to an equation which we can treat using our methods. This situation arises when displacements are independent of one direction, say x_3. When this occurs, the displacements (u_1,u_2,u_3) may be represented as a superposition of $(u_1,u_2,0)$ and $(0,0,u_3)$ where u_1 and u_2 satisfy a system independent of u_3 and u_3 satisfies (cf. eg. [8, p. 23])

$$\rho u_3'' = \partial_1\sigma_{31} + \partial_2\sigma_{32}, \quad ' = \partial/\partial t$$

where ρ is the mass density, $\partial_i = \partial/\partial x_i$ and the stresses σ_{3i} satisfy the constitutive relations

$$\sigma_{3i} = \mu_0\partial_i u_3(t) + \int_{-\infty}^{t} R_0(t-\tau)\partial_i u_3(\tau)d\tau$$

with $\mu_0 > 0$, $R_0(t) \in W^{3,1}_{loc}(\mathbb{R}^+)$. If we assume the material is initially quiescent and drop the index 3 we obtain, with $\Delta = \partial_1^2 + \partial_2^2$,

$$u''(t) = \mu\Delta u(t) + \int_0^t R(t-\tau)\Delta u(\tau)d\tau \qquad (4.1)$$

where $\mu = \mu_0/\rho$, $R = R_0/\rho$. Here we have suppressed the variables x_1 and x_2. Alternatively, if the material is not initially quiescent we assume the term $\int_{-\infty}^{0} R(t-\tau)\Delta u(\tau)d\tau$ is a known quantity which can be put in a nonhomogeneous term.

If now we let $X = L^2(\mathbf{R}^2)$, $D(A) = H^2(\mathbf{R}^2)$ with $Au = \mu\Delta u$, $B(t)u = F(t)Au = R(t)\Delta u$ we see that A generates a cosine family on X and that the hypotheses of Theorem 3.1 are satisfied. Thus we may compare (4.1) with the equation

$$v''(t) = [\mu\Delta + R'(0) - R^2(0)]v(t) + R(0)v'(t). \tag{4.2}$$

In particular, if one chooses initial data u_0 and u_1 for (4.1) which are in $L^2(\mathbf{R}^2)$, the difference $u(t) - v(t)$ of the solutions $u(t)$ and $v(t)$ of (4.1) and (4.2), respectively, is in $H^2(\mathbf{R}^2)$. As $H^2(\mathbf{R}^2)$ consists of continuous functions $u(t) - v(t)$ is continuous in t and $x \in \mathbf{R}^2$.

Assuming a plane wave, following [17, p. 694], one calculates that if the solution has a discontinuity along $x_1 \cos\theta + x_2 \sin\theta = 0$ at time $t=0$, this discontinuity along the bicharacteristic $x_1 = t\cos\theta$, $x_2 = t\sin\theta$ will have the decay rate $\exp R(0)t/2$.

One can get a similar problem involving three space dimensions by starting with the momentum equation for a homogeneous isotropic viscoelastic material, [8, p. 204],

$$\begin{aligned}\rho u_i''(t) &= (\lambda + \mu)\partial_i\partial_j u_j(t) + \mu\partial_j^2 u_i(t)\\ &\quad + \int_0^t [R_\lambda(t-\tau) + R_\mu(t-\tau)]\partial_i\partial_j u_j(\tau)d\tau\\ &\quad + \int_0^t R_\mu(t-\tau)\partial_j^2 u_i(\tau)d\tau, \quad i = 1,2,3\end{aligned}$$

or, in vector form,

$$\begin{aligned}\rho u''(t) &= (\lambda+2\mu)\,\mathrm{grad}\,\mathrm{div}\,u(t) - \mu\,\mathrm{curl}\,\mathrm{curl}\,u\\ &\quad + \int_0^t [R_\lambda(t-\tau) + 2R_\mu(t-\tau)]\mathrm{grad}\,\mathrm{div}\,u(\tau)d\tau\\ &\quad - \int_0^t R_\mu(t-\tau)\mathrm{curl}\,\mathrm{curl}\,u(\tau)d\tau.\end{aligned}$$

Then the dilatation $\theta = \mathrm{div}\,u$ satisfies

$$\rho\theta''(t) = (\lambda+2\mu)\Delta\theta(t) + \int_0^t [R_\lambda(t-\tau)+2R_\mu(t-\tau)]\Delta\theta(\tau)d\tau$$

while the rotation ω = curl u satisfies

$$\rho\omega''(t) = \mu\Delta\omega \quad + \int_0^t R_\mu(t-\tau)\Delta\omega(\tau)d\tau.$$

One can now examine each of these equations and determine the decay rates for plane waves as $\exp\{(R_\lambda(0)+2R_\mu(0))t/2$ and $\exp R_\mu(0)t/2\}$ on bicharacteristics for θ and ω respectively.

As a final note we remark that these results correspond to the results on decay of acceleration waves obtained in [1].

References

[1] B.D. Coleman and M.E. Gurtin, Waves in materials with memory, II. On the growth and decay of one-dimensional acceleration waves, Arch. Rational Mech. Anal. 19 (1965), 239-265.

[2] W. Desch and R. Grimmer, Propagation of singularities for integro-differential equations, J. Diff. Eq., 65 (1986), 411-426.

[3] W. Desch and R. Grimmer, Initial boundary value problems for integro-differential equations, J. Integral Eq. 10 (1985), 73-97.

[4] W. Desch, R. Grimmer and W. Schappacher, Some considerations for linear integrodifferential equations, J. Math. Anal. Appl. 104 (1984), 219-234.

[5] H.O. Fattorini, Second Order Linear Differential Equations in Banach Spaces, North Holland, Amsterdam, 1985.

[6] K.B. Hannsgen and R.L. Wheeler, Behavior of the solution of a Volterra equation as a parameter tends to infinity, J. Integral Eq. 7 (1984), 229-237.

[7] W.J. Hrusa and M. Renardy, On wave propagation in linear viscoelasticity, Q. Appl. Math. 43 (1985), 237-254.

[8] J.A. Hudson, The Excitation and Propagation of Elastic Waves, Cambridge University Press, 1980.

[9] R.K. Miller and R.L. Wheeler, Well-posedness and stability of linear Volterra integrodifferential equations in abstract spaces, Funkcial. Ekvac. 21 (1978), 279-305.

[10] B. Nagy, On cosine operator functions in Banach spaces, Acta Scientiarum Mathematicarum (Szeged) 36 (1974), 281-289.

[11] A. Narain and D.D. Joseph, Linearized dynamics for step jumps of velocity and displacement of shearing flows of a simple fluid, Rheol. Acta 21 (1982), 228-250, Corrig. Rheol. Acta 22 (1983), 519.

[12] J. Prüss, Positivity and regularity of hyperbolic Volterra equations in Banach spaces, preprint.

[13] M. Renardy, Some remarks on the propagation and nonpropagation of discontinuities in linearly viscoelastic liquids, Rheol. Acta 21 (1982), 251-254.

[14] C.C. Travis and G.F. Webb, An abstract second order semilinear Volterra integrodifferential equation, SIAM J. Math. Anal. 10 (1979), 412-424.

[15] C.C. Travis and G.F. Webb, Perturbation of strongly continuous cosine family generators, Colloquium Math. XLV (1981), 277-285.

[16] K. Tsuruta, Bounded linear operators satisfying second order integro-differential equations in Banach space, J. Int. Eq. 6 (1984), 231-268.

[17] E. Zauderer, Partial Differential Equations of Applied Mathematics, Wiley-Interscience, New York, 1983.

Acknowledgement

This work was done within the US- Austrian Cooperative Research Program between NSF and Fonds zur Förderung der Wissenschaftlichen Forschung (Project P-5691).

W. Desch and W. Schappacher
Institut für Mathematik
Universität Graz
A-8010 Graz
Austria

R. Grimmer
Department of Mathematics
Southern Illinois University
Carbondale
Illinois 62901
U.S.A.

H. ENGLER

Boundary value problems for weak solutions of a class of quasilinear hyperbolic integrodifferential equations

The author was supported by the National Science Foundation under Grant No. DMS 8601762.

0. TWO EXAMPLES FROM VISCOELASTICITY

Consider a one-dimensional body which in its reference configuration is homogeneous with unit density and occupies the interval $[0,L] \subset \underline{R}$ and which moves longitudinally. Let $x + u(x,t)$ be the position of the material particle x at time t; then the equations of motion are

$$u_{tt}(x,t) = \sigma_x(x,t) + f(x,t), \tag{0.1}$$

where f denotes volume forces, and σ is the Piola-Kirchhoff stress.

As constitutive relation we assume the following nonlinear generalization of Boltzmann's linear model (cf. [19] and [33])

$$\sigma(x,t) = p(u_x(x,t)) - \int_{-\infty}^{t} a(t-s)q(u_x(x,s))ds \tag{0.2}$$

with suitable functions p,q: $\underline{R} \to \underline{R}$ that vanish at the origin and with a positive integrable kernel $a(\cdot)$.

Setting $A = \int_0^\infty a(t)\,dt$, (0.2) can also be written in the form

$$\sigma(x,t) = g(u_x(x,t)) + \int_{-\infty}^{t} a(t-s)h(u_x(x,t),u_x(x,s))\,ds, \tag{0.2'}$$

where

$$h(\xi,\zeta) = q(\xi) - q(\zeta), \tag{0.3}$$

$$g(\xi) \quad = p(\xi) - A\cdot q(\xi). \tag{0.4}$$

In this formulation, p describes the instantaneous elastic response of the material, and g describes the equilibrium elastic response. For a solid, one expects g to be increasing at the origin, while $g \equiv 0$ will correspond to a material whose long-time behaviour resembles that of a liquid. To solve (0.1), (0.2) on $[0,L] \times [0,T]$, one has to prescribe the displacement history on $[0,L] \times (-\infty,0]$ and the initial velocity $u_t(\cdot,0)$ on $[0,L]$. In addition, boundary conditions have to be prescribed: if the motion of the material particles 0 and L is known, then $u(0,\cdot)$ and $u(L,\cdot)$ are given on $[0,T]$. If given traction forces act at the endpoints of the body, then the values of $-\sigma(0,\cdot)$ and $+\sigma(L,\cdot)$ are to be prescribed.

As a second example, we consider dynamic rectilinear shear flows of incompressible Non-Newtonian liquids, such as polymer melts and concentrated polymer solutions. The presentation below is based on [4] and on [19]; more details can be found in [15].

Let the particles of the material under consideration be labelled by Cartesian body coordinates $\underline{\xi} = (\xi^1,\xi^2,\xi^3) \in \underline{Q} \subset \underline{R}^3$, where Q is some reference domain, and let $\underline{y}(\underline{\xi},t) \in \underline{R}^3$ denote the spatial position of the particle $\underline{\xi}$ at time t. We want to study such motions for which either

$$\underline{y}(\underline{\xi},t) = (\xi^1,\xi^2,\xi^3 + u(\xi^1,t)) \tag{0.5}$$

with $\xi^1 \in \Omega = [0,L] \subset \underline{R}$, ξ^2, $\xi^3 \in \underline{R}$, corresponding to rectilinear shear flow between two plates of separation L, or

$$\underline{y}(\underline{\xi},t) = (\xi^1,\xi^2,\xi^3 + u(\xi^1,\xi^2,t)) \tag{0.6}$$

with $(\xi^1,\xi^2) \in \Omega \subset \underline{R}^2$, $\xi^3 \in \underline{R}$, corresponding to rectilinear flow in a pipe with cross-section $\Omega \subset \underline{R}^2$. Only forces that depend on the same variables as u and that act in the ξ^3-direction (as volume forces in Ω or as traction forces on $\partial\Omega$) are allowed. To describe the behaviour of the material, again balance laws (corresponding to (0.1)) and constitutive relations (corresponding to (0.2)) have to be combined, resulting in a nonlinear system of second order partial integrodifferential equations for three unknown functions. For a widely used class of such relations, the K-BKZ-model (proposed in [4]; see also [5], [18], [27]) and for motions of the form (0.5) and (0.6), the system of equations of motion can be reduced to a single second order equation for u which can be written as the system

$$u_{tt}(x,t) = \mathrm{div}_x \sigma(x,t) + f(x,t) \tag{0.7}$$

$$\sigma(x,t) = \int_{-\infty}^{t} a(t-s)p(|\nabla u(x,t) - \nabla u(x,s)|^2)(\nabla u(x,t) - \nabla u(x,s))ds, \tag{0.8}$$

if again the density is assumed to be one. Here $x = \xi^1$ or $x = (\xi^1, \xi^2)$, and the function $p(\cdot)$ and the kernel $a(\cdot)$ depend on the material. In typical physical situations, at the boundary of Ω for $0 \leq t \leq T$ either the displacement $u(\cdot,t)$ or the values of traction forces will be known, i.e. the values of $\nu(x)\cdot\sigma(x,t)$, where ν is the unit exterior normal vector field on $\partial\Omega$. To solve (0.7), (0.8) on $\Omega \times [0,T]$, one has to prescribe additionally the history of the deformation u for $t < 0$, the initial displacement $u(\cdot,0)$ and the initial velocity $u_t(\cdot,0)$. Depending on the kernel $a(\cdot)$, a discontinuity in t of u at $t = 0$ may still result in finite values for the stress σ, i.e. a jump in the displacement may be dynamically admissible. The systems (0.1), (0.2)' and (0.7), (0.8) are scalar quasilinear hyperbolic integrodifferential equations, and in this paper we show how to construct certain solutions for two basic boundary value problems for such equations. The reader is referred to [33] for a discussion of other viscoelastic models and for a comprehensive treatment of special problems such as the ones sketched above, as well as of the full three-dimensional equations of motion.

1. STATEMENT OF THE PROBLEM

Let $\Omega \subset \underline{R}^n$ be an open, bounded and connected set with piecewise Lipschitz boundary $\partial\Omega$ and unit exterior normal vector field $\nu: \partial\Omega \to S^{n-1}$ (defined outside a subset of $\partial\Omega$ with $(n-1)$-dimensional Hausdorff measure zero). Let $g:\underline{R}^n \to \underline{R}^n$, $h:\underline{R}^{2n} \to \underline{R}^n$, $a:(0,\infty) \to \underline{R}$, $f: \Omega \times [0,T] \to \underline{R}$ and $k:\partial\Omega \times [0,T \to \underline{R}$ be given functions. We want to study second order integrodifferential equations for the unknown function $u: \Omega \times [0,T] \to \underline{R}$ of the form

$$u_{tt}(x,t) = \mathrm{div}_x \sigma(x,t) + f(x,t) \text{ on } \Omega \times (0,T], \tag{1.1}$$

$$\begin{aligned} \sigma(x,t) &= g(\nabla u(x,t)) + \int_0^{\infty} a(s)h(\nabla u(x,t),\nabla u(x,t-s))ds \\ &= g(\nabla u(x,t)) + \int_{-\infty}^{t} a(t-s)h(\nabla u(x,t),\nabla u(x,s))ds, \end{aligned} \tag{1.2}$$

subject to initial conditions

$$u(\cdot,0) = u_0,\ u_t(\cdot,0) = u_1 \text{ in } \Omega, \tag{1.3}$$

$$u \equiv \psi \quad \text{on } \Omega \times (-\infty,0) \tag{1.4}$$

and boundary conditions on $\partial\Omega \times [0,T]$

$$u(x,t) = \varphi(x,t) \tag{1.5a}$$

or

$$\nu(x)\cdot\sigma(x,t) = k(x,t) \tag{1.5b}$$

for suitable data u_0, u_1, ψ and φ such that the integrals in (1.2) converge. We always assume that the function φ is defined on $\bar{\Omega} \times [0,T]$. Two examples for such equations were given in the previous section.

The system (1.1), (1.2) formally resembles a quasilinear wave equation. However, for nonlinear problems of this form only local existence of classical solutions for arbitrary smooth data is known (see e.g. [7], [10]). In general, classical solutions of the nonlinear problem must develop singularities in finite time ([24]), and techniques leading to local and global existence and uniqueness results for weak solutions of related one-dimensional problems (with $\Omega = \underline{R}$) have only recently been developed ([1], [13], [30]). On the other hand, in the study of wave propagation in visco-elastic materials in [8] it was shown implicitly that "regularizing effects" should be expected, if the kernel a(.) has a suitable singularity at 0.

The main purpose of this paper is to use this additional property of the equation to prove the existence of solutions.

We study weak solutions of (1.1), (1.2) (defined below) and present global existence results for data of arbitrary size for the two initial-boundary value problems (1.1) - (1.4), (1.5a) (the "Dirichlet problem) in Theorem 1 and for (1.1)-(1.4), (1.5b) (the "Neumann problem") in Theorem 2. These results are stated in Section 2, and sketches of their proofs are given in Section 3. In a related paper [15], further properties of these solutions are derived for the case of zero boundary data. We list these results in

Section 2, but do not give their proofs. We cannot claim uniqueness for the solutions that are constructed below.

The usual notation for Lebesgue and Sobolev spaces will be used. In particular, the norm of $L^2(\Omega,\underline{R})$ and $L^2(\Omega,\underline{R}^n)$ is abbreviated by $\|\cdot\|_0$, the scalar product on these spaces is written $\langle\cdot,\cdot\rangle$. The norm on the Hilbert space $H^1(\Omega)$ is given by $\|u\|_1^2 = \|u\|_0^2 + \|\nabla u\|_0^2$. $H_0^1(\Omega)$ is the closure of the space of test functions $u \in C_0^\infty(\Omega)$ under the norm $\|\nabla u\|_0$, equivalent to $\|\cdot\|_1$ and $H^{-1}(\Omega)$ is the set of all distributions u on Ω for which $|u(\varphi)| \leqq C\cdot \|\varphi\|_1$ with associated norm $\|\cdot\|_{-1}$, i.e. the dual space of $H_0^1(\Omega)$. We write $H_b^{-1}(\Omega)$ for the dual space of $H^1(\Omega)$ and shall use $\langle\cdot,\cdot\rangle$ both for the L^2-scalar product and for the product between elements of $H^{-1}(\Omega)$ and $H_0^1(\Omega)$. Also, $\langle\cdot,\cdot\rangle_{\partial\Omega}$, $\|\cdot\|_{0,\partial\Omega} = \|\cdot\|_{\partial\Omega}$ refer to the space $L^2(\partial\Omega)$. If H is any Hilbert space, then $C^{1/2}([0,T],H)$ is the usual space of H-valued Hölder-continuous functions on [0,T], and we write

$$W^{1/2}(0,T;H) = \{f \in L^1(0,T;H) \mid \int_0^T \int_0^t \|f(t)-f(s)\|_H \cdot (t-s)^{-3/2} ds\, dt < \infty\}$$

for the interpolation space $[L^1(0,T;H), W^{1,1}([0,T],H]_{1/2,1}$ (cf. [2]) with its natural norm. Then $W^{1/2}(0,T;H) \subset L^2(0,T;H)$; note that functions in $W^{1/2}(0,T;H)$ can have jumps and in general do not possess a trace at $t = 0$. Other spaces and norms are introduced as needed. Norms on function spaces are always denoted by double bars; single bars denote the absolute value or Euclidean norms of vectors or matrices. The letters c and C will be used for various constants whose value can change from line to line. The end of a proof is indicated by the symbol "□".

2. MAIN RESULTS

Throughout this paper, it will be assumed that

$$g(\xi) = \nabla_\xi G(\xi) \text{ for some } G \in C^2(\underline{R}^n,\underline{R}), \tag{2.1}$$

$$h(\xi,\zeta) = \nabla_\xi H(\xi,\xi) \text{ for some } H \in C^2(\underline{R}^{2n},\underline{R}), \tag{2.2}$$

and that

$$|D_\xi g(\xi)|,\ |D_\xi h(\xi,\zeta)|,\ |D_\zeta h(\xi,\zeta)| \leqq K < \infty \tag{2.3}$$

for some constant K, for all ξ, $\zeta \in \underline{R}^n$.

We can assume without loss of generality that

$$G(0) = 0,\ h(0,0) = g(0) = 0,\ H(\xi,\xi) = 0 \text{ for all } \xi \in \underline{R}^n. \tag{2.4}$$

Concerning the kernel $a(\cdot)$, it will be assumed that

$$a(\cdot) \in W^{1,1}_{loc}(0,\infty;\underline{R}),\ a \geqq 0,\ \int_0^\infty (a(t) + \min(1,t)|a'(t)|)dt < \infty\ . \tag{2.5}$$

For $a(\cdot)$ as above and $0 < T < \infty$, we define $A_T(\cdot)$ by $A_T(t) = \int_t^{t+T} a(s)ds$ and denote the set of all measurable functions ψ: $\Omega \times (-\infty,0) \to \underline{R}$ for which

$$\|\psi\|^2_{a,T} = \int_{-\infty}^0 \|f(\cdot,t)\|^2_1 \cdot A_T(-t)dt < \infty \tag{2.6}$$

by $L^2_{a,T}(-\infty,0;\ H^1(\Omega))$, with semi-norm $\|\cdot\|_{a,T}$. This set contains $L^\infty(-\infty,0;H^1(\Omega))$ as a subspace; if $a(t)$ vanishes for $t > T_0$, then $L^2_{a,T}(-\infty,0;H^1(\Omega))$ can be identified with a set of measurable functions on $(-T_0,0) \times \Omega$. The set $L^2_{a,T}(-\infty,0;H^1_0(\Omega))$ is defined similarly. We also abbreviate

$$A = \int_0^\infty a(t)\ dt < \infty\ .$$

For all results below, it will further be assumed that for some constant $\lambda > 0$ and for all $z,\xi,\zeta \in \underline{R}^n$

$$z^T\cdot(D_\xi g(\xi) + \int_0^\infty a(t)dt\cdot D_\xi h(\xi,\zeta))\cdot z \geqq \lambda.|z|^2, \tag{2.7}$$

$$H(\zeta + \xi,\zeta) \geqq \lambda\cdot|\xi|^2. \tag{2.8}$$

Then (2.4) and (2.8) imply that

$$h(\xi,\xi) = 0 \text{ for all } \xi \in \underline{R}^n. \tag{2.9}$$

The kernel a will be assumed to satisfy

$$a'(t) \leqq b(t) - \gamma(t) \text{ for a.e. } t > 0 \tag{2.10}$$

for some $b \in L^1(0,\infty;\underline{R}^+)$ and some non-negative and non-increasing function γ with $\int_0^1 \gamma(s)\,ds = \infty$. Thus $a(t)$ must go to $+\infty$ as $t \downarrow 0$, but it is not required that a is decreasing or convex near 0.

All results can also be shown if (1.2) is replaced by the more general equation

$$\sigma(x,t) = g(\nabla u(x,t)) + \sum_{i=1}^{N} \int_0^\infty a_i(s) h_i(\nabla u(x,t), \nabla u(x,t-s))\,ds, \tag{2.11}$$

with g and the (a_i,h_i) satisfying suitable modifications of the assumptions above.

By a <u>weak solution</u> of the Dirichlet problem (1.1) - (1.4), (1.5a) we understand a function $u \in L^2(0,T;H^1(\Omega))$ such $u(.,t) - \varphi(\cdot,t) \in H_0^1(\Omega)$ for a.e. t and such that for all test functions $\beta \in C_0^\infty(\Omega \times [0,T))$

$$\int_0^T \{\langle u,\beta_{tt}\rangle + \langle\sigma,\nabla\beta\rangle - \langle f,\beta\rangle\, dt = \langle u_0,\beta_t(\cdot,0)\rangle - \langle u_1,\beta(\cdot,0)\rangle \tag{2.12}$$

where u is continued as ψ on $\Omega \times (-\infty,0)$ and where (1.2) holds a.e. on $\Omega \times [0,T]$.

Similarly, a weak solution of the "Neumann problem" (1.1) - (1.4), (1.5b) is a function $u \in L^2(0,T;H^1(\Omega))$ for which (2.12) holds for all test functions $\beta \in C_0^\infty(\bar{\Omega} \times [0,T))$ with the term $\int_0^T \langle k,\beta\rangle_{\partial\Omega}\, dt$ added to the right hand side.

<u>THEOREM 1</u>. For any set of data $\psi \in L^\infty_{loc}(-\infty,0;H_0^1(\Omega)) \cap L^2_{a,T}(-\infty,0;H_0^1(\Omega))$, $\varphi \in W^{1,2}([0,T],H^1(\Omega)) \cap W^{2,1}([0,T],L^2(\Omega))$, $u_0 = \varphi(\cdot,0) \in H^1(\Omega)$, $u_1 = \varphi_t(\cdot,0) \in L^2(\Omega)$ and $f \in L^1(0,T;L^2(\Omega))$, there exist a weak solution u of (1.1) - (1.4), (1.5a). In addition, $\sigma \in L^\infty(0,T;L^2(\Omega,\underline{R}^n))$, the mapping $t \to (u(.,t),u_t(\cdot,t))$ is weakly continuous with values in $H^1(\Omega) \times L^2(\Omega)$ for $t \in [0,T]$, $u(\cdot,0) = u_0$, $u_t(\cdot,0) = u_1$ in this sense, and $u_{tt} - f$ is in $L^\infty(0,T;H^{-1}(\Omega))$.

<u>THEOREM 2</u>. For any set of data $\psi \in L^\infty_{loc}(-\infty,0;H_0^1(\Omega)) \cap L^2_{a,T}(-\infty,0;H_0^1(\Omega))$,

$u_o \in H_o^1(\Omega)$, $u_1 \in L^2(\Omega)$, $f \in L^1(0,T;L^2(\Omega))$, and $k \in W^{1/2}$ $(0,T;L^2(\partial\Omega))$ there exist a weak solution u of (1.1) - (1.4), (1.5b). The quantities σ,u, and u_t have the same regularity properties as in Theorem 1, and u_{tt} - f is an element of $L^\infty(0,T;H_b^{-1}(\Omega))$.

Recall that $W^{1/2}(0,T;L^2(\partial\Omega))$ is defined at the end of Section 1.

It follows from the proofs given below that solutions can be found such that the quantities

$$\eta(x,t) = \frac{1}{2}|u_t(x,t)|^2 + G(\nabla u(x,t)) + \int_{-\infty}^{t} a(t-s)H(\nabla u(x,t),\nabla u(x,s))ds,$$

$$q(x,t) = -u_t(x,t)\cdot\{g(\nabla u(x,t)) + \int_{-\infty}^{t} a(t-s)h(\nabla u(x,t),\nabla u(x,s))ds\},$$

$$\Sigma(x,t) = - \int_{-\infty}^{t} a'(t-s)H(\nabla u(x,t),\nabla u(x,s))ds$$

are in $L^1(0,T;L^1(\Omega))$ resp. in $L^1(0,T;L^1(\Omega,\underline{R}^n))$, and

$$\eta_t + \operatorname{div} q + \Sigma \leqq u_t\cdot f \tag{2.13}$$

in the sense of distributions on $\Omega \times (0,T)$. Moreover, $q.\nu = 0$ on those parts of $\partial\Omega \times [0,T]$ where φ is t-independent (in case of the Dirichlet problem) resp. where k vanishes (in the case of the Neumann problem). In (2.13), η can be interpreted as the density of the kinetic plus the potential energy, q is the energy flux, $u_t\cdot f$ is the power supplied by body forces, is a dissipation term due to history effects, and the inequality holds due to possible additional dissipation effects that might come from shocks. For systems of conservation laws (in one space dimension), (2.13) is known as an "entropy inequality" and is used to single out physically meaningful solutions, cf. [13], [34].

Several existence results for global small smooth solutions of the special case (0.1), (0.2), both on intervals Ω and for $\Omega = \underline{R}$, have been given previously ([9], [20], [21], [29], [35]). In all these results, nonlinear equations were discussed, assuming that the kernel a(·) is bounded near zero. These results were then extended to the case of singular kernels in

[22]. In [11], breakdown of classical solutions of (1.1), (1.2) for large data was shown to occur in quite general situations, again for smooth and bounded kernels and the case of one space dimension. For a related type of abstract nonlinear equations with singular kernels, weak solutions similar to the ones found here were constructed in [28].

These existence results can be applied to the two examples given above. Define

$$G(\xi) = P(\xi) - A\cdot Q(\xi), \quad H(\xi,\zeta) = Q(\xi) + (\zeta-\xi)q(\zeta) - Q(\zeta),$$

where $P' = p$, $Q' = q$, and $P(0) = Q(0) = 0$.

Then the existence theorems hold if we assume that for some $K \geqq \lambda > 0$ and for all $\xi \in \underline{\mathrm{R}}$

$$\lambda \leqq p'(\xi),\ q'(\xi) \leqq K.$$

Since the integral kernel $a(\cdot)$ is assumed to be integrable near zero, jumps both for the displacement and for the velocity across $t = 0$ are dynamically admissible, and this is indeed allowed by the existence result.

The assumption that $p'(\cdot)$ is bounded from above and below has the mechanical interpretation that the instantaneous elastic modulus of the material under consideration should be strictly positive. However, the equilibrium elastic modulus, which is related to $D_\xi g = p'(\cdot) - A\cdot q'(\cdot)$, need not satisfy any sign conditions, since the existence proof below does not rely on linearization about a rest state.

For the second example outlined above, the kernel is frequently assumed to be a positive finite or infinite linear combination of decaying exponentials, and (2.10), that is the assumption that $a(0)$ is infinite, is suggested by theoretical considerations ([14]) and by experiments ([26], [31]). The eixstence theorems above then can be applied if additionally

$$0 < \lambda \leqq p(\alpha) + 2\alpha p'(\alpha) \leqq K \text{ for all } \alpha \geqq 0$$

for suitable constants λ, K. This assumption is <u>not</u> supported by observations, which are better described by functions p that decay at infinity ([26], [31]).

We next give additional properties of the solutions constructed in Theorems 1 and 2. Proofs of these results are contained in [15] in the special case of the boundary condition (1.5a) with $\varphi \equiv 0$.

Finite speed of propagation into rest states:

Let $\Omega' \subset \Omega$ be open, $\Omega' \neq \Omega$, and let $c > 0$. For $x \in \Omega'$, we define

$$D(x) = \inf\{r > 0 \mid \exists\theta \in W^{1,\infty}([0,r],\Omega), |\theta'(\cdot)| \leq 1, \theta(0) \in \partial\Omega' \cap \Omega, \theta(r) = x\}, \tag{2.14}$$

$$\underline{C}(\Omega',c) = \{(x,t) \in \Omega' \times [0,T] \mid c.t < D(x)\}.$$

Thus $(x,t) \in \underline{C}(\Omega',c)$ if and only if x cannot be reached up to time t by a particle that travels in Ω, starting at time 0 anywhere on $\partial\Omega' \cap \Omega$ and moving with a maximal speed of c. Set

$$c^* = \sup_{\xi \neq 0} \frac{|g(\xi)+A\cdot h(\xi,0)|}{\sqrt{2G(\xi)+2AH(\xi,0)}} \in (0,\infty).$$

Let u be a weak solution of (1.1) and (1.2) satisfying (2.13), and let $\Omega' \subset \Omega$ be open. If $u = 0$ a.e. on $\Omega' \times (-\infty,0)$, $u_0 = 0$ a.e. on Ω', and $f = 0$ and $\varphi = 0$ resp. $k = 0$ a.e. on $\underline{C}(\Omega',c^*) \cap \partial\Omega \times [0,T]$, then $u(x,t) = 0$ a.e. on $\underline{C}(\Omega',c^*)$.

The result shows that perturbations of the zero rest state (which is trivially a solution of (1.1), (1.2)) can only propagate with the finite speed c^*, and they have to travel in Ω. In the case of example (0.1), (0.2), we obtain as a bound for the speed of propagations of the rest state

$$c^* = \sup_{\xi \in \underline{R}} \frac{|p(\xi)|}{\sqrt{2P(\xi)}} \leq \sup_{\xi \in \underline{R}} \sqrt{p'(\xi)}.$$

This estimate is independent of the kernel a and of $q(\cdot)$.

In the special case where $g(\xi) = 0$ for all $\xi \in \underline{R}^n$ (which corresponds to viscoelastic liquids), any time-independent function $\psi_0 \in H^1(\Omega)$ is a solution of (1.1), (1.2) with $f \equiv 0$ and therefore a rest state, due to (2.9). In this case the speed of propagation of such rest states can be estimated by

$$c^* = \sup_{\xi \neq \xi} \sqrt{A}. \frac{|h(\xi,\zeta)|}{\sqrt{2H(\xi,\zeta)}} \in (0,\infty).$$

Let $\psi_0 \in H_0^1(\Omega)$, and assume that $u(x,t) = \psi_0(x)$ a.e. on $\Omega' \times (-\infty,0)$, $u_0 = \psi_0$, $u_1 = 0$ a.e. on Ω', $f = 0$ a.e. on $\underline{C}(\Omega',c^*)$, and $\varphi = 0$ resp. $k = 0$ a.e. on $\underline{C}(\Omega',c^*) \cap \partial\Omega \times [0,T]$. Then $u(x,t) = \psi_0(x)$ a.e. on $\underline{C}(\Omega',c^*)$.

For the example given by (0.7), (0.8), g obviously vanishes. Defining $P(\alpha) = \int_0^\alpha p(s)ds$, any local perturbation of a rest state then propagates with a speed bounded by

$$c^* = \sqrt{A} \sup_{\alpha>0} \frac{\sqrt{\alpha p(\alpha)}}{\sqrt{P(\alpha)}} \leq \sup_{\alpha>0} \sqrt{A} \cdot \sqrt{p(\alpha) + 2\alpha p'(\alpha)}.$$

For linear problems, Laplace transform methods were used in [22] to study wave propagation properties for solutions of the one-dimensional version of (1.1) - (1.2) on all $\underline{R}$, and in [32] non-homogeneous boundary value problems on a half line were studied. Other propagation results are proven in [17] with abstract techniques and in [6] with an energy argument.

<u>Additional regularity properties</u>:

Assume that for some t_0, $\varepsilon > 0$ and some $0 < \alpha < \frac{1}{2}$

$$\gamma(t) \geq \varepsilon t^{-1-2\alpha} \text{ for } 0 < t < \delta,$$

where $\gamma(\cdot)$ is as in (2.10), and that there exist functions $g_0:\underline{R} \to \underline{R}$, $h_0:\underline{R}^2 \to \underline{R}$ such that for all $\xi \in \underline{R}^n$ with $|\xi| = 1$ and all $s,t \in \underline{R}$

$$g(s\xi) + A \cdot h(s\xi,t\xi) = (g_0(s) + h_0(s,t))\xi,$$

where as above $A = \int_0^\infty a(t)dt$. Let u be a weak solution of (1.1), (1.2), and assume that $\psi \equiv 0$, $\varphi \equiv 0$ resp. $k \equiv 0$, and $f \in L^2(0,T;H^{-1+\alpha}(\Omega))$. Then u is in $H^\alpha([0,T],H^1(\Omega)) \cap L^2(0,T;H^{1+\alpha}(\Omega))$, $u_t \in H^\alpha([0,T],L^2(\Omega)) \cap L^2(0,T;H^\alpha(\Omega))$, and $u_{tt} \in H^\alpha([0,T],H^{-1}(\Omega)) \cap L^2(0,T;H^{-1+\alpha}(\Omega))$. Here H^α etc. are the usual fractional order Sobolev spaces.

Regularizing effects of integrodifferential equations have been studied in greater detail for linear model problems in [22]. The results given there indicate that actually the solution becomes instanteously much more regular, namely $u(\cdot,t) \in H^{\beta+2}$ for $t > 0$, if $u_0 \in H^{\beta}(\underline{R})$, and show also that no instantaneous smoothing will occur if $a(\cdot)$ has a weaker (e.g. logarithmic) singularity. Unlike the situation for the heat equation, solutions are never arbitrarily smooth for $t > 0$, due to the presence of stationary singularities that the solution inherits from the data.

Other authors ([12], [16]) have studied regularity properties of solutions of related linear equations in the case that a has a non-integrable power type singularity at zero.

These regularity properties immediately apply to the examples (0.1), (0.2) (since the problem is one-dimensional) and (0.7), (0.8). In the one-dimensional case, solutions are therefore in $H^{1+\alpha}([0,L] \times [0,T])$ and thus in $C^{\alpha/2}(\Omega)$ for a.e. $t > 0$. We note that solutions of linear wave equations with the same data need not be continuous.

Asymptotic properties:

Results on the asymptotic behaviour of solutions can be shown if additionally

$$a'(t) + \delta\cdot a(t) \leqq 0 \quad \text{a.e. on } (0,\infty) \text{ for some } \delta > 0,$$

i.e. if $a(t)e^{\delta t}$ is non-increasing in t, and if φ resp. k vanish identically.

Assume first that

$$g(\xi)\cdot\xi \geqq \varepsilon\cdot|\xi|^2 \text{ for some } \varepsilon > 0, \text{ for all } \xi \in \underline{R}^n.$$

Then only the zero solution will be a rest state of (1.1), (1.2) under the boundary condition (1.5a), and one can show that for some $\varepsilon > 0$

$$\|u_t(\cdot,t)\|_0 + \|u(\cdot,t)\|_1 \leqq C(e^{-\varepsilon t}C_0 + \int_0^t e^{\varepsilon(s-t)} \|f(\cdot,s)\|_0 ds),$$

where C_0 depends on the initial data u_0, u_1, and ψ. Thus, if $f \in L^1(0,\infty;L^2(\Omega))$, the solution will go to zero. In the case of the Neumann condition (1.5b), the average displacement $\bar{u}(t) = \frac{1}{|\Omega|} \cdot \int_\Omega u(x,t)dx$ can be computed directly from the corresponding average volume force $\bar{f}(\cdot)$, and the quantity $u(\cdot,t) - \bar{u}(t)$ satisfies a similar estimate.

Next consider the case

$$G(\cdot) \equiv 0 \text{ on } \underline{R}^n.$$

Then any t-independent function is a possible rest state for (1.1), (1.2). In the case of the boundary condition (1.5a), and if $f \in L^1(0,\infty;L^2(\Omega))$, one can show that some $u_\infty \in H_0^1(\Omega)$ exists such that

$$\|u_t(\cdot,t)\|_0 + \|u(\cdot,t)-u_\infty\|_1 \leqq C(e^{-\varepsilon t}C_0 + \int_0^t e^{\varepsilon(s-t)}\|f(\cdot,s)\|_0 ds)$$

with C_0 depending on the initial data and ε depending only on the equation. A similar estimate holds in the case of (1.5b) for $u(\cdot) - \bar{u}(\cdot)$, if $f - \bar{f} \in L^1(0,\infty,L^2(\Omega))$, with $u_\infty \in H^1(\Omega)$ and $\bar{u}_\infty = 0$.

Applied to the example given in (0.1), (0.2), the zero solution is exponentially stable, if $a' + \delta a \leqq 0$, and if for some $\varepsilon > 0$

$$p(\xi)\xi \geqq A\cdot q(\xi)\xi + \varepsilon.|\xi|^2$$

for all $\xi \in \underline{R}$, corresponding to a strictly positive elastic equilibrium modulus. In the case of negative equilibrium elastic modulus (which is permitted in all other results above), one cannot expect to find stable rest states, as was shown in [18]. If $p(\cdot) \equiv A\cdot q(\cdot)$, i.e. $G(\cdot) \equiv 0$, the second part of these results applies, and the solution will converge to some equilibrium u_∞ for volume forces $f \in L^1(0,\infty,L^2(\Omega))$. Then u_∞ must satisfy the equation

$$-A_1\cdot(q(u_\infty'))' = \int_0^\infty f(\cdot,t)dt + u_1 - \int_0^\infty \bar{a}(s)q(\psi_x(\cdot,-s))_x ds$$

in $H^{-1}(0,L)$, where $\bar{a}(t) = \int_t^\infty a(s)\,ds$ and $A_1 = \int_0^\infty \bar{a}(t)dt = \int_0^\infty t\cdot a(t)dt < \infty$. Thus the rest state does not depend on the initial displacement u_0.

In the case of example (0.7), (0.8), the assumption on the kernel has the physical interpretation that $\frac{1}{\delta}$ is the maximal relaxation time of the material ([14]). It is not clear whether an explicit formula for the rest state can be found, except in the case of a linear equation.

3. PROOFS OF THE EXISTENCE RESULTS

We use a viscosity approximation to construct solutions with the desired properties, finding first solutions u^ε of approximating equations, then deriving various *a priori* estimates for the u^ε, and finally showing that an appropriate subsequence converges in a sufficiently strong topology.

Throughout this section, the x-dependence of all functions will be suppressed, where no confusion can arise. Also, without loss of generality, it can be assumed that

$$A = \int_0^\infty a(t)dt = 1.$$

Let $\psi \in L^2_{a,T}(-\infty,0;H^1(\Omega))$ be given as in the assumptions of Theorem 1 or Theorem 2. For $w \in L^2(0,T;L^2(\Omega,\underline{R}^n))$, continued for $t < 0$ by $\nabla\psi$, we define

$$\begin{aligned} F(w)(x,t) &= g(w(x,t)) + \int_{-\infty}^t a(t-s)h(w(x,t),w(x,s))ds \\ &= \int_{-\infty}^t a(t-s)\{g(w(x,t)) + h(w(x,t),w(x,s))\}ds \end{aligned} \tag{3.1}$$

on $\Omega \times (0,T)$. Then F maps into $L^2(0,T;L^2(\Omega,\underline{R}^n))$ and is Lipschitz-continuous, as a straight-forward calculation shows. Some further properties of F are listed in the following

<u>LEMMA 1</u>. Let $u \in W^{1,2}([0,T],L^2(\Omega,\underline{R}^n))$ and $v,w \in L^2(0,T;L^2(\Omega,\underline{R}^n))$.

a) Then for a.e. $t \in [0,T]$

$$\frac{d}{dt}\{\int_{-\infty}^{t} a(t-s)\cdot\int_{\Omega} H(u(x,t),u(x,s))dx\,ds + \int_{\Omega} G(u(x,t))dx\}$$

$$= \int_{-}^{t} a'(t-s)\cdot \int_{\Omega} H(u(x,t),u(x,s))dx\,ds + \langle F(u)(\cdot,t),u_t(\cdot,t)\rangle.$$

b) Let $\gamma(\cdot)$ be as in (2.10). Then for some constant C (depending on $a(\cdot)$, ψ, and the properties of g and h)

$$\operatorname*{ess\,sup}_{0\leq t\leq T} \|F(v)(\cdot,t)\|_0 \leq C\cdot(\operatorname*{ess\,sup}_{0\leq t\leq T} \|v\|_0 + 1),$$

$$\int_0^T\int_0^t \gamma(t-s)\,\|F(v)(t) - F(v)(s)\|_0^2\,ds\,dt$$

$$\leq C\cdot(\int_0^T\int_0^t \gamma(t-s)\,\|v(t) - v(s)\|_0^2\,ds\,dt + \operatorname*{ess\,sup}_{0\leq t\leq T} \|v\|_0^2 + 1),$$

whenever the right hand side is finite.

c) There exist μ, $\delta > 0$, depending only on $a(\cdot)$, g, and h such that

$$\int_0^T e^{-\mu t}\langle F(v)(t) - F(w)(t),v(t) - w(t)\rangle dt \geqq \delta\cdot\int_0^T\|v(t) - w(t)\|_0^2 dt.$$

Proof. Assertion a) certainly holds if additionally $a'(\cdot)$ is integrable near zero. Noting that $\int_{\Omega} G(u(\cdot,t),u(\cdot,s)) \leq C\cdot|t-s|$, since $u_t \in L^2(0,T;L^2)$, and using (2.5), it follows in the general case by approximation.

The first estimate in part b) follows again by a direct calculation. To establish the second estimate, we set $D_0 = \int_0^T\int_0^t \gamma(t-s)\,\|v(t)-v(s)\|_0^2\,ds\,dt$, $D_1 = \operatorname*{ess\,sup}_{0\leq t\leq T} \|v(t)\|_0$, and write for $0 \leq s \leq t \leq T$

$$F(v)(s) - F(v)(t) = g(v(t)) - g(v(s))$$

$$+ \int_0^s a(\tau)\cdot\{h(v(t),v(t-\tau)) - h(v(s),v(t-\tau))\}d\tau$$

$$+ \int_0^s a(\tau)\cdot\{h(v(s),v(t-\tau)) - h(v(s),v(s-\tau))\}d\tau$$

$$+ \int_{-\infty}^{0} a(t-\tau)\cdot\{h(v(t),\nabla\psi(\tau)) - h(v(s),\nabla\psi(\tau))\}d\tau$$

$$+ \int_{s}^{t} a(\tau)\cdot h(v(t),v(t-\tau))d\tau$$

$$+ \int_{-\infty}^{0} \{a(t-\tau) - a(s-\tau)\}\cdot h(v(s),\nabla\psi(\tau))d\tau$$

$$= E_0(t,s) + E_1(t,s) + E_2(t,s) + E_3(t,s) + E_4(t,s) + E_5(t,s).$$

Then

$$\int_0^T \int_0^t \gamma(t-s)\ \|E_i(t,s)\|_o^2 \ ds\ dt \leq C\cdot D_o$$

for $i \in \{0,1,2,3\}$. To show corresponding estimates for E_4 and E_5, let $\delta > 0$ be arbitrary, and abbreviate $t(\delta) = \min(0,t-\delta)$. We then have

$$\int_0^T \int_0^{t(\delta)} \gamma(t-s)\|E_4(t,s)\|_o^2 \ ds\ dt \leq C\cdot\int_0^T \int_0^{t(\delta)} \gamma(t-s)(\int_s^t a(\tau)d\tau)^2 \ dsdt.D_1^2$$

$$\leq C\cdot(1 + 2\cdot(\int_0^T a(t)dt)^3)\cdot D_1^2,$$

using the definition of $\gamma(\cdot)$ and an integration by parts. Sending δ to 0 we obtain a uniform bound. Similarly, since a is non-increasing up to an absolutely continuous function,

$$\int_0^T \int_0^t \gamma(t-s)\|E_4(t,s)\|_o^2 \ ds\ dt$$

$$\leq C\cdot\int_0^T \int_0^t \gamma(t-s)\cdot((\int_{-\infty}^0 a(t-\tau)-a(s-\tau)d\tau)^2 + 1)\ ds\ dt \cdot D_1^2$$

$$\leq C\cdot(1 + \int_0^T \int_0^t \gamma(t-s)\cdot(\int_s^t a(\tau)d\tau)^2 ds\ dt)\cdot D_1^2,$$

and this quantity is bounded by the previous argument.

Together, this establishes part b).

For the proof of part c), we set $\xi(x,t) = v(x,t) - w(x,t)$ and recall that $\int_0^\infty a(t)dt = 1$. Then

$$\int_0^T e^{-\mu t} \langle F(v)(t) - F(w)(t), v(t) - w(t)\rangle \, dt =$$

$$= \int_0^T e^{-\mu t} \int_0^\infty a(s)\langle g(v(t))+h(v(t),v(t-s))-\{g(w(t))+h(w(t),v(t-s))\}, \xi(t)\rangle ds dt$$

$$+ \int_0^T e^{-\mu t} \int_0^\infty a(s)\langle h(w(t),v(t-s))-h(w(t),w(t-s)),\xi(t)\rangle \, ds \, dt$$

$$\geqq \lambda \cdot \int_0^T e^{-\mu t} \|\xi(t)\|_o^2 \, dt - K \cdot \int_0^T e^{-\mu t} \int_0^\infty a(s)\|\xi(t-s)\|_o \cdot \|\xi(t)\|_o ds \, dt$$

$$\geqq \int_0^T e^{-\mu t} \{\tfrac{\lambda}{2} - C \cdot \int_0^{T-t} e^{-\mu s} a(s) ds\} \cdot \|\xi(t)\|_o^2 \, dt,$$

with K,λ from (2.3) and (2.8) and a suitable $C > 0$. Choosing μ sufficiently large, part c) follows.

The proofs of the existence theorems are given in several steps; we shall give the arguments for Theorem 1 in detail and then outline the proof of Theorem 2.

PROOF OF THEOREM 1. For the purpose of this proof, we abbreviate $X(T) = L^2(0,T;H_0^1(\Omega))$ and $X^*(T) = L^2(0,T;H^{-1}(\Omega))$ and equip these spaces with their natural norms $\|\cdot\|$ resp. $\|\cdot\|_*$.

(i) Construction of approximating solutions

Let $\varepsilon > 0$ be fixed, let φ and f be as in the assumptions of Theorem 1, and let $\underline{S} = \underline{S}^\varepsilon: L^2(0,T;L^2(\Omega,\underline{R}^n)) \to X(T)$ be defined as the operator mapping $\underline{h} \in L^2(0,T;L^2(\Omega,\underline{R}^n))$ to the unique solution $v \in X(T)$ of

$$v_t - \varepsilon\Delta v = f + \text{div}\,(\underline{h} + \varepsilon\nabla\varphi_t) - \varphi_{tt}$$

in the sense of distributions on $\Omega \times (0,T)$ with $v(\cdot,0) = 0$.

The operator $\underline{S}$ is well-defined (see [25]), and there is an estimate

$$\| (\underline{Sh}_1)_t - (\underline{Sh}_2)_t \|_* + \| \underline{Sh}_1 - \underline{Sh}_2 \| \leq C \cdot \| \underline{h}_1 - \underline{h}_2 \|_{L^2}$$

with $C = C(\varepsilon)$ independent of T, if T remains in a bounded set. Thus $\underline{Sh} \in C([0,T],H)$.

Define next $R: X(T) \to L^\infty(0,T;H_0^1(\Omega))$ by

$$Rv(x,t) = \int_0^t v(x,s)ds.$$

We want to solve the equation

$$v_t - \varepsilon\Delta v = f + \mathrm{div}(F(\nabla Rv + \nabla\varphi) + \varepsilon\nabla\varphi_t) - \varphi_{tt} \tag{3.2}$$

in $X(T)$ with initial condition $v(\cdot,0) = 0$; i.e. v is to be found in $X(T)$ as a solution of the fixed point equation

$$v = \underline{S}(F(Rv + \nabla\varphi)).$$

Then the Lipschitz property of F implies that the operator $\underline{M}$, defined by the right hand side of this equation, satisfies

$$\|\underline{M}(v) - \underline{M}(w)\|_{X(T)} \leq C^* \cdot \int_0^T \|v - w\|_{X(t)}\, dt$$

with some generic constant C^*. Thus for any $k \geq 1$

$$\|\underline{M}^k(v) - \underline{M}^k(w)\|_{X(T)} \leq \frac{(C^* \cdot T)^k}{k!} \|v - w\|_{X(T)},$$

and for sufficiently large k, the k-th iterate $\underline{M}^k$ is a contraction. Its unique fixed point v^ε will also be the unique fixed point of $\underline{M}$. Setting $u^\varepsilon = Rv^\varepsilon + \varphi$, we obtain a unique solution in $W^{1,2}([0,T];H^1(\Omega)) \cap C^1([0,T],L^2(\Omega))$ of the approximating equation

$$u^\varepsilon_{tt}(x,t) - \varepsilon\cdot\Delta u^\varepsilon_t(x,t) = \mathrm{div}_x\sigma^\varepsilon(x,t) + f(x,t) \text{ on } \Omega \times (0,T], \tag{3.3}$$

in the sense of distributions, where σ^ε is defined as in (1.2), with initial conditions (1.3) and (1.4) and boundary condition (1.5a).

A priori estimates. The superscript ε will be dropped in this paragraph. We can take the scalar product of (3.3) with the admissible test function $u_t - \varphi_t \in L^2(0,T;H_o^1(\Omega))$. Integrating over $[0,t]$ and using part a) of Lemma 1 results in the identity

$$\frac{1}{2}\|u_t(t) - \varphi_t(t)\|_o^2 + \int_{-\infty}^{t} a(t-s)\int_\Omega \{G(\nabla u(\cdot,t))+H(\nabla u(\cdot,t),\nabla u(\cdot,s))\}dxds$$

$$+ \varepsilon\cdot\int_0^t \|\nabla u_t(s)\|_o^2 ds - \int_0^t\int_{-\infty}^{s}\int_\Omega a'(s-\tau)H(\nabla u(\cdot,s),\nabla u(\cdot,\tau))dx\, d\tau\, ds$$

$$= \int_0^\infty\int_\Omega a(s)\{G(\nabla u_o(\cdot)) + H(\nabla u_o(\cdot),\nabla\psi(\cdot,-s))\}dx\, ds$$

$$- \int_0^t \langle F(\nabla u)(s),\nabla\varphi_t(s)\rangle\, ds + \int_0^t \langle f(s)-\varphi_{tt}(s),u_t(s)-\varphi_t(s)\rangle ds$$

$$+ \varepsilon\int_0^t \langle\nabla u_t(s),\nabla\varphi_t(s)\rangle)\, ds. \tag{3.4}$$

Now (2.3), (2.4), (2. 7), (2.8) imply that for any ξ, $\zeta \in \underline{R}^n$

$$G(\xi) + H(\xi,\zeta) \geqq c\cdot|\xi|^2 - C\cdot|\zeta|^2,$$

$$|g(\xi)| + |h(\xi,\zeta)| \leqq C\cdot(|\xi| + |\zeta|\ ). \tag{3.5}$$

with suitable constants c, C. Using moreover (2.2) and (2.5), we deduce from (3.4) the estimate

$$\frac{1}{2}\cdot\|u_t(t)\|_o^2 + \varepsilon.\int_0^t \|u_t(s)\|_1^2\, ds + c\cdot\|u(t)\|_1^2 +$$

$$+ \int_0^t\int_0^s (-a')_+\ (s-\tau)\,\|u(s) - u(\tau)\|_1^2\, d\tau\, ds$$

$$\leqq C\cdot(1+\int_0^t((1+a(t-s))\,\|u(s)\|_1^2 + (\ \|f(s)\|_o + \|\varphi_{tt}(s)\|_o)\cdot\ \|u_t(s)\|_o)ds)$$

with some constants c, C, that depend only on the data, on the kernel a, on the constants in (3.5), and on T. A standard application of Gronwall's Lemma then implies that for some universal constant $\bar{C}$, independent of ε,

$$\sup_{[0,T]} \left(\|u_t(t)\|_0 + \|u(t)\|_1 \right) \leq \bar{C}, \tag{3.6}$$

$$\varepsilon \cdot \|u_t\|^2 \leq \bar{C}, \tag{3.7}$$

$$\int_0^T \int_0^t (-a')_+(t-s) \|u(t) - u(s)\|_1^2 \, ds \, dt \leq \bar{C}. \tag{3.8}$$

Let $w = w^\varepsilon = u_{tt}^\varepsilon - \varepsilon\Delta u_t^\varepsilon - f(t) \in L^2(0,T;H_b^{-1}(\Omega)) \subset L^2(0,T;H^{-1}(\Omega))$. Then part b) of Lemma 1 implies that additionally

$$\sup_{[0,T]} \|w(t)\|_{-1} \leq \bar{C}, \tag{3.10}$$

$$\int_0^T \int_0^t \gamma(t-s) \|w(t) - w(s)\|_{-1}^2 \, ds \, dt \leq \bar{C}. \tag{3.11}$$

Passing to the limit. Using (3.6) and (3.7), we can pick a sequence $\varepsilon_n \to 0$ and corresponding sequences $u^n = u^{\varepsilon_n}$, $w^n = w^{\varepsilon_n}$ such that

$$u^n \to u^* \qquad \text{weak - * in } L^\infty(0,T;H^1(\Omega)), \tag{3.12}$$

$$u_t^n \to u_t^* \qquad \text{weak - * in } L^\infty(0,T;L^2(\Omega)), \tag{3.13}$$

$$w^n \to w^* \qquad \text{weak - * in } L^\infty(0,T;H^{-1}(\Omega)). \tag{3.14}$$

Since $\varepsilon_n \Delta u_t^n \to 0$ strongly in $X^*(T)$ due to (3.12), w^* must equal $u_{tt}^* - f$ in the sense of distributions on $\Omega \times (0,T)$.

We now claim that in fact for this sequence

$$w^n \to u_{tt}^* - f \qquad \underline{\text{strongly}} \text{ in } L^2(0,T;H^{-1}(\Omega)), \tag{3.15}$$

$$u^n \to u^* \qquad \underline{\text{strongly}} \text{ in } L^2(0,T;H^1(\Omega)), \tag{3.16}$$

$$u_t^n \to u_t^* \qquad \underline{\text{strongly}} \text{ in } L^2(0,T;L^2(\Omega)), \tag{3.17}$$

as $n \to \infty$.

We first prove (3.15). Set $\bar{w}^n = u^n_{tt} - f = w^n + \varepsilon_n \Delta u^n$, and define the shift semigroup $J_s: X^*(T) \to X^*(T)$ for $s \geqq 0$ by

$$J_s v(t) = \begin{cases} v(t+s), & \text{if } 0 \leq t \leq T-s \\ 0 \;, & \text{if } T-s < t \leq T \end{cases} . \tag{3.18}$$

The $(J_s)_{s\geqq 0}$ form a C_0-semigroup of contractions on $X^*(T)$ and on $X(T)$, equipped with their standard norms. Then for all n

$$\int_0^T \gamma(s) \| J_s w^n - w^n \|^2_* \, ds = \int_0^T \int_0^t \gamma(t-s) \| w^n(t) - w^n(s) \|^2_{-1} \, ds \, dt$$

$$+ \int_0^T \int_{T-s}^T \gamma(\tau) d\tau \, \| w^n(s) \|^2_{-1} ds \leq C.$$

We apply Lemma 2 (i) below and note that by (3.7) $\| w^n - \bar{w}^n \|_* \leqq \sqrt{\varepsilon_n} \cdot C$. Then

$$\| J_s \bar{w}^n - \bar{w}^n \|_* \leqq \omega(s) \tag{3.19}$$

uniformly in n, for some $\omega:[0,1] \to [0, \;)$, with $\omega(s) \downarrow 0$ as $s \downarrow 0$. In addition, from (3.6) and the assumptions on f, we infer that

$$\sup_{s,t} \left\| \int_s^t \bar{w}^n(\tau) d\tau \right\|_0 \leqq \sup_{s,t} \{ \| u_t(t) - u_t(s) \|_0 + \int_s^t \| f(\tau) \|_0 d\tau \} \leqq C$$

and thus

$$\sup_h \{ \left\| \int_0^h J_\tau \bar{w}^n \, d\tau \right\|_{L^2(0,T;L^2(\Omega))} + \left\| \int_0^h J_\tau \bar{w}^n \, d\tau \right\|_{W^{1,2}([0,T];H^{-1}(\Omega))} \} \leqq C$$

uniformly in n. We now use part (ii) of Lemma 2 with $Y = L^2(0,Y;L^2(\Omega)) \cap W^{1,2}([0,T],H^{-1}(\Omega))$ to conclude that the $\bar{w}^n$ and thus also the w^n converge to w^* strongly in $L^2(0,T;H^{-1}(\Omega))$.

We next show (3.16), using the monotonicity property given in Lemma 1c). Noting that all u^n have the same boundary values and thus $u^n - u^m \in X(T)$, one can assert that for suitable δ, μ,

$$\delta \cdot \|u^n - u^m\|^2 \quad \int_0^T e^{-\mu t} \langle F(\nabla u^n)(t) - F(\nabla u^m)(t), \nabla u^n(t) - \nabla u^m(t) \rangle dt$$

$$= \int_0^T e^{-\mu t} \langle w^n(t) - w^m(t), u^n(t) - u^m(t) \rangle dt$$

$$\leq C \cdot \|w^n - w^m\|_* \cdot \|u^n - u^m\| .$$

Since $\|w^n + f - u^*_{tt}\|_* \to 0$, the $u^n - \varphi$ form a Cauchy-sequence in $X(T)$, and their strong limit in $L^2(0,T;H^1(\Omega))$ must equal u^*. This proves (3.16). From (3.15) and (3.16), we deduce (3.17) by integration by parts.

Using the Lipschitz property for F, one sees now that indeed u^* is a weak solution of (1.1), (1.2) and that $u^* - \varphi \in L^\infty(0,T;H^1_0(\Omega))$. Since $u^* \in L^\infty(0,T;H^1(\Omega)) \cap W^{1,\infty}([0,T];L^2(\Omega))$ and $u^*_{tt} - f \in L^\infty([0,T];H^{-1}(\Omega))$, $u^*(\cdot)$ is weakly continuous with values in V^* and u^*_t is weakly continuous with values in H, and the initial values are attained in this sense. □

PROOF OF THEOREM 2: In this proof, we use the notation $X(T) = L^2(0,T;H^1(\Omega))$ and $X^*(T) = L^2(0,T;H_b^{-1}(\Omega))$ with corresponding norms $\| \cdot \|$ and $\|\cdot\|_*$. Also, $\|\cdot\|_{-1}$ will here denote the norm on $H_b^{-1}(\Omega) = (H^1(\Omega))'$.

Let $f \in L^1(0,T;L^2(\Omega))$, $k \in W^{1/2}(0,T;L^2(\partial\Omega))$ and $u_1 \in L^2(\Omega)$ be as in the assumptions of the Theorem. To construct approximating solutions, let $\varepsilon > 0$, and define the operator $S = S^\varepsilon$: $L^2(0,T;L^2(\Omega,\underline{R}^n)) \to X(T)$ by $\underline{S}(\underline{h}) = v$, where $v \in X(T)$ satisfies

$$\int_0^T \{\varepsilon\langle\nabla\beta,\nabla v\rangle - \langle\beta_t,v\rangle\}dt = \int_0^T \{\langle\beta,f\rangle - \langle\nabla\beta,\underline{h}\rangle + \langle\beta,k\rangle_{\partial\Omega}\}dt + \langle\beta(\cdot,0),u_1\rangle$$

for all test functions $\beta \in C_0^\infty(\bar{\Omega} \times [0,T))$. Formally, classical solutions of this relation will solve $v_t - \varepsilon\Delta v = f + \operatorname{div} \underline{h}$ with initial condition $v(\cdot,0) = u_1$ and boundary condition $\varepsilon\cdot\partial_\nu v + \nu\cdot\underline{h} = k$. More precisely, let $\Gamma: L^2(\partial\Omega) \to H_b^{-1}(\Omega)$ be the adjoint of the trace operator from $H^1(\Omega)$ to $L^2(\partial\Omega)$ and let *div:* $L^2(\Omega,\underline{R}^n) \to H_b^{-1}(\Omega)$ be defined by $\langle \mathit{div}\, \underline{h}, v\rangle = -\langle\underline{h},\nabla v\rangle$ ($H^1 - H_b^{-1}$ - pairing on the left hand side, L^2-scalar product on the right-hand side). Then $\underline{S}(v)$ solves $v_t - \varepsilon\cdot \mathit{div}\, \nabla v = f + \mathit{div}\, \underline{h} + \Gamma k$ in $X^*(T)$.

Using the methods of [25] one can show that $\underline{S}$ is well-defined and that

$$\|(\underline{S}h_1)_t - (\underline{S}h_2)_t\|_* + \|\underline{S}h_1 - \underline{S}h_2\| \leq C\cdot \|h_1 - h_2\|_{L^2}$$

as in the previous proof. Define next $R : X(T) \to L^\infty(0,T;H^1(\Omega))$ by

$$Rv(x,t) = u_o + \int_0^t v(x,s)ds.$$

We want to solve the fixed point equation

$$v = \underline{S}(F(\nabla Rv))$$

in $X(T)$. By the same argument as in the proof above, some iterate of the operator defined by the right hand side of this equation will be a strict contraction in $X(T)$, giving a unique fixed point v^ε. We set $u^\varepsilon = Rv^\varepsilon$ and obtain a unique solution in $W^{1,2}([0,T];H^1(\Omega)) \cap C^1([0,T], L^2(\Omega))$ of the approximating equation (3.3) (in the sense of distributions) with approximating boundary condition

$$\varepsilon\cdot\partial_\nu u_t + \nu\cdot\sigma^\varepsilon = k \tag{3.20}$$

in the usual weak sense. Also, u^ε satisfies the initial conditions (1.3) and (1.4) by construction.

To deduce <u>a priori estimates</u>, we can take the scalar product of the weak form of the equation for $u = u^\varepsilon$ with the test function u_t. Integrating over $[0,t]$ now results in the identity

$$\begin{aligned}
&\tfrac{1}{2}\|u_t(t)\|_o^2 + \int_{-\infty}^t a(t-s)\int_\Omega \{G(\nabla u(\cdot,t)) + H(\nabla u(\cdot,t),\nabla u(\cdot,s))\}dx\,ds \\
&+ \varepsilon\cdot\int_0^t \|\nabla u_t(s)\|_o^2\,ds - \int_0^t\int_{-\infty}^s\int_\Omega a'(s-\tau)H(\nabla u(\cdot,s),\nabla u(\cdot,\tau))dx\,d\tau\,ds \\
&= \int_0^\infty\int_\Omega a(s)\{G(\nabla u_o(\cdot)) + H(\nabla u_o(\cdot),\nabla\psi(\cdot,-s))\}dx\,ds + \tfrac{1}{2}\|u_1\|_o^2 \\
&\quad + \int_0^t \langle f(s),u_t(s)\rangle\,ds + \int_0^t \langle k(s),u_t(s)\rangle_{\partial\Omega}\,ds.
\end{aligned} \tag{3.21}$$

Here the last term on the right hand side requires some attention: Let $H = L^2(\partial\Omega)$, and define

$$B(p,q) = \int_0^T \langle p,q_t\rangle_H dt = - \int_0^T \langle p_t,q\rangle_H dt$$

on $C_0^\infty([0,T],H) \times C^\infty((0,T),H)$. Then B can be extended to arguments from $L^1(0,T;H) \times W^{1,\infty}([0,T],H)$ and from $W_0^{1,1}([0,T],H) \times L^\infty(0,T;H)$ as a continuous bilinear form, where elements from $W_0^{1,1}([0,T],H)$ vanish at the end points of $[0,T]$ by definition. Applying a standard interpolation result for bilinear forms (see [3]), we thus can also define B on $W^{1/2}(0,T;H) \times C^{1/2}([0,T],H)$ with the estimate

$$|B(p,q)| \leq C\cdot \|p\|_{W^{1/2}}\cdot \|q\|_{C^{1/2}}.$$

For any $v \in L^\infty(0,T;H^1(\Omega)) \cap W^{1,\infty}([0,T],L^2(\Omega))$, we have $v \in C^{1/2}([0,T],L^2(\partial\Omega))$ by a standard trace estimate (cf. [25]), and thus for the term above

$$\left|\int_0^t \langle k,u_t\rangle_{\partial\Omega} ds\right| \leq C\cdot\|k\|_{W^{1/2}}\cdot \|u\|_{C^{1/2}([0,t],L^2(\partial\Omega))}$$

$$\leq C\cdot\|k\|_{W^{1/2}}\cdot \sup_{[0,t]} \left(\|u(\cdot)\|_1 + \|u_t(\cdot)\|_0\right).$$

By the same arguments as in the proof above, we then deduce from (3.21)

$$\frac{1}{2}\cdot \|u_t(t)\|_0^2 + \varepsilon\cdot\int_0^t \|u_t(s)\|_1^2\, ds + c\cdot \|u(t)\|_1^2 +$$

$$+ \int_0^t \int_0^s (-a')_+(s-\tau)\, \|u(s) - u(\tau)\|_1^2\, d\tau\, ds$$

$$\leq C\cdot(1 + \int_0^t ((1 + a(t-s))\, \|u(s)\|_1^2 + \|f(s)\|_0 \|u_t(s)\|_0) ds)$$

$$+ \|k\|_{W^{1/2}}\cdot \sup_{[0,t]} \left(\|u(\cdot)\|_1 + \|u_t(\cdot)\|_0\right).$$

Gronwall's Lemma implies again that for some universal constant $\bar{C}$

$$\sup_{[0,T]} (\|u_t(t)\|_0 + \|u(t)\|_1) \leq \bar{C}, \tag{3.22}$$

$$\varepsilon \cdot \|u_t\|^2 \leq \bar{C}, \tag{3.23}$$

$$\int_0^T \int_0^t (-a')_+(t-s) \, \|u(t) - u(s)\|_1^2 \, ds \, dt \leq \bar{C} \tag{3.24}$$

and by Lemma 1

$$\sup_{[0,T]} \|w(t)\|_{-1} \leq \bar{C}, \tag{3.25}$$

$$\int_0^T \int_0^t \gamma(t-s) \, \|w(t) - w(s)\|_{-1}^2 \, ds \, dt \leq \bar{C}, \tag{3.26}$$

where $w = w^\varepsilon = u_{tt}^\varepsilon - \varepsilon \cdot \mathit{div} \, \nabla u_t^\varepsilon - f - \Gamma k = \mathit{div} \, F(\nabla u^\varepsilon)$.

In order to pass to the limit, we extract a sequence $\varepsilon_n \to 0$ such that the u^n, u_t^n and w^n converge in the weak - * sense as in (3.12) - (3.14), with $H^{-1}(\Omega)$ everywhere replaced by $H_b^{-1}(\Omega)$. Again, the weak limit of the w^n must equal $u_{tt}^* - f - \Gamma k \in X^*(T)$.

To prove that in fact u^*, u_t^* and u_{tt}^* are strong limits as in (3.15) - (3.17), the same argument can be used as above:

Starting with (3.15), it follows from (3.25) that time shifts of the w^n give equicontinuous functions on $X^*(T)$, and (3.23) implies that integrals of time shifts of the w^n are uniformly bounded in a suitable space that is compactly embedded in $X^*(T)$. Thus the w^n converge strongly by Lemma 2. Next, (3.16) again follows from the monotonicity property of the integral operator F, given in Lemma 1 c). Finally, (3.17) follows simply from an integration by parts. We leave the details to the reader.

All remaining assertions again follow as in the previous proof and we obtain a weak solution of (1.1) - (1.4), (1.5b) as desired. □

LEMMA 2. Let $Y \subset X$ be Banach spaces with compact imbedding and norms $\|\cdot\|_X = \|\cdot\|$, $\|\cdot\|_Y$, let X be reflexive, and let $(J_s)_{s \geq 0}$ be a C_0-semigroup of contractions on X with infinitesimal generator A.

(i) Let $\gamma:(0,2] \to \underline{R}$ be non-negative and non-decreasing, such that $\int_0^1 h\cdot\gamma(h)dh < \infty$. Then there exists a constant $\bar{C} > 0$ such that for $0 < s \leq 1$ and $x \in X$

$$\|J_s x - x\|^2 \cdot (\int_s^1 \gamma(2\tau)d\tau) \leq \bar{C} \text{ whenever } \int_0^1 \gamma(s)\cdot\|J_s x - x\|^2 \, ds + \|x\| \leq 1.$$

(ii) Let ω: $[0,1] \to [0,\infty)$ be non-decreasing, with $\omega(s) \downarrow 0$ as $s \downarrow 0$, and let $C > 0$. Then the set of all $v \in X$ with

$$\|J_s v^n - v^n\|_X \leq \omega(s), \quad \sup_s \|\int_0^s J_\tau v^n \, d\tau\|_Y \leq C$$

is relatively compact in X.

Proof. (i) Let $Y = D(A)$, equipped with the graph norm. Define

$$K(h,x) = \inf \{\|z_1\| + h\cdot(\|Az_2\| + \|z_2\|) | z_2 \in D(A), \ z_1 + z_2 = x\}$$

(cf. [2], [3]). Let $x \in X$ satisfy the assumption above, and set

$$x^h = \frac{1}{h}\int_h^{2h} J_s x \, ds \text{ for } 0 < h \leq 1,$$

then obviously $Ax^h = \frac{1}{h}(J_{2h}x - J_h x)$ and thus

$$K(h,x) \leq \frac{1}{h}\int_h^{2h} \|J_s x - x\| \, ds + \|J_h(J_h x - x)\|_X + h.$$

Thus

$$\int_0^1 \gamma(2h)\cdot K(h,x)^2 \, dh \leq \int_0^1 \gamma(2h)\cdot \frac{2}{h}\int_h^{2h} \|J_s x - x\|^2 \, ds \, dh$$
$$+ 2. \int_0^1 \gamma(2h)\|J_h(J_h x - x)\|^2 \, dh + C.$$

Since $\gamma(2h) \leq \gamma(h)$, the second integral is finite. Since $\gamma(\cdot)$ is non-increasing, the first integral can be estimated as follows:

$$\int_0^1 \gamma(2h)\cdot \frac{1}{h}\int_h^{2h} \|J_s x - x\|^2 \, ds\, dh \leq C + \int_0^1 \int_{s/2}^{s} \frac{\delta(2h)}{h}\, dh\cdot \|J_s x - x\|^2 \, ds$$

$$\leq C + \int_0^1 \log 2 . \gamma(s)\cdot \|J_s x - x\|^2 \, ds$$

$$\leq C.$$

Therefore $\int_0^1 \gamma(2h)\cdot K(h,x)^2 \, dh \leq C_1$. Since $K(\cdot,x)$ is non-decreasing in its first argument, this implies that for $0 < \delta \leq 1$

$$K(\delta,x)^2\cdot(\int_\delta^1 \gamma(2h)\, dh) \leq \int_\delta^1 \gamma(2h)\cdot K(h,x)^2 \, dh \leq C_1.$$

For fixed $s > 0$, choose now $y \in D(A)$ such that

$$\|x - y\|_X + s\cdot \|y\|_{D(A)} \leq 2K(s,x),$$

thus

$$\|J_s x - x\|_X \leq \|J_s x - J_s y\|_X + \|J_s y - y\|_X + \|y - x\|_X$$

$$\leq 6K(s,x)$$

and therefore

$$\|J_s x - x\|^2 \cdot(\int_s^1 \gamma(2\tau) d\tau) \leq 36\cdot K(s,x)^2 \cdot \int_s^1 \gamma(2\tau) d\tau \leq 36\cdot C_1 = \bar{C}.$$

(ii) Without loss of generality, let $(v^n)_{n\geq 1}$ be a sequence satisfying the assumptions above and converging weakly to $0 \in X$. For $h > 0$, let

$$v^{n,h} = \frac{1}{h}\int_0^h J_\tau v^n \, d\tau \in X.$$

Then $\|v^{n,h}\|_Y \leq \frac{C}{h}$ and thus $v^{n,h} \to 0$ strongly in X for any fixed h. On the other hand, $\|v^{n,h} - v^n\|_X \leq \omega(h)$ as $h \downarrow 0$, uniformly in n. Thus

$$\limsup_{n \to \infty} \|v^n\|_X \leq \limsup_{n \to \infty} \|v^n - v^{n,h}\|_X + \limsup_{n \to \infty} \|v^{n,h}\|_X \leq \omega(h).$$

The right hand side can be made arbitrarily small, and part (ii) follows. □

References

[1] H.D. Alber: Local existence of weak solutions to the quasi-linear wave equation for large initial values. Math. Z. 190 (1985), 249-276.

[2] H. Behrens & P.L. Butzer: Semigroups of Operators and Approximation. Springer, Berlin, New York, Heidelberg, 1967.

[3] J. Bergh & J. Lofstrom: Interpolation Spaces. Springer, Berlin, New York, Heidelberg, 1976.

[4] B. Bernstein, E.A. Kearsley & L.J. Zapas: A study of stress relaxation with finite strain. Trans. Soc. Rheology 7 (1963), 391-410.

[5] R.B. Bird, R.C. Armstrong & O. Hassager: Dynamics of Polymeric Liquids. John Wiley, New York, 1977.

[6] G. Canadas: Speed of propagation of solutions of a linear integro-differential equation with nonconstant coefficients. SIAM J. Math. Anal. 16 (1985), 143-147.

[7] C. Chen, W. V. Wahl: Das Rand-Anfangswertproblem fur quasilineare Wellengleichungen in Sobolevraumen niedriger Ordnung. J. Reine Angew. Math. 337 (1982), 77-119.

[8] B.D. Coleman & M.E. Gurtin: Waves in materials with memory, II. On the growth and decay of one-dimensional acceleration waves. Arch. Rat. Mech. Anal. 19 (1965), 239-265.

[9] C.M. Dafermos & J.A. Nohel: A nonlinear hyperbolic Volterra equation in viscoelasticity. Amer. Math. J. Supplement (1981), 87-116.

[10] C.M. Dafermos & W. Hrusa: Energy methods for quasilinear hyperbolic initial-boundary value problems. Applications to elastodynamics. Arch. Rat. Mech. Anal. 87 (1985), 267-292.

[11] C.M. Dafermos: Development of singularities in the motion of materials with fading memory. Arch. Rat. Mech. Anal. 91 (1986), 193-205.

[12] G. Da Prato & M. Ianelli: Linear integro-differential equations in Banach spaces. Redn. Sem. Mat. Univ. Padova 62 (1980), 191-206.

[13] R. DiPerna: Compensated Compactness and General Systems of Conservation Laws. Trans. Amer. Math. Soc. 292 (1985), 383-420.

[14] M. Doi & S.F. Edwards: The Theory of Polymer Dynamics. Clarendon Press, Oxford, 1986.

[15] H. Engler: Weak solutions of a class of quasilinear hyperbolic integro-differential equations describing viscoelastic materials. Submitted to Arch. Rat. Mech. Anal.

[16] R.C. Grimmer & A.J. Pritchard: Analytic resolvent operators for integral equations in Banach spaces. J. Differential Eq. 50 (1983), 234-259.

[17] R.C. Grimmer & M. Zeman: Wave propagation for linear integro-differential equations in Banach space. J. Differential Eq. 54 (1984), 274-282.

[18] M.E. Gurtin: Thermodynamics and the energy criterion for stability. Arch. Rat. Mech. Anal. 52 (1973), 93-103.

[19] M.E. Gurtin: An Introduction to Continuum Mechanics. Academic Press, New York, 1981.

[20] W. Hrusa: A nonlinear functional differential equation in Banach space with applications to materials with fading memory. Arch. Rat. Mech. Anal. 84 (1983), 99-137.

[21] W. Hrusa & J.A. Nohel: The Cauchy problem in one-dimensional nonlinear viscoelasticity. J. Differential Eq. 59 (1985), 388-412.

[22] W. Hrusa & M. Renardy: On wave propagation in linear viscoelasticity. Quart. Appl. Math. 43 (1985), 237-253.

[23] W. Hrusa & M. Renardy: On a class of quasilinear partial integro-differential equations with singular kernels. J. Differential Eq. 64 (1986), 195-220.

[24] S. Klainerman & A. Majda: Formation of singularities for wave equations including the nonlinear vibrating string. Comm. Pure Appl. Math. 33 (1980), 241-263.

[25] O.A. Ladyzenskaya, V.A. Ural'tseva & N.N. Solonnikov: Linear and Quasilinear Equations of Parabolic Type. Amer. Math. Soc., Providence, 1968.

[26] H.M. Laun: Description of the non-linear shear behaviour of a low density polyethylene melt by means of an experimentally determined strain dependent memory function. Rheologica Acta 17 (1978), 1-15.

[27] A.S. Lodge: Body Tensor Fields in Continuum Mechanics. Academic Press, New York, 1974.

[28] S.-. Londen: An existence result on a Volterra equation in a Banach space. Trans. Amer. Math. Soc. 235 (1978), 285-304.

[29] R.C. MacCamy: A model for one-dimensional nonlinear viscoelasticity. Quart. Appl. Math. 35 (1977), 21-33.

[30] J.A. Nohel, R.C. Rogers, A. Tzavaras: Weak solutions for a nonlinear system in viscoelasticity. MRC Technical Summary Report # 2976, Madison, Wisconsin 1987.

[31] A.C. Papanastasiou, L.E. Scriven & C.W. Macosko: An integral constitutive equation for mixed flows: Viscoelastic characterization. J. Rheology 27 (1983), 387-410.

[32] J. Pruss: Positivity and regularity of hyperbolic Volterra equations in Banach spaces. Preprint (1986).

[33] M. Renardy, W.E. Hrusa & J.A. Nohel: Mathematical Problems in Viscoelasticity. Longman Group (formerly Pitman), London, 1987.

[34] J. Smoller: Shock Waves and Reaction - Diffusion Equations, Springer, Berlin, New York, Heidelberg, 1982.

[35] O. Staffans: On a nonlinear hyperbolic Volterra equation. SIAM J. Math. Anal. 11 (1980), 793-812.

Hans Engler
Department of Mathematics
Georgetown University
Washington DC 20057
USA.

A. FAVINI

Implicit integrodifferential equations

1. INTRODUCTION AND NOTATION

We shall concern ourselves with strict solutions (in a sense to be stated precisely) of the linear integrodifferential degenerate problem

$$\begin{cases} \frac{d}{dt}(M(t)u(t)) + L(t)u(t) = \int_0^t K(t-s)u(s)ds + f(t), \ 0 \leq t \leq \tau, \\ M(t)u(t)|_{t=0} = w_0; \end{cases} \tag{1.1}$$

here w_0 belongs to a complex Banach space X, f is a continuous function from $[0,\tau]$ into X and $M(t)$, $L(t)$, $K(t)$, $0 \leq t \leq \tau$, are three families of closed linear operators from a (complex) Banach space Y into X.

Degeneration may arise because either $M(t)$ has no inverse or its inverse is unbounded near $t = 0$.

There are a lot of papers about the regular situation $M(t) \equiv I$, the identity operator, [besides the numerous works by R. Grimmer, we want to quote [1, 2, 7, 10, 11, 12, 14]], but, to our knowledge, (1.1) has not been studied in literature.

The present paper deals with the "parabolic" case and gives some conditions on L,M, K, f and w_0 in order to obtain a unique local solution u of (1.1) such that $L(\cdot)u(\cdot)$ and $d/dt(M(\cdot)u(\cdot))$ are Hölder-continuous; we refer to [1,2] for this type of results in the non degenerate case.

All this is achieved as an application of certain general statements concerning the operational (nonlinear) equation

$$BMu + Lu = G(u), \tag{1.2}$$

where B is a closed linear operator from the complex Banach space E into itself, L, M are closed linear operators from a Banach space F into E and G acts suitably from D(L) into E. It is always assumed that the domain D(L) of L is embedded in the domain D(M) of M (and thus that $D(L(t)) \subseteq D(M(t))$

with continuous embedding for all t) and that L has a bounded inverse (L(t) has a bounded inverse for each t). To reach our goal, we shall use:

1) A variation of the well-known Da Prato-Grisvard method [3], which we have developed in [4,5], for the linear equation

$$BMu + Lu = h; \tag{1.3}$$

2) A maximal regularity result for the solution u of (1.3), in the sense that $h \in (E;D(B))_{\theta,\infty} = V$, the real interpolation space between E and D(B), implies Lu, $BMu \in V$;

3) A fixed-point argument for obtaining from 2) an existence result for (1.2), [see [6], where there are treated nonlinear implicit differential equations, too].

The paper is organized as follows. In Section 2 we recall some results on (1.2), (1.3) from [5,6], as well as their application to (1.1) with $K(\cdot) \equiv 0$.

Section 3 contains the main results of this work and gives regularity and compatibility conditions on f and w_0 entailing solvability for (1.1).

In Section 4 we show that two general classes of problems can be handled by means of our technique.

The first example concerns operators M, L, K defined by suitable sesquilinear forms on a Hilbert space V and it is outlined how some concrete degenerate integrodifferential equations fall within this theory.

In the second one we see that this happens also when M(t) = I, L(t) is an elliptic differential operator of order 2m, $m \in \mathbb{N}$, satisfying Kato-Tanabe-type conditions (and thus the domain of L(t) in this case depends upon t) and K(t) is another differential operator of order $k \leq 2m$: see [1] for similar problems.

The following notation will be used. If E is a complex Banach space, its norm shall be denoted by $\|\cdot;E\|$. If F is another Banach space, L(E;F) denotes the space of all bounded linear operators from E into F, with the usual norm. If E = F, we shall use L(E) for L(E;E).

If $T \in L(E,F)$, R(T) is the range of T.

We will need the following spaces of functions:

If $\tau > 0$, X is a Banach space, then

$$C[0,\tau;X] = \{u:[0,\tau] \to X;\ u \text{ continuous}\},\ \|u:C[0,\tau;X]\| = \sup_{0\leq t\leq\tau} \|u(t);X\|\ ;$$

$$C_0[0,\tau;X] = \{u \in C[0,\tau;X];\ u(0) = 0\};$$

$$C^{(k)}[0,\tau;X] = \{u:[0,\tau] \to X;\ u \text{ is k-times continuously differentiable}\},$$

$k = 1,2,\ldots$; if $0 < \theta < 1$,

$$C^{\theta}[0,\tau,X] = \{u \in C[0,\tau,X];[u]_{\theta} = \sup_{\substack{0\leq t,s\leq\tau \\ t\neq s}} (\|u(t)-u(s);X\| / |t-s|^{\theta}) < \infty$$

$$\|u;C^{\theta}[0,\tau;X]\| = \|u;C[0,\tau;X]\| + [u]_{\theta};$$

$$C_0{}^{\theta}[0,\tau;X] = \{u \in C^{\theta}[0,\tau,X]:u(0) = 0\}.$$

It is well known [8] that $C_0{}^{\theta}[0,\tau;X] = (E;D(B))_{\theta,\infty}$, where $E = C_0[0,\tau;X]$ and

$$D(B) = \{u \in C^{(1)}[0,\tau;X]:u(0) = u'(0) = 0\},\ Bu = u'.$$

2. SOLUTION OF (1.2) AND (1.3)

First, consider equation (1.3). To this end, we shall assume the following:

H1: If $z \in \mathbb{C}$ satisfies $|\pi-\arg z| \leqq \varphi < \pi/2$, or $z = 0$, then $B-z$ is densely defined with bounded inverse $(B-z)^{-1} \in L(E)$ and

$$\|(B-z)^{-1};L(E)\| \leqq C(1 + |z|)^{-1},$$

z in this sector.

H2: L, M are closed linear operators from F into E, $D(L) \subseteq D(M)$, L has a bounded inverse, there exists $(zM+L)^{-1} \in L(E;F)$

$\forall z$: $|\arg z| \leqq \pi-\varphi + \varepsilon$, $\varepsilon > 0$, and

$$\|L(zM+L)^{-1};L(E)\| \leqq C(1 + |z|)^{\alpha},$$

where $0 \leqq \alpha < 1$.

Let us denote by Γ the path in $\mathbb{C}$ defined by $z = \tau \exp(\pm i(\pi-\varphi+\varepsilon/2))$, with $\tau \geqq a_0 > 0$, a_0 suitably small, and by $z = a_0 \exp(i\omega)$, $|\omega| \leqq \pi-\varphi + \varepsilon/2$. Then the third hypothesis reads

H3. $D(B)$ is invariant under $(zT+I)^{-1}$, $T = ML^{-1}$, $z \in \Gamma$, and if $[B;(zT+I)^{-1}]$ denotes the commutator $B(zT+I)^{-1} - (zT+I)^{-1}B$, then $[B;(zT+I)^{-1}]$ has a bounded extension as an operator au $L(E)$ and in $L(V)$, $V = (E;D(B))_{\theta,\infty}$; further, there exist $0 \leqq \sigma < 1$, $C > 0$ such that

$$\max \{\|[B;(zT+I)^{-1}]; L(E)\|, \|[B;(zT+I)^{-1}];L(V)\|\} \leqq C(1+|z|)^{\sigma}, \ z \in \Gamma.$$

One has the following existence-uniqueness-regularity result [5]:

THEOREM 2.1. Under H1-3, let $\alpha < \theta < 1$. Then for all $h \in V$ and real s sufficiently large, the equation

$$(B+s)Mu + Lu = h \tag{2.1}$$

has exactly one solution u such that $Lu, BMu \in (E;D(B))_{\theta-\alpha,\infty}$.

In most applications, we are allowed to take $s = 0$ in (2.1) and thus in this case the preceding result concerns Equation (1.3). Henceforth we shall always suppose $s = 0$, and hence (1.3) is solved for any $h \in V$.

APPLICATION. Let $L(t)$, $M(t)$, $0 \leqq t \leqq \tau$, be two families of closed linear operators from Y into X such that

i) $L(t)$ is invertible for all t, $0 \leqq t \leqq \tau$,

ii) $D(L(t)) \subseteq D(M(t))$ for all $0 \leqq t \leqq \tau$;

iii) $t \to M(t)L(t)^{-1} = T(t)$ is continuous from $[0,\tau]$ into $L(X)$;

iv) $t \to L(t)^{-1}$ is continuous from $[0,\tau]$ into $L(X;Y)$.

We shall use the following definiton of strict solution for the problem

$$\begin{cases} \dfrac{d}{dt}(M(t)u(t)) = -L(t)u(t) + f(t),\ 0 \leq t \leq \tau \\ M(t)u(t)|_{t=0} = w_0 \in X. \end{cases} \tag{2.2}$$

DEFINITION 2.1. Let f be a continuous function from $[0,\tau]$ into X. A strict solution u for (2.2) is an element u of $C[0,\tau;Y]$ such that $u(t) \in D(L(t))$, $0 \leq t \leq \tau$, $M(\cdot)u(\cdot) \in C^{(1)}[0,\tau;X]$, $L(\cdot)u(\cdot) \in C[0,\tau;X]$, and (2.2) holds.

In order to treat (2.2) by means of Theorem 2.1, we observe that if $T(\cdot) \in C^{(1)}[0,\tau;L(X)]$, $w_0 = M(0)u_0$, $u_0 \in D(K(0))$, and we put $\varphi(t) = v_0 + tv_1$ where $v_0 = L(0)u_0$ and v_1 shall be precised later on, then $w(t) = L(t)u(t) - \varphi(t)$ transforms (2.2) into the problem of finding $w \in C_0[0,\tau;X]$ such that

$$\frac{d}{dt}(T(t)w(t)) + w(t) = h(t),\ 0 \leq t \leq \tau, \tag{2.3}$$

$h(t) = f(t) - (I + T(t))\,\varphi(t) - T(t)\,v_1$. With the aid of these preliminaries, we can demonstrate

THEOREM 2.2. Assume i) - iv) and suppose that $T(\cdot) \in C^{(1)}[0,\tau;L(X)]$ and there are $0 \leq \alpha,\ \sigma < 1$, $\alpha < \theta < 1$ such that, if $R(z,t) = (zT(t) + I)^{-1}$,

$$\text{H4:}\ \begin{cases} \|R(z,t);L(X)\| \leq C(1+|z|)^{\alpha},\ 0 \leq t \leq \tau,\ z \in \Sigma\colon |\arg z| \leq \pi/2 + \eta,\ \eta > 0 \\ \|\frac{\partial}{\partial t} R(z,t);L(X)\| \leq C(1+|z|)^{\sigma},\ 0 \leq t \leq \tau,\ z \in \Sigma, \\ \|\frac{\partial}{\partial t} R(z,t) - \frac{\partial}{\partial s} R(z,s);L(X)\| \leq C(1+|z|)^{\sigma}|t-s|^{\theta},\ 0 \leq t,\ s \leq \tau,\ z \in \Sigma. \end{cases}$$

Then (2.2) has a unique strict solution u such that $L(\cdot)u(\cdot)$, $(d/dt)(M(\cdot)u(\cdot)) \in C^{\theta-\alpha}[0,\tau;X]$ for any $f \in C^{\theta}[0,\tau;X]$ and all $w_0 \in X$ satisfying $w_0 = T(0)v_0 \in R(T(0))$ and $f(0) - (I+T'(0))v_0 \in R(T(0))$.

Proof. First of all, the hypotheses imply that $T'(\cdot) \in C^{\theta}[0,\tau;L(X)]$; on the other hand, if $f(0) - (I+T'(0))v_0 = T(0)v_1$, take this v_1 in the definition of φ. Then $h \in C_0^{\theta}[0,\tau;X]$ and it is easy to verify that all assumptions in Theorem 2.1 hold if $V = C_0^{\theta}[0,\tau;X]$ (see §1). Hence we have unique solvability of

$$\begin{cases} (\frac{d}{dt} + s)(T(t)w_1(t)) + w_1(t) = h_1(t), \ 0 \leqq t \leqq \tau, \\ T(0)w_1(0) = 0, \end{cases}$$

where $h_1(t) = e^{-st}h(t)$. Then $w(t) = e^{st}w_1(t)$ satisfies (2.3). #

It follows from Theorem 2.1 that under the stated hypotheses with $\alpha = 0$ there is a bounded linear operator S from V into itself such that Sf, $\forall f \in V$, verifies BTSf + Sf = f.

Let us give some conditions which allow to solve (1.2); denote $\alpha_0 = \|S, L(V)\|$, and suppose that the nonlinear operator G acts from $D(G) \supset D(L)$ into E. Then we have (see [6]):

Theorem 2.3. Assume the hypotheses in Theorem 2.1 for B, L, M, and $\alpha = 0$. If $D(L) \subseteq D(G)$ and there $\beta > 0$ such that $\alpha_0\beta < 1$ and

$$\|G(L^{-1}h_1) - G(L^{-1}h_2); V\| \leqq \beta\| h_1-h_2; V\|, \ h_1,h_2 \in V,$$

then (1.2) has a unique solution u such that Lu, BMu $\in$ V.

Proof. The proof is an easy application of Banach fixed point Theorem. Note that if v, w are two solutions of $BTx + x = G(L^{-1}x)$ with v,w, BTv, BTw $\in$ V, then, by Theorem 2.1, $v = SG(L^{-1}v)$, $w = SG(L^{-1}w)$, and thus

$$\|v-w;V\| \leqq \alpha_0\beta\| v - w;V\| . \#$$

3. SOLUTION OF (1.1)

In this section we shall always suppose that L(t) and M(t) have the properties indicated in Theorem 2.2; we then shall make some hypotheses on K and f which enable us to apply the general result given in Theorem 2.3.

Formally, by means of the technique that we have used for the Application in §2, if $(s,t) \to K(t,s)L(s)^{-1}$ is well-defined and continuous from

$$\Delta = \{(s,t): 0 \leqq s \leqq t \leqq \tau\}$$

into L(X), (for sake of simplicity), then (1.1) can be transformed into the problem

$$\frac{d}{dt}(T(t)w(t)) + w(t) = F(w)(t),\ 0 \leq t \leq \tau,$$

$$T(t)w(t)|_{t=0} = 0,$$

where

$$F(w)(t) = \int_0^t K(t-s)L(s)^{-1}w(s)ds + \int_0^t K(t-s)L(s)^{-1}\{v_0+sv_1\}ds - T(t)\{v_0+tv_1\}$$
$$- T(t)v_1 - v_0 - tv_1 + f(t),\ 0 \leq t \leq \tau.$$

Before we list the needed assumptions on K(t) as connected with L(t), we specify what we mean by a strict solution for (1.1).

DEFINITION 3.1. Assume $(s,t) \to K(t-s)L(s)^{-1}$ continuous from Δ into L(X), with K(t) a closed linear operator from Y into X for any $t \in [0,\tau]$.

Then $u \in C([0,\tau;Y]$ is a strict solution of (1.1) if $u(t) \in D(L(t))$ for each $t \in [0,\tau]$, $L(\cdot)u(\cdot) \in C[0,\tau;X]$, $M(\cdot) \in C^{(1)}[0,\tau;X]$, and (1.1) holds.

The hypotheses on K(t), L(t) an as follows:

H.5 There exists a Banach space $Y_1 \subset Y$ (continuous embedding) such that

$$D(L(t)) \subset Y_1 \subset D(K(t')),\ \forall t,t' \in [0,\tau],$$

and K(t) is continuous from $[0,\tau]$ into $L(Y_1;X)$.

H.6 There is C > 0 such that

$$\|L(t')^{-1} - L(t'')^{-1};\ L(X;Y_1)\| \leq C|t'-t''|^{\theta},$$

where $0 < \theta < 1$, for all t', t'' $\in [0,\tau]$.

We are now ready to prove

THEOREM 3.1. Assume H4, 5, 6, with $\alpha = 0$, $0 < \theta < 1$, $f \in C^{\theta}[0,\tau;X]$, $w_0 = T(0)v_0, f(0) - (I+T'(0))v_0 (= T(0)v_1) \in R(T(0))$.

If $\tau > 0$ is sufficiently small, then (1.1) has a unique strict solution on $[0,\tau]$ such that $L(\cdot)u(\cdot)$, $(d/dt)\ (M(\cdot)u(\cdot)) \in C^{\theta}[0,\tau;X]$.

Proof. If we want to apply Theorem 2.3, it suffices to show that the aforesaid map F acts from $C_0^{\theta}[0,\tau;X] = V$ into itself and

$$\|F(w_1)-F(w_2);\ V\| \leq \beta\ \|w_1 - w_2;V\|,$$

$\beta > 0$ small.

Clearly, $F(w)$, $w \in V$, is continuous on $[0,\tau]$ and

$$\|F(w)(t) - F(w)(t');\ X\| \leq (1) + (2) + (3) + (4) + (5) + (6),$$

where

$$(1) = \left\|\int_0^t K(s)L(t-s)^{-1}w(t-s)ds - \int_0^{t'} K(s)L(t'-s)^{-1}w(t'-s)ds;X\right\|,$$

$$(2) = \left\|\int_0^t K(s)L(t-s)^{-1}\{v_0+(t-s)v_1\}ds - \int_0^t K(s)L(t'-s)^{-1}\{v_0+(t'-s)v_1\}ds;X\right\|,$$

$$(3) = \|T'(t)(v_0+tv_1) - T'(t')(v_0+t'v_1);X\|,$$

$$(4) = \|(T(t) - T(t'))v_1;\ X\|,\quad (5) = |t-t'|\ \|v_1;X\|,\ (6) = \|f(t)-f(t');X\|.$$

Suppose $0 < t' < t$. Then, by H5.6,

$$\begin{aligned}(1) \leq &\int_0^{t'} \|K(s)[L(t-s)^{-1}-L(t'-s)^{-1}]w(t-s);X\| + \\ &+ \int_0^{t'} \|K(s)L(t'-s)^{-1};\ L(X)\|\ \|w(t-s)-w(t'-s);X\|\, ds\ + \\ &+ \int_{t'}^{t} \|K(s)L(t-s)^{-1};\ L(X)\|\,\|w(t-s);X\|\, ds \leq \\ \leq\ &C_1 t'|t-t'|^{\theta}(\tau^{\theta}\ \|w;V\|) + C_2\tau|t-t'|^{\theta}\ \|w;V\| + C_3|t-t'|\tau^{\theta}\ \|w;V\| \leq \\ \leq\ &C_4(w)|t-t'|^{\theta}.\end{aligned}$$

Also (2)-(6) are estimated in the same way. Finally, if $w_1,w_2 \in V$, then $(F(w_1)-F(w_2))(t) =$

$$= \int_0^t K(t-s)L(s)^{-1}\{w_1(s) - w_2(s)\}ds$$

implies, with $0 \leqq t' < t \leqq \tau$,

$$\|F(w_1)(t)-F(w_2)(t);X\| \leqq C\tau \ \|w_1-w_2;\ V\| \ , \ \|F(w_1)(t)-F(w_2)(t)-F(w_1(t') +$$

$$F(w_2)(t');X\| \leqq \int_0^{t'} \|K(s)L(t'-s)^{-1};L(X)\| \, \|w_1(t-s)-w_2(t-s)-w_1(t'-s) +$$

$$+ w_2(t'-s);X\| \, ds \ +$$

$$+ \int_0^{t'} \|K(s)[L(t-s)^{-1} - L(t'-s)^{-1}];\ L(X)\| \ \|w_1(t-s)-w_2(t-s);X\| \, ds \ +$$

$$+ \int_{t'}^{t} \|K(s)L(t-s)^{-1};\ L(X)\| \ \|w_1(t-s)-w_2(t-s);X\| \, ds \leqq$$

$$\leqq \ ct'(t-t')^{\theta}\| \ w_1-w_2;V\| + c't'\tau^{\theta}(t-t')^{\theta} \ \|w_1-w_2;V\| + c''\tau^{\theta}(t-t') \ \|w_1-w_2;V\|$$

$$\leqq \ c_1\tau(t-t')^{\theta} \ \ \|w_1-w_2;\ V\| \, .$$

Hence Theorem 2.3 applies immediately. #

4. Applications

Application 1. Let $V \subset W \subset H$ be separable complex Hilbert spaces, with continuous and dense embeddings, so that, if V', W' denote the antidual spaces of V and W, respectively, and H is identified with its own antidual, we also have the continuous and dense embeddings $H \subset W' \subset V'$. We denote by $[f,v]$ the scalar product of $\in V'$ and $v \in V$ in this antiduality between V and V'.

For all $t \in [0,\tau]$, $\tau > 0$, let $a_0(t;u,v)$, u, $v \in V$, $a_1(t,x,y)$, $x,y \in W$, be two sesquilinear forms on V and W, respectively, such that

$$|a_0(t;u,v)| \leq C_1 \|u;V\| \|v;V\|, \quad 0 \leq t \leq \tau, \ u,v \in V,$$

$$\text{Re } a_0(t;u,u) \geq C_2 \|u;V\|^2, \quad 0 \leq t \leq \tau, \ u \in V,$$

K_1: $$|a_1(t;x,y)| \leq C_3 \|x; W\| \|y;W\|, \quad 0 \leq t \leq \tau, \ x,y \in W,$$

$$a_1(t;u,u) \geq 0, \quad 0 \leq t \leq \tau, \ u \in V,$$

for certain positive constants C_i, $i = 1,2,3$;

$a_0(t;u,v), u,v \in V, a_1(t;x,y), x,y \in W$, are twice continuously differentiable in $t \in [0,\tau]$ and their derivatives with respect to t satisfy

K_2: $$|a_0^{(k)}(t;u,v)| \leq C_4 \|u;V\| \|v;V\|, \quad u,v \in V, \ k = 1,2,$$

$$|a_1^{(k)}(t;x,y)| \leq C_5 |a_1(t;x,y)|, \quad x,y \in W, \ k = 1,2, C_4, C_5 > 0.$$

Notice that in view of K_2 there is a positive number C_6 such that

$$|a_0(t;u,v)-a_0(s;u,v)| \leq C_6|t-s| \|u;V\| \|v;V\|, 0 \leq t, \ s \leq \tau, u,\tau \in V. \tag{4.1}$$

Denote by $L(t)$, $M(t) \in L(V;V')$ the bounded operators defined by $a_0(t;u,v)$ and the restriction of $a_1(t;x,y)$ to V, respectively.

In virtue of K_1, K_2, using the classical results in [9], it is easy to recognize (see also [4;pp. 38-39]) that $zM(t) + L(t)$ has a bounded inverse from V' into V for all z, Re $z \geq 0$, and $\|(zM(t)+L(t))^{-1};L(V';V)\| \leq C_2^{-1}$. Hence, if $T(t) = M(t)L(t)^{-1}$, $R(z,t) = (zT(t) + I)^{-1}$, then $\|R(z,t);L(V')\|$ is uniformly bounded on z, Re $z \geq 0$ and $t \in [0,\tau]$.
The assumptions on $a_1(t,x,y)$ and the fact that V' is the antidual space of V permit to deduce that

$$\|R(z,t)M'(t)u; V'\| \leq C \|R(z,t)M(t)u;V'\|,$$

for all $u \in V$, $t \in [0,\tau]$ and any z in a sector containing Re $z > 0$. Moreover, (see again [4; pp. 38-39]) $\| \frac{\partial^k}{\partial t^k} R(z,t);L(V')\|$ is bounded on such a sector times $[0,\tau]$, $k = 1,2$. This in its turn clearly implies that

$$\|\frac{\partial}{\partial t} R(z,t) - \frac{\partial}{\partial s} R(z,s);L(V')\| \leq C\,|t-s|,\ t,s \in [0,\tau],\ C > 0.$$

Our further hypotheses concern the operator $K(t)$:

For all $t \in [0,\tau]$, $K(t)$ is the linear operator associated with a sesquilinear form $a_2(t;u,v)$, $u,v \in V$, such that

K_3: $|a_2(t;u,v)| \leq C_7 \|u;V\|\,\|v;V\|$, $u,v \in V$,

$$|a_2(t;u,v) - a_2(s;u,v)| \leq C_8|t-s|^a \|u;V\|\,\|v;V\|,\ u,v \in V,$$

with $0 < a \leq 1$, C_7, $C_8 > 0$.

K_3 implies that $t \to K(t)$ is continuous from $[0,\tau]$ into $L(V;V')$; therefore we can apply Theorem 3.1. if we take $X = V'$, $Y = Y_1 = V$.

Notice that H6 is satisfied since (4.1) implies

$$\|L(t)u-L(s)u;V'\| \leq C_6|t-s|\ \|u;V\|$$

and the estimates $C'' \|u;V\| \leq \|L(t)u;V'\| \leq C'' \|u;V\|$, C', $C'' > 0$, permit to deduce [see [13; pp. 144-145]] that

$$\|L(t)^{-1}f-L(s)^{-1}f;\ V\| \leq k\ |t-s|\ \|f;V'\|,\ k > 0,\ t,s \in [0,\tau],\ f \in V'.$$

Summarizing these observations, we have obtained

<u>THEOREM 4.1.</u> Assume K_1, K_2, K_3. Then (1.1) has a unique strict solution u on $[0,\tau]$, sufficiently small, such that $L(\cdot)u(\cdot),(d/dt)(M(\cdot)u(\cdot)) \in C^\theta[0,\tau;V']$, $0 < \theta < 1$, for any $f \in C^\theta[0,\tau;V']$ and any $w_0 \in V'$ such that $[w_0,v] = a_1(0;u_0,v)$, $[f(0),v]-a_0(0;u_0,v)-a'_1(0;u_0,v_1)$, with suitable u_0, $u_1 \in V$ and all $v \in V$.

In fact, the conditions on w_0 translate the ones given in Theorem 3.1.

The preceding regularity assumptions become less restrictive if the sesquilinear form $a_1(t;x,y)$ is independent of t.

In fact, then it suffices to assume that

$a_0(t;u,v)$ is continuously differentiable with respect to t and there are C, C', C" > 0 such that

K_4: $|a_0^{(k)}(t;u,v)| \leq C \|u;V\| \|v;V\|$, $0 \leq t \leq \tau$, $u,v \in V$, $k = 0,1$,

$\text{Re } a_0(t;u,u) \geq C' \|u;V\|^2$, $0 \leq t \leq \tau$, $u \in V$,

$|a_0'(t;u,v)-a_0'(s;u,v)| \leq C''|t-s|^{\mu} \|u;V\| \|v;V\|$, $0 \leq s,t \leq \tau$, $u,v \in V$,

with $0 < \mu \leq 1$.

K_4 implies that

$$\|L'(t)-L'(s);L(V,V')\| \leq C'' \; |t-s|^{\mu}, \; 0 \leq t,s \leq \tau.$$

On the other hand, from

$$(zM+L(t))^{-1}-(zM+L(s))^{-1}=(zM+L(t))^{-1}(L(s)-L(t))(zM+L(s))^{-1}$$

and

$$\frac{\partial}{\partial t} R(z,t) - \frac{\partial}{\partial s} R(z,s) = \{L'(t)L(t)^{-1}R(z,t) - L'(s)L(s)^{-1}R(z,s)\} -$$

$$- \{R(z,t)L'(t)L(t)^{-1}R(z,t) - R(z,s)L'(s)L(s)^{-1}R(z,s)\},$$

we deduce that there is k > 0 for which

$$\|\frac{\partial}{\partial t} R(z,t) - \frac{\partial}{\partial s} R(z,s);L(V')\| \leq k \; |t-s|^{\mu}, \; 0 \leq t, \; s \leq \tau.$$

Hence,

THEOREM 4.2. Let $a_1(x,y)$ be a continuous sesquilinear form on W such that $a_1(x,x) \geqq 0$ for all $x \in W$; assume K_3 and K_4, $0 < \mu < 1$, $f \in C^{\mu}[0,\tau;V']$, $w_0 \in V'$, $[w_0,v] = a_1(u_0,v)$, $[f(0),v] - a_0(0;u_0,v) = a_1(u_1,v)$, for certain $u_0,u_1 \in V$ and all $v \in V$.

Then (1.1) has a unique local solution u such that $L(\cdot)u(\cdot)$, $(d/dt)(M(\cdot)u(\cdot)) \in C^{\mu}[0,\tau;V']$.

If $\mu = 1$, the preceding conclusion holds for any $\theta \in (0,1)$ instead of μ, that is, $f \in C^{\theta}[0,\tau;V']$ implies $L(\cdot)u(\cdot)$, $(d/dt)\,(M(\cdot)u(\cdot)) \in C^{\theta}[0,\tau;V']$, $0 < \theta < 1$.

EXAMPLE Let Ω be an open set in R^n, $n \geqq 1$, with a smooth boundary, and define

$$a_0(t;u,v) \quad \sum_{i,j=1}^{n}\int_\Omega (a_{ij}(t,x)\ \partial u/\partial x_i \cdot \partial\bar{v}/\partial x_j + c(t,x)u\bar{v})dx,\ t \in [0,\tau],$$

$$u,v \in H_0^1(\Omega) \quad = V \subset L^2(\Omega) = H,$$

$$a_2(t;u,v) = \sum_{i,j=1}^{n}\int_\Omega (c_{ij}(t,x)\partial u/\partial x_i \cdot \partial\bar{v}/\partial x_j + d(t,x)u\ \bar{v})dx,\ t \in [0,\tau],$$

$$u,v \in V,$$

with coefficients sufficiently regular in t and

$$\sum_{i,j=1}^{n} a_{ij}(t,x)z_i\ \bar{z}_j \geqq \gamma \sum_{i=1}^{n} |z_i|^2\ \forall z_i \in \mathbb{C},\ c(t,x) \geqq 0,\ \gamma > 0.$$

If $a_1(u,v)$ is given by

$$a_1(u,v) = \int_\Omega m(x)\ u\ \bar{v}\ dx,\ m(x) \geqq 0 \text{ for all } x \in \Omega,$$

or by

$$a_1(u,v) = \sum_{i,j=1}^{n}\int_\Omega (b_{ij}(x)\partial u/\partial x_i \cdot \partial\bar{v}/\partial x_j)dx,$$

where $\sum_{i,j=1}^{n} b_{ij}(x)z_i\ \bar{z}_j \geqq 0\quad \forall z_i \in \mathbb{C},$

then Theorem 4.2 works.

Application 2 Let us consider the problem

$$\partial u(t,x)/\partial t+A(t,x,D)u(t,x) = \int_0^t K(t-s,x,D)u(s,x)ds+f(t,x), 0 \leq t \leq \tau, x \in \Omega,$$

$$u(0,x) = u_0(x),\ x \in \Omega, \tag{4.2}$$

$$B_j(t,x,D)u(t,x) = 0,\ 0 \leq t \leq \tau,\ x \in \partial\Omega,\ j = 1,\dots,m.$$

Here, Ω is a bounded region in R^n which is of class C^{2m},

$$A(t,x,D) = \sum_{|\alpha|\leq 2m} a_\alpha(t,x)D^\alpha$$

is strongly elliptic and satisfies the conditions in [13; p. 129, p. 140],

$$B_j(t,x,D) = \sum_{|\beta|\leq m_j} b_{j\beta}(t,x)D^\beta,\ j = 1,\dots,m,$$

is a set of normal boundary differential operators with the properties listed in [13, p. 140] and

$$K(t,x,D) = \sum_{|\gamma|\leq q} c_\gamma(t,x)D^\gamma,\ q \leq 2m$$

is a differential operator whose coefficients $c_\gamma(t,x)$ are continuous functions on $[0,\tau] \times \bar{\Omega}$.

It is well known [13; pp. 140-144] that if $1 < p < \infty$ and if the coefficients are suitably regular, then the operators $L(t)$ defined by

$$D(L(t)) = \{u \in W_p^{2m}(\Omega);\ B_j(t,x,D) = 0,\ x \in \partial\Omega,\ j = 1,\dots,m\},$$

$$(L(t)u(x) = A(t,x,D)u(x),\ u \in D(L(t)),$$

satisfy the assumptions in Theorem 2.2, with $\alpha = \sigma = 0$, $\theta = 1$, $X = Y = L^p(\Omega)$. Further, if we take $W_p^{2m}(\Omega)$ as Y_1, Lemma 5.3.4 in [13, p. 142] ensures that H6 is verified.

On the other hand, the conditions $0 \leq q \leq 2m$ and

$$\max_{|\gamma|\leq q} \sup_{x\in\Omega} |c_\gamma(t',x)-c_\gamma(t'',x)| \leq k_1|t'-t''|^a, \ 0 < a \leq 1, \ k_1 > 0,$$

imply that H5 holds, and thus Theorem 3.1 allows us to deduce the following

Theorem 4.3. Assume the preceding assumptions on the coefficients fulfilled; let $u_0 \in D(L(0))$ and $f(t) = f(t,\cdot) \in C^\theta[0,\tau;L^p(\Omega)]$, $0 < \theta < 1$, $f(0) - (I + ((d/dt)L(t)^{-1}|_{t=0}))L(0)\, u_0 \in D(L(0))$.

Then the abstract problem in $L^p(\Omega)$ connected with (4.2) has a unique strict local solution u such that $u'(\cdot), L(\cdot)u(\cdot) \in C^\theta[0,\tau;L^p(\Omega)]$.

References

[1] Acquistapace, P: Existence and maximal regularity for linear parabolic integrodifferential equations, in "Integrodifferential Equations and Applications, Trento, 1984 Proceedings", J. Integral Eqs. 10, Special Conference Issue (1985), 5-43.

[2] Da Prato, G: Abstract differential equations, maximal regularity, and linerarization, Proceed. Symp. Pure Math. 45 (1986), Part 1, 359-370.

[3] Da Prato, G., Grisvard, P: Sommes d'opérateurs linéaires et équations différentielles opérationnelles, J. Math. Pures Appl. 54 (1975), 305-387.

[4] Favini, A: Degenerate and singular evolution equations in Banach space, Math. Ann. 273 (1985), 17-44.

[5] Favini, A., Plazzi, P: On some abstract degenerate problems of parabolic type 1: the linear case, (to appear in Nonlinear Anal.).

[6] Favini, A., Plazzi, P: On some abstract degenerate problem of parabolic type 2: the nonlinear case, (to appear in Nonlinear Anal.).

[7] Friedman, A., Shinbrot, M: Volterra integral equations in Banach space, Trans. Amer. Math. Soc. 126 (1967), 131-179.

[8] Grisvard, P: Spazi di tracce e applicazioni, Rend. Matem. 5 (1972), 657-929.

[9] Lions, J.L., Magenes, E: Problèmes aux limites non homogènes et applications, vol. 1, Paris: Dunod 1968.

[10] Lunardi, A: Laplace transform methods in integrodifferential equatins, in "Integrodifferential equations and applications, Trento, 1984, Proceedings", J. Integral Equations, 10 (1985), 185-211.

[11] Lunardi, A., Sinestrari, E: C^{α}-Regularity for non-autonomons linear integrodifferential equations of parabolic type, J. Diff. Equations 63 (1986), 88-116.

[12] Prüss, J: On resolvent operators for linear integro-differential equations of Volterra type, J. Integral Equations 5 (1983), 211-236.

[13] Tanabe, H: Equation of evolution, London: Pitman 1979.

[14] Tanabe, H: Volterra integro-differential equations of parabolic type of higher order in t, preprint.

Angelo Favini
Dipartimento di Matematica
Universita di Bologna
40126 Bologna
Italy.

J.A. GOLDSTEIN, S. OHARU & T. TAKAHASHI

Semilinear Hille-Yosida theory

1. INTRODUCTION

The Hille-Yosida theory of linear problems puts linear initial value problems of the form $du(t)/dt = Au(t)$ $[t \geqq 0]$, $u(0) = x$ in one-to-one correspondence with (C_0) semigroups $T = \{T(t) : t \geqq 0\}$, which in turn are in one-to-one correspondence with their generators A. Ever since the founding of the theory of nonlinear semigroups in the nineteen sixties, there has been great interest in obtaining necessary and sufficient conditions for the generation of the parameter nonlinear semigroups acting on arbitrary Banach spaces.

The current theory of nonlinear semigroups has at least two shortcomings. First, only certain initial value problems of interest are covered by the existing contraction (or quasi-contraction) semigroup theory. A standard example not covered by quasi-contractive theory is the Korteweg-de Vries equation. Second, even for contraction semigroups, conditions which are both necessary and sufficient are known only for Hilbert space and special Banach spaces. Nonreflexive spaces with "bad" unit sphere geometry arise naturally in the applications (e.g. L^1 for a single conservation law, and supremum norm spaces for the Hamilton-Jacobi equation).

Thus nonlinear semigroup theory, while it does a lot, is presently inadequate, both as a theory and for the applications.

Recently S. Oharu and T. Takahashi [5] initiated a Hille-Yosida type theory for a class of semilinear problems of the form

$$du(t)/dx = Au(t) + Bu(t) \quad [t \geqq 0],\ u(0) = x. \tag{1}$$

Here A is a linear generator of a (C_0) semigroup T on an arbitrary Banach space X, B maps C, a convex set in X, to X and is locally Lipschitz continuous, $x \in C$, and solutions are sought with values in C. By a mild solution of (1) is meant a strongly continuous solution u of the integral equation

$$u(t) = T(t)x + \int_0^t T(t-s)Bu(s)ds, \quad t \geqq 0.$$

Standard techniques give rise to the existence of a unique local (in t) mild solution of (1). It will be a global solution if we can find for it suitable a priori bounds which will prevent blow-up in a finite time. We shall "measure" the size of the solution by some nonnegative proper convex l.s.c. functional p on C (which may or may not agree with the norm).

The following example illustrates such a functional p which is not a norm. Consider the nonlinear wave equation

$$w_{tt}(= \partial^2 w/\partial t^2) = \Delta w - |w|^{q-1}w \quad (t \geqq 0,\ x \in \mathbf{R}^n).$$

This becomes a first order system for $u = (w, w_t)$, and the energy functional

$$p(w,v) = \|v\|^2_{L^2(\mathbf{R}^n)} + \|\nabla w\|^2_{L^2(\mathbf{R}^n)} + \frac{2}{q+1}\|w\|^{q+1}_{L^{q+1}(\mathbf{R}^n)}$$

is conserved for strong solutions, i.e. $(d/dt)p(w,w_t) \equiv 0$.

In Section 2 we present the Oharu-Takahashi semilinear generation theorem. Beginning in Section 3 we survey the new work of Goldstein, Oharu, and Takahashi [2]. We discuss groups of operators, approximation theory, and differentiability of the nonlinear semigroup solution S(t)x as a function of both t and x. We conclude with an example covered by this theory but not by the standard nonlinear semigroup theory.

2. THE SEMIGROUP GENERATION THEOREM

Let $p : X\ [0,\infty]$ be a proper convex l.s.c. functional which is finite on a given convex set C. Assume $C_r = \{x \in C : p(x) \leqq r\}$ is closed for each $r > 0$. Let $B : C \to X$ be locally Lipschitzian in the sense that for each $r > 0$ there is a number $\omega(r)$ such that

$$\|Bx - By\| \leqq \omega(r)\ \|x-y\|$$

holds for all $x,y \in C_r$. Finally let A generate a (C_0) semigroup T on X (which we may assume to be contractive without loss of generality).

When the above hypotheses hold we write $(A,B,C,p) \in \Omega$.

By a locally Lipschitzian semigroup on C we mean a strongly continuous one parameter semigroup $S = \{S(t) : t \geq 0\}$ of operators from C to C satisfying: for each $r > 0$, $\tau > 0$ there is a constant $\omega(r,\tau)$ such that

$$\|S(t)x - S(t)y\| \leq \exp\{\omega(r,\tau)t\} \quad \|x-y\|$$

holds for $x,y \in C_r$ and $0 \leq t \leq \tau$.

THEOREM 1 [5]. Let $(A,B,C,p) \in \Omega$, and let $a \geq 0$, $b \geq 0$. Then [I] holds if and only if [II] holds.

[I] There is a locally Lipschitzian semigroup S on C such that for all $x \in C$ and $t \geq 0$,

$$S(t)x = T(t)x + \int_0^t T(t-s)BS(s)x \, ds, \tag{I.1}$$

$$p(S(t)x) \leq e^{at}(p(x) + bt). \tag{I.2}$$

[II] For each $r > 0$ there is a $\lambda_0(r) > 0$ such that for all $x \in C_r$ and $\lambda \in (0,\lambda_0(r))$ there is an $x_\lambda \in \mathrm{Dom}(A) \cap C$ satisfying

$$x_\lambda - (A + B)x_\lambda = x, \tag{II.1}$$

$$p(x_\lambda) \leq (1 - a\lambda)^{-1}(p(x) + b\lambda). \tag{II.2}$$

Note that (II.2) is the analogue of "dissipativity" (which is a uniqueness type condition), and (II.1) says that $I - \lambda(A+B)$ has big range (which is an existence type condition).

We give a brief outline of the proof. Assume [II] and $\|T(t)\| \leq 1$. Let $x_0^n = x$, $h_n = \tau/n = \lambda$ and solve

$$h_n^{-1}(x_k^n - x_{k-1}^n) = (A+B)x_k^n. \tag{2}$$

In the <u>stability condition</u> (from (II.2))

$$p(x_k^n) \leqq (1 - ah_n)^{-k}(p(x) + kh_n b) \tag{3}$$

we let $k \to \infty$, $n \to \infty$ in such a way that $kh_n \to \tau$. Then $(x_k^n)_{k=0}^n$ defines a step function u^n which converges to a continuous function u. The limit of (3) is (I.2). From (2) we get

$$x_k^n = (I - h_n A)^{-k} x + \sum_{j=1}^{k} h_n (I - h_n A)^{-(k-j+1)} B x_j^n,$$

the limit of which is (I.1).

For the converse, assume [I], let $R > 0$ be large, and let ω_R be the Lipschitz seminorm of B on C_R. Let $h > 0$ and let

$$A_h x = h^{-1}(T(h) - I)x, \quad B_h x = h^{-1} \int_0^h T(h-s)BS(s)x \, ds$$

(so that $A_h k \to Ax$, $B_h x \to Bx$ as $h \to 0$ for $x \in \mathrm{Dom}(A) \cap C$).

One uses the Banach fixed point theorem (or contraction mapping principle) to find a solution $y = x(\lambda,h)$ of

$$y = h(\lambda+h)^{-1}x + \lambda(\lambda+h)^{-1}S(h)y$$

for $x \in C_r$ and $0 < \lambda < \lambda_0(r)$. Then

$$y = (I - \lambda A_h)^{-1}x + \lambda(I - \lambda A_h)^{-1}B_h\, y,$$

$$p(y) \leqq (1 - a_h\lambda)^{-1}(p(x) + b_h \lambda)$$

for $a_h = a(1 - ah)^{-1}$, $b_h = b(1 - ah)^{-1}$. Once we show that

$$x_\lambda = \lim_{h \downarrow 0} y = \lim_{h \downarrow 0} x(\lambda,h)$$

exists, [II] readily follows.

The existence of the limit x_λ follows with the aid of the <u>ball measure of noncompactness</u> β (cf. e.g. [1], [4]). For V a bounded subset of

X, $\beta(V)$ is the infimum of $\varepsilon > 0$ such that V is contained in a finite union of balls of radius ε. For $\delta > 0$ small we let

$$V_\lambda = \{x(\lambda,h) : 0 < h < \delta\}, \quad W_\lambda = \{(I-\lambda A_h)^{-1}B_h x(\lambda,h) : 0 < h < \delta\}.$$

Using the local Lipschitz condition of B, one can establish $\beta(W_\lambda) \leqq 2\omega_R\beta(V_\lambda)$. But $K = \{(I - \lambda A_h)^{-1}x : 0 < h < \delta\}$ has compact closure, so

$$\beta(V_\lambda) \leqq \beta(K + \lambda W_\lambda) \leqq \beta(K) + \lambda\beta(W_\lambda) \leqq 0 + 2\omega_R\lambda\beta(V_\lambda),$$

which implies that $\beta(V_\lambda) = 0$ for λ small enough. But this implies that $\bar{V}_\lambda$ is compact, and now it is routine to pass to the limit of $x(\lambda,h)$ as $h \to 0$. We omit the rest of the details.

3. GROUPS OF OPERATORS

The rest of this article sketches the paper [2] by Goldstein, Oharu, and Takahashi. (This paper is not yet finished.)

Let $(A,B,C,p) \in \Omega$, and assume in addition that A generates a (C_0) group $T = \{T(t) : t \in \mathbb{R}\}$. Then Theorem 1 applies to the nonlinear semigroups $S_\pm$ determined by $\pm(A + B)$. We can define a group S on C by

$$S(t) = S_+(t) \text{ or } S_-(-t),$$

according as $t \geqq 0$ or $t \leqq 0$. To show that S is a group it suffices to show that

$$(d/dt)S_+(t)S_-(t)x = 0$$

for all $t \geqq 0$ and all x in a dense subset of C. Our first proof of this was achieved by approximating B by smooth operators B_n, deducing regularity of $S_{n\pm}$ (generated by $\pm(A + B_n)$, and passing to the limit as $n \to \infty$ with the aid of a version of the Trotter-Neveu-Kato approximation theorem. The paper [2] contains a simpler proof, using the variation of parameters formula and the linearity of A.

The generation theorem for groups is as follows.

THEOREM 2. Let A generate a (C_0) group T, let $(\pm A, \pm B, C, p) \in \Omega$, and let $a, b \geqq 0$. Then $[\tilde{I}]$ is equivalent to $[\tilde{\tilde{II}}]$.

$[\tilde{I}]$ There is a locally Lipschitzian group S on C such that for all $x \in C$ and all $t \in \mathbb{R}$, (I.1) and $(I.\tilde{2})$ hold:

$$p(S(t)x) \leqq e^{a|t|}(p(x) + b|t|). \tag{I.$\tilde{2}$}$$

$[\tilde{\tilde{II}}]$ For each $r > 0$ there is a $\lambda_0(r) > 0$ such that for all $x \in C_r$ and $\lambda \in (-\lambda_0(r), \lambda_0(r))$ there is an $x_\lambda \in \mathrm{Dom}(A) \cap C$ satisfying (II.1) and $(II.\tilde{2})$:

$$p(x_\lambda) \leqq (1 - a|\lambda|)^{-1}(p(x) + b|\lambda|). \tag{II.$\tilde{2}$}$$

4. THE APPROXIMATION THEOREM

THEOREM 3. For each $n = 0,1,2,\ldots$ let $(A_n, B_n, C_n, p_n) \in \Omega$, and let $a \geqq 0$, $b \geqq 0$.

(a) Suppose $(I - \alpha A_n)^{-1}x \to (I - \alpha A_0)^{-1}x$ for each $x \in X$ and $\alpha > 0$.

(b) For $R > 0$ there is an $R' > 0$ such that for each $x_0 \in C_0$ with $p_0(x_0) \leqq R$ there is an $x_n \in C_n$ such that $x_n \to x_0$ and $\sup_n p_n(x_n) \leqq R'$.

(c) For $x_n \in C_n$ with $x_n \to x_0 \in C_0$ and $\sup_n p_n(x_n) < \infty$, it follows that $B_n x_n \to B_0 x_0$.

(d) $\{B_n\}$ is uniformly approximable by globally Lipschitzian functions in the following sense. For each n and each $r > 0$ there is a globally Lipschitzian function L_n which agrees with B_n on $C_{nr} = C_n \cap \{x \in X; p_n(x) \leqq r\}$. The smallest Lipschitz constant ω_{nr} of L_n (or of B_n on C_{nr}) satisfies $\sup_n \omega_{nr} < \infty$ for each $r > 0$.

(e) Suppose that [II] of Theorem 1 holds for (A_n, B_n, C_n, p_n) with $\lambda_0(r)$ depending on $r > 0$ but not on n.

Then if $x_n \in C_n$, $x_n \to x_0 \in C_0$, and $\sup_n p_n(x_n) < \infty$, it follows that $S_n(t)x_n \to S_0(t)x_0$ as $n \to \infty$ for each $t \geqq 0$.

The idea of the proof is to use Kisyński's sequence space approach [3]. Let $\mathcal{X} = \{x = (x_n)_0^\infty \subset X : x_n \in X, x_n \to x_0\}$ be the convergent sequences in X, normed by $|x| = \sup_n \|x_n\|$. Define $\mathcal{A}$ by $\mathcal{A}x = y$ means $x,y \in \mathcal{X}$ and $A_n x_n = y_n$ for all $n \geq 0$. Then (by (a)) $\mathcal{A}$ generates a (C_0) contraction semigroup on $\mathcal{X}$. Define $\mathcal{B}$ by: $\mathcal{B}x = y$ means $x,y \in \mathcal{X}$, $x_n \in C_n$ and $B_n x_n = y_n$ for all n. Define $p: \mathcal{C} \to [0,\infty]$ by $p(x) = \sup_n p_n(x_n)$ for $x \in \mathcal{C} = \{y \in \mathcal{X} : y_n \in C_n$ for all $n \geq 0\}$.

The hypotheses of Theorem 3 implies that $(\mathcal{A},\mathcal{B},\mathcal{C},p)$ belong to Ω (relative to the space $\mathcal{X}$). Then Theorem 3 follows from Theorem 1, remembering that a member of $\mathcal{X}$ is a sequence converging to its zeroth element. Thus if $x \in \mathcal{X}$ and $\mathcal{S}$ is the semigroup generated by $\mathcal{A} + \mathcal{B}$, then $\mathcal{S}(t) = (S_n(t)x_n)_0^\infty$ and $S_n(t)x_n \to S_0(t)x_0$ as $n \to \infty$.

5. REGULARITY OF S

It is of great interest to know when $S(t)x$ is differentiable as a function of t as well as a function of x.

We say that $(A,B,C,p) \in \Omega'$ if $(A,B,C,p) \in \Omega$ and, in addition, B is Fréchet differentiable on C relative to p in the sense that for each $r > 0$ and $x \in C_r$ there is a bounded linear operator $B'(x) \in L(X)$ and a function w from C_r cross a neighbourhood of zero to X such that $w(x,0) = 0$ and

$$B(x+v) = B(x) = B'(x)v + |v|w(x,y)$$

for $|v|$ small with $x + v \in C_r$ and $|w(x,v)| \to 0$ as $v \to 0$. We further demand that $B'(S(\cdot,x))$ is strongly measurable and locally bounded in $L(X)$ for each $x \in C$.

THEOREM 4. Let $(A,B,C,p) \in \Omega'$. Then for all $x \in C$ and all $t \geq 0$, $y \to S(t)y$ is Fréchet differentiable at x. It is continuously Fréchet differentiable, if, in addition, B is continuously Fréchet differentiable on C relative to p (i.e. $B'(S(\cdot,x))$ is strongly continuous for each $x \in C$).

The Fréchet differentiability and the t-differentiability of $S(t)x$ are related by the following calculation. For $x \in C \cap \mathrm{Dom}(A)$, $t \geq 0$, $h > 0$, and $v = S(h)x-x$, using obvious notation we have

$$h^{-1}[S(t+h)x - S(t)x] = h^{-1}[S(t)(x+v) - S(t)x]$$

$$= S'(t,x)(h^{-1}v) + |h^{-1}v|w(t,x,v).$$

Letting $h \to 0$ leads to the following result.

THEOREM 5. Let $(A,B,C,p) \in \Omega'$ and suppose [I] (or [II]) holds. Then for each $x \in C \cap \mathrm{Dom}(A)$, the function $S(\cdot)x$ is strongly continuously differentiable and satisfies

$$(d/dt)S(t)x = S'(t,x)(A + B)x = (A + B)S(t)x$$

for $t \geqq 0$.

The paper [2] contains further regularity results. The local Lipschitz conditions on B are relaxed in the paper of Oharu and Takahashi [6].

We close with one simple example of an $(A,B,C,p) \in \Omega$ which is not governed by the Crandall-Liggett theorem or other standard results in semigroup theory. This will illustrate the role of the constant b in Theorem 1.

For $1 \leqq q < \infty$ let $X_q = L^q(\mathbb{R})$ and let $X_\infty = C_0(\mathbb{R})$. Fix $q \in [1,\infty]$ and let X_q be the corresponding space of real functions. Of concern is the parabolic equation

$$\partial u/\partial t = \partial^2 u/\partial x^2 + F(u)$$

for $x \in \mathbb{R}$, $t \geqq 0$ where $F \in C^1(\mathbb{R})$ (or more generally F is locally Lipschitzian). Assume that the ordinary differential equation $y' = F(y)$ has the property that if y is a solution and $y(0) \geqq 0$, then $y(t) \geqq 0$ for all $t \geqq 0$. We shall apply Theorem 1 with $A = d^2/dx^2$, $X = X_q$, $C = \{u \in X_q \cap C_0(\mathbb{R}) : u(x) \geqq 0$ for all $x \in \mathbb{R}\}$, $(Bu)(x) = F(u(x))$, and $p(u) = \|u\|_\infty$.

If $F(x) - F(y) \leqq \omega|x-y|$ holds for some real ω and all real x,y, then standard nonlinear semigroup theory implies that [II] holds with $a = \omega$ and $b = 0$. We suppose instead that $F(0) = 0$ and $F \in L^\infty(\mathbb{R})$ but F' is not bounded above (e.g. $F(y) = \sin(ky^2)$ for k a positive constant). Thus $F - \omega I$ is not dissipative for each $\omega \in \mathbb{R}$.

We must find $u \in \mathrm{Dom}(A) \cap C$ (i.e. $u \in C^1(\mathbb{R})$, $u' \in AC(\mathbb{R})$, $u'' \in X_q$, $u \geqq 0$ on $\mathbb{R}$) such that

$$u - \lambda(u'' + F(u)) = h$$

where $h \in X_q$ and $\lambda > 0$ are given. Let F_n be a globally Lipschitzian function, bounded by $\|F\|_\infty$, which agrees with F on $[-n,n]$. We can solve $u_n - \lambda(u_n'' + F(u_n)) = h$ uniquely in $\mathrm{Dom}(A) \cap C$. The maximum principle implies $\|u_n\|_\infty \leqq \|h\|_\infty + \lambda\|F\|_\infty$. We can then deduce uniform bounds on $\|u_n''\|_\infty$ and then on $\|u_n'\|_\infty$. Using compactness we can pass to the limit along a suitable subsequence and obtain the desired solution u. The final conclusion is that [II] holds with $a = 0$ and $b = \|F\|_\infty$.

This work was partially supported by a USNSF grant, which is gratefully acknowledged.

References

[1] Deimling, K., Ordinary Differential Equations in Banach Spaces, Lecture Notes in Math. No. 596, Springer, Berlin-Heidelberg-New York, 1977.

[2] Goldstein, J.A., S. Oharu and T. Takahashi, Semilinear Hille-Yosida theory: The approximation theorem and groups of operators, in preparation.

[3] Kisyński, J., A proof of the Trotter-Kato theorem on approximation of semigroups, Colloq. Math. 18 (1967), 181-184.

[4] Martin, Jr., R.H., Nonlinear Operators and Differential Equations in Banach Spaces, Wiley, New York, 1976.

[5] Oharu, S. and T. Takahashi, Locally Lipschitz continuous perturbations of linear dissipative operators and nonlinear semigroups, Proc. Amer. Math. Soc. 100 (1987), 187-194.

[6] Oharu, S. and T. Takahashi, Characterization of nonlinear semigroups associated with semilinear evolution equations, to appear.

Jerome A. Goldstein
Department of Mathematics
Tulane University
New Orleans
Louisiana 70118
USA.

Shinnosuke Oharu
Department of Mathematics
Faculty of Science
Hiroshima University
Hiroshima 730
Japan.

Tadayasu Takahashi
National Aerospace Laboratory
Chofu
Tokyo
Japan

R. GRIMMER & W. SCHAPPACHER

Integrodifferential equations in Banach spaces with infinite memory

I. INTRODUCTION

Many problems in viscoelasticity and in heat conduction with memory involve equations with infinity memory, (cf. [3], [8], [12]). We are interested in formulating such problems in a way so that under reasonable assumptions the problem is well-posed with respect to initial data and history. Equations in Banach spaces which involve delay are known to involve special difficulties with regard to well-posedness (cf. [9]). Frequently, in the study of integrodifferential equations in Banach spaces it is assumed that the material under study is initially quiescent so that the history can be taken to be zero. Alternatively, the history is incorporated into the inhomogeneity in the problem. Each of these methods avoids confronting the problem of continuity with respect to the history, however. Our approach here will be to show that a certain equation is adjoint to another integrodifferential equation which is known to be well-posed by other methods. The form of our equation with infinite memory appears somewhat nonstandard but a close examination of the equations occurring in viscoelasticity and heat-conduction with memory indicates that our equation is precisely the appropriate one.

We note that our results which relate equations with unbounded operators are extensions of results of Burns and Herdman [2] which do not apply to our infinite dimensional problem but do relate the finite dimensional system studied in [1] with that in [11]. Additional work dealing with adjoint theory for equations in finite dimensional spaces may be found in [13], cf. also the references therein.

Consider now the linear Volterra equation

$$x'(t) = Ax(t) + \int_0^t B(t-s)x(s)ds + f(t), \quad t \geq 0 \tag{1.1}$$

in a Banach space $(X, \|\cdot\|)$. We assume

(Hi) A is the infinitesimal generator of a linear C_o-semigroup S(.) on X.

As a result, D(A) endowed with the graph norm

$$|x|_A = \|x\| + \|Ax\|$$

becomes a Banach space which is denoted by X_A.

(Hii) For each $t \geqq 0$, the operator B(t) is a continuous linear operator $X_A \to X$.

(Hiii) For each $x \in X_A$, the map $t \to B(t)x$ is measurable.

(Hiv) There is a constant ω so that $e^{-\omega t}|B(t)|_{X_A,X}$ is integrable.

REMARK. In view of (Hii) we may decompose B(t) as

$$B(t)x = F(t)Ax + K(t)x$$

where the closed graph theorem implies both F(.) and K(.) are bounded linear operators on X.

A variety of techniques have been employed to obtain conditions which ensure the well-posedness of the problem (1.1). One particularly useful technique is to consider the pair (x_o,f) as the initial state of the system in $X \times F$ where F is a space of functions defined on $[0,\infty) \to X$. This technique seems to have been first used for integrodifferential equations by Miller [11] where (1.1) was considered with $X = \mathbb{R}^n$.

In finite dimensions the usual existence and uniqueness results show that this flow is indeed a C_o-semigroup. In infinite dimensional spaces the usual technique is the reverse of that for finite dimensions. That is, one shows that the formal generator is, in fact, really the generator of a C_o-semigroup and then recovers the well-posedness of (1.1) from the well-posedness of the equation

$$(x,g)' = \mathcal{A}(x,g)$$

$$(x(0),g(0)) = (x_0,f).$$

For our purposes here it is convenient to choose for F the space

$$F = L^2_\omega(0,\infty;X) = \{f \,|\, e^{-\omega t} f(t) \in L^2(0,\infty;X)\}.$$

On F the left shift semigroup $f(.) \to f(t+.)$ has generator D_s with $D(D_s) = W^{1,2}_\omega(0,\infty;X) = \{u \in L^2_\omega(0,\infty;X) \,|\, u' \in L^2_\omega(0,\infty;X)\}$, and

$$D_s f(s) = \frac{d}{ds} f(s).$$

If we define δ on $D(D_s)$ by $\delta f = f(0)$ and B by $(Bx)(t) = B(t)x$ for all $x \in D(A)$ and $t \geq 0$, we can write $\mathcal{A}$ in the matrix form

$$\mathcal{A} = \begin{pmatrix} A & \delta \\ B & D_s \end{pmatrix} \tag{1.2}$$

with domain $D(\mathcal{A}) = D(A) \times D(D_s)$ in $X \times F$. For additional details regarding this approach we refer to the references [6] and [7]. In particular, one has (cf. [6], [7]):

THEOREM 1.1. Let A be the infinitesimal generator of a C_0-semigroup on X and assume that for any $x \in D(A)$ the map $t \to B(t)x$ belongs to $W^{1,2}_\omega(0,\infty;X)$. Then $\mathcal{A}$ generates a C_0-semigroup $S(t)$ on $X \times F$. It thus follows that (1.1) is well-posed with respect to the data $(x_0,f) \in X \times F$.

2. THE ADJOINT OF $\mathcal{A}$

Given a Banach space Y, we denote the pairing between Y and its dual Y* by $\langle .,. \rangle_Y$. So the pairing between $Z = X \times L^2(0,\infty;X)$ and its dual is

$$\langle \binom{x}{\phi}, \binom{z}{\psi} \rangle_Z = \langle x,z \rangle_X + \int_0^\infty \langle \phi(s),\psi(s) \rangle_X \, ds.$$

Our main result is

THEOREM 2.1. The adjoint A^* of A is given by

$$D(A^*) = \{\binom{x}{\psi} \in Z^* \mid z = \psi(0),\ \psi \in W^{1,2}(0,\infty;X^*),$$
$$z + \int_0^\infty F^*(s)\psi(s)ds \in D(A^*)\},$$

$$A^*\begin{pmatrix}\psi(0)\\ \\ \psi\end{pmatrix} = \begin{pmatrix}(A^*(\psi(0) + \int_0^\infty F^*(s)\psi(s)ds) + \int_0^\infty K^*(s)\psi(s)ds\\ \\ -\psi'\end{pmatrix}$$

Proof. Suppose that for any pair $\binom{x}{\phi} \in D(A) \times D(D_s)$ we have

$$\left\langle \begin{pmatrix} A & \delta \\ B & D_s\end{pmatrix}\begin{pmatrix} x \\ \phi\end{pmatrix}, \begin{pmatrix} z \\ \psi\end{pmatrix}\right\rangle_Z = \left\langle \begin{pmatrix} x \\ \phi\end{pmatrix}, \begin{pmatrix} z^* \\ \psi^*\end{pmatrix}\right\rangle_Z$$

i.e.,

$$\langle Ax + \phi(0),\ z\rangle_X + \langle Bx+D_s\phi,\psi\rangle_{L^2} = \langle x,z^*\rangle_X + \langle \phi,\psi^*\rangle_{L^2}. \tag{2.1}$$

To begin with, take $x = 0$ and $\phi \in D(D_s)$ with $\phi(0) = 0$. Then

$$\langle D_s\phi,\psi\rangle_{L^2} = \langle \phi,\psi^*\rangle_{L^2}$$

and so ψ^* must be the negative of the weak derivative of ψ. Hence $\psi \in W_\omega^{1,2}(0,\infty;X)$ and

$$\psi^* = -\psi' \tag{2.2i}$$

Sobolev embedding implies that $\psi(0)$ is well defined. If ϕ has compact support but $\phi(0) \neq 0$, we obtain

$$\langle \phi(0),z\rangle_X + \langle D_s\phi,\psi\rangle_{L^2} = - \langle \phi,\psi'\rangle_{L^2}$$

and thus $\qquad z = \psi(0).$ $\qquad$ (2.2ii)

Next, we consider (2.1) with $\phi = 0$. Then

$$\langle Ax,z\rangle_X + \langle Bx,\psi\rangle_{L^2} = \langle x,z^*\rangle_X$$

and from the form of B, we obtain

$$\langle Ax,z\rangle_X + \int_0^\infty \langle F(s)Ax + K(s)x,\psi(s)\rangle_X ds = \langle x,z^*\rangle_X.$$

Thus for any $x \in D(A)$

$$\langle Ax,z + \int_0^\infty F^*(s)\psi(s)ds\rangle_X = \langle x,z^* - \int_0^\infty K^*(s)\psi(s)ds\rangle_X$$

which is equivalent to

$$z + \int_0^\infty F^*(s)\psi(s)ds \in D(A^*)$$

and

$$A^*(z + \int_0^\infty F^*(s)\psi(s)ds) = z^* - \int_0^\infty K^*(s)\psi(s)ds. \tag{2.3}$$

Combining (2.2)-(2.3) yields the claim.

We next investigate the associated abstract Cauchy problem.

THEOREM 2.2. The abstract Cauchy problem associated with the operator A^* is the infinite delay equation

$$\frac{dz}{dt}(t) = A^*(z(t) + \int_{-\infty}^t F^*(t-s)z(s)ds) + \int_{-\infty}^t K^*(t-s)z(s)ds. \tag{2.4}$$

$$z(0) = z,\ z(\theta) = \psi_0(-\theta),\ \theta < 0,\ z \in X^*,\ \psi_0 \in L^2(0,\infty;X^*).$$

Proof. Consider the equation

$$\frac{d}{dt}\begin{pmatrix} z(t) \\ \psi(t,\cdot)\end{pmatrix} = A^*\begin{pmatrix} z(t) \\ \psi(t,.)\end{pmatrix},$$

$$z(0) = z,\ \psi(0,\theta) = \psi_0(\theta).$$

The first component reads

$$\frac{d}{dt} z(t) = A^*(z(t) + \int_0^{\infty} F^*(\theta)\psi(t,\theta)d\theta) + \int_0^{\infty} K^*(\theta)\psi(t,\theta)d\theta \tag{2.5}$$

while for the second component, we get

$$\frac{\partial}{\partial t} \psi(t,\theta) = - \frac{\partial}{\partial \theta} \psi(t,\theta), \qquad \psi(t,0) = z(t).$$

Integrating this equation yields

$$\psi(t,\theta) = \begin{cases} z(t-\theta) & \text{if } t-\theta > 0 \\ \psi_o(\theta-t) & \text{if } t-\theta < 0 \end{cases}$$

This representation with (2.5) gives

$$\begin{aligned} \frac{dz}{dt}(t) &= A^*(z(t) + \int_0^t F^*(\theta)z(t-\theta)d\theta + \int_t^{\infty} F^*(\theta)\psi_o(\theta-t)d\theta) \\ &+ \int_0^t K^*(\theta)z(t-\theta)d\theta + \int_t^{\infty} K^*(\theta)\psi_o(\theta-t)d\theta \\ &= A^*(z(t) + \int_0^t F^*(\theta)z(t-\theta)d\theta + \int_{-\infty}^0 F^*(t-s)\psi_o(-s)ds) \\ &+ \int_0^t K^*(\theta)z(t-\theta)d\theta + \int_{-\infty}^0 K^*(t-s)\psi_o(-s)ds. \end{aligned}$$

If we set $z(s) = \psi_o(-s)$ for $s \leqq 0$, we thus arrive at the infinite delay equation

$$\frac{dz}{dt}(t) = A^*(z(t) + \int_{-\infty}^t F^*(t-s)z(s)ds) + \int_{-\infty}^t K^*(t-s)z(s)ds$$

$$z(s) = \psi_o(-s) \text{ for } s \leqq 0.$$

<u>COROLLARY 2.3</u>. If X is reflexive, (2.4) is well-posed with respect to the data $(z,\psi_o) \in X^* \times L^2_\omega(0,\infty;X^*)$.

3. APPLICATIONS

EXAMPLE 3.1. The simplest model in viscoelasticity is the one dimensional case on the real line. The equations of interest are

$$\rho u_{tt}(t,x) = \sigma_x(t,x) \tag{3.1}$$

and

$$\sigma(t,x) = G(0)u_x(t,x) + \int_{-\infty}^{t} G'(t-s)u_x(s,x)ds. \tag{3.2}$$

Here u is the displacement, ρ is the mass density, σ is the stress and G is the relaxation function. Eq. (3.1) is the linear momentum equation while (3.2) gives the constitutive relation between stress and strain.

Physical considerations lead to $\rho > 0$ and $G(0) > 0$. We assume for simplicity that $\rho = 1$, or equivalently, notation is changed so that $G(t)/\rho$ becomes $G(t)$. Then, upon placing $v = u_t$ and $w = G^{1/2}(0)u_x$ and $G_1(t) = G(t)/G^{1/2}(0)$ one obtains, with $\gamma = G^{1/2}(0)$,

$$\begin{pmatrix} v \\ w \end{pmatrix}_t = \begin{pmatrix} 0 & \gamma\partial_x \\ \gamma\partial_x & 0 \end{pmatrix}\left(\begin{pmatrix} v \\ w \end{pmatrix} + \int_{-\infty}^{t} \begin{pmatrix} 0 & 0 \\ 0 & G_1'(t-s) \end{pmatrix}\begin{pmatrix} v(s) \\ w(s) \end{pmatrix}ds.\right) \tag{3.3}$$

Now, take $X = L^2(\mathbb{R}) \times L^2(\mathbb{R})$ and $D(A) = H^1(\mathbb{R}) \times H^1(\mathbb{R})$ with A given by

$$\begin{pmatrix} 0 & \gamma\partial_x \\ \gamma\partial_x & 0 \end{pmatrix}.$$

It is easy to see that $A^* = -A$. Thus, both A and A^* are dissipative and so by a corollary of the Lumer-Phillips Theorem, A and A^* generate a C_0-semigroup.

The equation associated with (3.3) is

$$\begin{pmatrix} v \\ w \end{pmatrix}_t = -\begin{pmatrix} 0 & \gamma\partial_x \\ \gamma\partial_x & 0 \end{pmatrix}\begin{pmatrix} v \\ w \end{pmatrix} - \int_{0}^{t} \begin{pmatrix} 0 & 0 \\ 0 & G_1'(t-s) \end{pmatrix}\begin{pmatrix} 0 & \gamma\partial_x \\ \gamma\partial_x & 0 \end{pmatrix}\begin{pmatrix} v(s) \\ w(s) \end{pmatrix}ds. \tag{3.4}$$

It is clear now from our result that (3.3) with initial data (v(0),w(0)) prescribed in $L^2(\mathbb{R}) \times L^2(\mathbb{R})$ and (v(s),w(s)) prescribed in

$L^2(R^-,L^2(R)) \times L^2(R^-,L^2(R))$ is well-posed.

It is interesting to note that (3.4) is the system formulation of the equation

$$u_{tt} = G(0)u_{xx} + \int_0^t G'(t-s)u_{xx}(s)ds$$

obtained by taking $w = -u_t$ and $v = G^{1/2}(0)u_x$. Also, we note that in certain problems it may be more reasonable to prescribe velocity v and stress σ as initial and/or boundary data. If the material being studied is a viscoelastic solid, $\int_0^\infty |G'(s)|ds < G(0)$, ([3, p. 11]), and the relation between σ and u_x may be inverted, (cf. [5, p. 66]).

It is possible to write a system in velocity and stress without this technique, however. Consideration of the example just presented indicates that the initial data for strong solutions of (3.3) consist of pairs (v,w) with $v(s), w(s) \in H^1(R^-, L^2(R))$. Thus, we see we can write

$$v_t = \sigma_x,$$

$$\sigma_t = (G(0)v + \int_{-\infty}^t G'(t-s)v(s)ds)_x$$

and proceed much as before.

<u>EXAMPLE 3.2</u>. Further examples can be obtained from the area of thermoviscoelasticity. We examine a one dimensional problem studied earlier by Leugering [10], cf. also Christensen [3, pp. 77-90]).

$$\rho u_{tt} = \sigma_x$$

$$\sigma = G(0)u_x + \int_{-\infty}^t G'(t-s)u_x(s)ds - b\theta$$

$$\rho\theta_t = \frac{k}{T_0}\theta_{xx} - m(0)\theta - \int_{-\infty}^t m'(t-s)\theta(s)ds - bu_{tx}.$$

Here ρ, u, and σ are as before while θ is the temperature difference from a base temperature T_0. Also, it is assumed that G(0), b, k, m(0), ρ and T_0

are all positive and that $G'(t)$ and $m'(t)$ are in $H^1(R^+)$.

It is convenient to let $\rho = 1$ or, equivalently, divide by ρ and redefine the coefficients by $G = G/\rho$, etc. One then obtains after letting $v = u_t$ and $w = G^{1/2}(0)u_x$,

$$\begin{pmatrix} v \\ w \\ \theta \end{pmatrix}_t = \left(\begin{pmatrix} 0 & \gamma\partial_x & -b\partial_x \\ \gamma\partial_x & 0 & 0 \\ -b\partial_x & 0 & \frac{k}{T_0}\Delta\ -m(0) \end{pmatrix}\begin{pmatrix} v \\ w \\ \theta \end{pmatrix} \right. \tag{3.5}$$

$$\left. \int_{-\infty}^{t} \begin{pmatrix} 0 & 0 & 0 \\ 0 & G_1'(t-s) & 0 \\ 0 & 0 & 0 \end{pmatrix} \begin{pmatrix} v(s) \\ w(s) \\ \theta(s) \end{pmatrix} ds \right) - \int_{-\infty}^{t} \begin{pmatrix} 0 & 0 & 0 \\ 0 & 0 & 0 \\ 0 & 0 & m'(t-s) \end{pmatrix} \begin{pmatrix} v(s) \\ w(s) \\ \theta(s) \end{pmatrix} ds.$$

We take $X = L^2(R)^3$ and $D(A) = H^1(R) \times H^1(R) \times H^2(R)$. One proceeds by showing that A and A* are dissipative. The result now follows as before.

ACKNOWLEDGEMENT. The research of R.G. was partially supported by the National Science Foundation under grant DMS-8502325. This research was completed while R.G. was a guest at the Institute of Mathematics, University of Graz. Support for this visit was provided through the Austrian-U.S. Cooperative Research Program administered by F.W.F. and N.S.F. It is a pleasure to thank the Institute of Mathematics, University of Graz, for the hospitality shown during the stay of R.G.

The work of W.S. was partially supported by Fonds zur Förderung der Wissenschaftlichen Forschung, Austria, Grant P-5691.

Bibliogrpahy

[1] V. Barbu and S.I. Grossman: Asymptotic behavior of linear integro-differential systems, Trans. Amer. Math. Soc., 171 (1972), 277-288.

[2] J.A. Burns and T.L. Herdman: Adjoint semigroup theory for a class of functional differential equations, SIAM J. Math. Anal. 7 (1976), 729-745.

[3] R.M. Christensen: The Theory of Viscoelasticity: An introduction, 2nd ed., Academic Press, New York, 1982.

[4] B.D. Coleman and M.E. Gurtin: Equipresence and constitutive equations for rigid heat conductors, Z. Angew. Math. Phys., 18 (1967), 199-208.

[5] C. Corduneanu: Integral Equations and Stability of Feedback Systems, Academic Press, New York, 1973.

[6] W. Desch and W. Schappacher: On relatively bounded perturbations of linear C_0-semigroups, Ann. Sc. Norm. Sup. Pisa, Ser IV. 11 (1984), 327-341.

[7] W. Desch and W. Schappacher: The semigroup approach to integro-differential equations, J. Integral Equations 10 (1985), 99-110.

[8] M.E. Gurtin and A.C. Pipkin: A general theory of heat conduction with finite wave speeds, Arch. Rational Mech. Anal. 31 (1968), 113-126.

[9] K. Kunisch and W. Schappacher: Necessary conditions for partial differential equations with delay to generate C_0-semigroups, J. Differential Eq. 50 (1983), 49-79.

[10] G. Leugering: A generation result for a class of linear thermovisco-elastic material, in "Dynamical Problems in Mathematical Physics", Brosowski and Martensen (eds.), Verlag Peter Lang, Frankfurt, 1983.

[11] R.K. Miller: Linear Volterra integrodifferential equations as semi-groups, Funkcial. Ekvac., 17 (1974), 39-55.

[12] R.K. Miller: An integrodifferential equation for rigid heat conductors with memory, J. Math. Anal. Appl. 66 (1978), 313-332.

[13] O. Staffans: Extending initial and forcing function semigroups generated by a functional equation, SIAM J. Math. Anal. 16 (1985), 1034-1048.

R. Grimmer
Department of Mathematics
Southern Illinois University
Carbondale
Illinois 62901
USA.

W. Schappacher
Institut für Mathematik
Universität Graz
Elisabethstrasse 16
A-8010 Graz
Austria.

K.B. HANNSGEN & R.L. WHEELER

Existence and decay estimates for boundary feedback stabilization of torsional vibrations in a viscoelastic rod

This work was partially supported by the Air Force Office of Scientific Research under Grant AFOSR-86-0085.

1. INTRODUCTION

In forthcoming work with Yuriko Renardy [2], we show that a simple boundary feedback scheme can improve the exponential rate of decay of torsional vibrations in a uniform linear viscoelastic rod of circular cross section. Analytically, the problem reduces to the study of the roots of a characteristic equation

$$\Delta(s) \equiv \Delta(s;k) = 0 \quad (\mathcal{R}\, s > s_*, \quad 0 \leqq k < \infty), \tag{1.1}$$

where k is the feedback gain parameter and

$$U(x,s) \equiv \int_0^\infty e^{-st}u(x,t)dt$$

is the formal Laplace transform, with respect to time, of the solution $u(x,t)$ of the boundary value problem for the angular displacement of the rod. One expects that if (1.1) has no solution in $\{\mathcal{R}\, s \geqq \sigma_1\}$, then $u(x,t)e^{-\sigma_1 t}$ should be bounded, in some sense, as $t \to \infty$. In this article we sketch the derivation of the analytic problem (1.1) and then show how one can establish the correspondence between root location and the existence of a weak solution u with the expected rate of growth or decay for $t \to \infty$.

2. THE FORMAL SOLUTION

We start with the mathematical problem to be solved; various physical constants have been scaled out. We refer the reader to [2] for a detailed derivation and further references. Consider, then, the initial-boundary value problem

$$u_t(x,t) = \int_0^t A(t-\tau)u_{xx}(x,\tau)d\tau + F(x,t) \quad (0 \leqq x \leqq 1,\ 0 \leqq t < \infty), \tag{2.1}$$

$$u(x,0) = u_0(x) \quad (0 < x < 1), \tag{2.2}$$

$$u(0,t) = 0 \qquad (0 \leqq t < \infty), \tag{2.3}$$

$$Iu_t(1,t) + \int_0^t A(t-\tau)u_x(1,\tau)d\tau \tag{2.4}$$

$$= f(t) - kH(t-\varepsilon)[u(1,t-\varepsilon)-u_0(1)] \quad (0 \leqq t < \infty).$$

The unknown $u(x,t)$ represents the angular displacement at position x along the rod at time t. Equation (2.1) is the force-momentum balance in the rod, u_0 in (2.2) gives the initial displacement. By (2.3), the left end of the rod is held fixed. Equation (2.4) is the feedback condition at the right end. $A(t)$ is the stress relaxation function; we assume

$$\begin{aligned} &A(t) = E + a(t) \text{ with } E > 0 \text{ and } a(t) \text{ completely monotonic} \\ &(\text{that is, } (-1)^j a^{(j)}(t) > 0,\ j = 0,1,2,\ldots) \text{ on } \underline{R}^+ \equiv (0,\infty), \\ &0 < a(0+) \leqq \infty), \text{ and } \int_0^\infty e^{\eta t}a(t)dt < \infty \text{ for some } \eta > 0. \end{aligned} \tag{2.5}$$

The functions F and f include external forces and contributions from the initial history ($t < 0$) of the rod. With this interpretation, the following hypothesis is reasonable. (Here and below, H^0 is the Hilbert space $L^2(0,1)$. An underlined symbol is interpreted as an element or operation in H^0. We write $\underline{F}(t) = \underline{F}(\cdot,t)$, etc.) We have

$$\begin{aligned} &f \in L^1_{loc}(\underline{R}^+),\ \underline{F} \in L^1_{loc}(\underline{R}^+,H^0). \text{ The Laplace transforms} \\ &\hat{f}(s) \text{ and } \hat{\underline{F}}(s) \text{ have analytic extensions such that } s\hat{f}(s) \\ &\text{and } s\hat{\underline{F}}(s) \text{ are analytic and bounded in each half-plane} \\ &\{\mathrm{R}s > -\eta'\},\ \eta' < \eta. \end{aligned} \tag{2.6}$$

Equation (2.4) is an integrated form of the force-momentum relation

$$Iu_{tt}(1,t) + \frac{d}{dt}\int_0^t A(t-\tau)u_x(1,\tau)d\tau = f_1(t) - ku_t(1,t-\varepsilon)H(t-\varepsilon). \tag{2.4'}$$

(In (2.4) $f(t) = \int_0^t f_1(\tau)d\tau + Iu_t(1,0+)$.) Here the first term allows for a mass concentration ("tip mass"), due, for example, to a feedback device, with moment of inertia $I \geqq 0$. The second term comes from the shear stress. On the right, in addition to f_1, we have the feedback force, parametrized by k. H is the Heaviside function, and $\varepsilon \geqq 0$ represents a possible delay between the observation of the velocity u_t and the application of the force. In related problems with purely elastic bodies, such a delay can cause a reasonance that leads to unbounded vibrations for arbitrarily small ε [1]. In the present study it is shown that this resonance phenomenon can occur for arbitrarily small ε only if

$$k \geqq \sqrt{A(0+)}\,\frac{1-C}{1+C},$$

where $C = \exp\left(-\frac{A'(0+)}{A(0+)^{3/2}}\right) \geqq 0$, and hence not at all when $A(0+) = \infty$ and ε is sufficiently small.

By taking Laplace transforms formally in (2.1) - (2.4), one arrives at the two-point boundary value problem (for $U = \hat{u}$, with complex parameter s)

$$\hat{A}(s)U_{xx}(x,s) - sU(x,s) = -[\hat{F}(x,s) + u_0(x)], \tag{2.7}$$

$$U(0,s) = 0, \tag{2.8}$$

$$I[sU(1,s)-u_0(1)] + \hat{A}(s)U_x(1,s) = -ke^{-\varepsilon s}[U(1,s)-s^{-1}u_0(1)]+\hat{f}(s). \tag{2.9}$$

Using the standard solution techniques for such problems, one deduces the representation

$$\begin{aligned} U(x,s) = &-\frac{1}{\Delta(s)}\int_0^1 G(x,y,s)[u_0(y) + \hat{F}(y,s)]dy \\ &+ \frac{\sinh\beta x}{\Delta(s)}[(ks^{-1}e^{-\varepsilon s} + I)u_0(1) + \hat{f}(s)], \end{aligned} \tag{2.10}$$

where the Green's function G is defined by

$$G(x,y,s) = -\sinh \beta x[\alpha^{-1}(ke^{-\varepsilon s} + Is)\sinh \beta(1-y) + \cosh \beta(1-y)]$$

$$(0 \leqq x < y \leqq 1), \tag{2.11}$$

$$G(x,y,s) = G(y,x,s) \quad (0 \leqq y < x \leqq 1),$$

and

$$\Delta(s) = [Is + ke^{-\varepsilon s}]\sinh\beta + \alpha\cosh \beta$$

$$= \frac{1}{2} e^{\beta}[Is + ke^{-\varepsilon s} + \alpha] \left(1 - e^{-2\beta} \frac{Is + ke^{-\varepsilon s} - \alpha}{Is + ke^{-\varepsilon s} + \alpha}\right). \tag{2.12}$$

In these formulas, the parameters $\alpha = \alpha(s)$ and $\beta = \beta(s)$ are given formally by $\alpha = \sqrt{s\hat{A}(s)}$, $\beta = \frac{s}{\alpha}$. More precisely, note that, by (2.5) and the representation theroem due to Bernstein, there is a nondecreasing function μ on $[\eta,\infty)$ such that $a(t) = \int_\eta^\infty e^{-xt}d\mu(x)$, without loss of generality, we assume that $\eta \in \text{supp } d\mu$. Then

$$\hat{A}(s) = \frac{E}{s} + \int_\eta^\infty \frac{d\mu(x)}{x+s} \quad (s \notin (-\infty, -\eta],\ s \neq 0). \tag{2.13}$$

It follows that $\mathcal{I}(s\hat{A}(s)) > 0$ $(\mathcal{I}s > 0)$ and that $\hat{A}$ has exactly one zero s_* in $(-\eta,0)$. We can thus define α by $\alpha^2 = s\hat{A}(s)$, $\mathcal{R}\alpha > 0$ $(s \notin (-\infty,s_*])$ and let $\beta = \frac{s}{\alpha}$. The properties of $\hat{A}$, α and β used below follow more or less directly from (2.13). In particular, $\beta(s) = E^{-\frac{1}{2}}s + o(s)$ $(s \to 0)$, so the only singularities of U in $\{\mathcal{R}s > s_*\}$ occur at points where $\Delta(s) = 0$. The location of these points, as a function of k, is investigated numerically and analytically in [2].

3. EXISTENCE

To illustrate how existence and exponential decay can be established, we restrict attention to the special case $I = \varepsilon = 0$; in this case $\Delta(s) \neq 0$ $(\mathcal{R}s \geqq 0)$.

When $A'(0+) > -\infty$, one finds that

$$s^2 \hat{A}(s) - sA(0+) - A'(0+) \to 0 \quad (s \to \infty), \tag{3.1}$$

uniformly in each half-plane $\{\mathcal{R}s \geqq \sigma_0\}$, $-\infty < \sigma_0 < \infty$, and it follows from standard methods of complex analysis that $\Delta(s)$ has a sequence of zeros asymptotic to the line

$$\mathcal{R}\, s = \sigma_* \equiv \frac{1}{2}\left[\frac{A'(0+)}{A(0+)} + \sqrt{A(0+)}\,\log\left|\frac{k - \sqrt{A(0+)}}{k + \sqrt{A(0+)}}\right|\right].$$

(There is no asymptotic line when $k = \sqrt{A(0+)}$; let $\sigma_* = -\infty$ in that case.) Suppose now that $\Delta(s) \neq 0$ ($\mathcal{R}s \geqq \sigma_1 > \max\{\sigma_*, s_*\}$). Using (3.1) and (2.12), we get

$$\frac{1}{|\Delta(s)|} \leqq Q\left|\frac{e^{-\beta}}{\alpha}\right| \quad (\mathcal{R}s \geqq \sigma_1) \tag{3.2}$$

for some constant Q. (Actually $\alpha \to \sqrt{A(0+)}$ and $\beta \sim \dfrac{s}{\sqrt{A(0+)}}$ $(s \to \infty, \mathcal{R}\,s \geqq \sigma_1)$ in this case.) To estimate (2.10), assume that

$$u_0 \in AC[0,1] \text{ and } u_0(0) = 0. \tag{3.3}$$

Then

$$\int_0^1 G(x,y,s)u_0(y)dy = G_1(x,1,s)u_0(1) - \int_0^1 G_1(x,y,s)u'(y)dy, \tag{3.4}$$

where $G_1(x,y,s) = \int_x^y G(x,z,s)dz = O(\frac{e^{\beta}}{\beta})(s \to \infty, \mathcal{R}\,s \geqq \sigma_1)$. Combining this with (2.6), (2.10) and (3.2), we get

$$|U(x,s)| \leqq \frac{Q_1}{(1 + |s|)} \quad (\mathcal{R}s \geqq \sigma_1). \tag{3.5}$$

It follows that the function $s \to \underline{U}(\cdot,s)$ belongs to the Hardy space $\mathcal{H}^2(\{\mathcal{R}s > \sigma_1\}, H^o)$, so $\underline{U}$ is indeed the transform of a function $\underline{u} \in L^2(\underline{R}^+, H^o)$ with $t \to e^{-\sigma_1 t}\underline{u}(t) \in L^2(\underline{R}^+, H^o)$. Similarly, there exists ϕ, with $t \to e^{\sigma_1 t}\phi(t) \in L^2(\underline{R}^+)$ and $\hat{\phi}(s) = U(1,s)$.

Now let $\mathcal{D} = H^2 \cap H^1_0$, where H^2 and H^1_0 are the usual Sobolev spaces in H^o. Let $\underline{L}$: $\mathcal{D} \to H^o$ be the operator $\dfrac{d^2}{dx^2}$, and let $\underline{L}^* : H^o \to H^{-2}$ be its adjoint. $\underline{U}(\cdot,s)$ can then be characterized as the L^2 solution of (2.7) such that

$\underset{\sim}{U}(\cdot,s)-\hat{\phi}(s)\underset{\sim}{x} \in \mathcal{D}$ ($\underset{\sim}{x}$ = identity function, $\underset{\sim}{x}(x) = x$). Thus

$$\underset{\sim}{U}(\cdot,s)-\hat{\phi}(s)\underset{\sim}{x} = \underset{\sim}{A}^{-1}(s)(\underset{\sim}{L}-\beta^2\underset{\sim}{I})^{-1}[s\hat{\phi}(s)\underset{\sim}{x} - \underset{\sim}{\hat{F}}(s) - \underset{\sim}{u}_o].$$

Taking Fourier series with respect to the basis $\{\underset{\sim}{e}_n(x)\}_{n=1}^{\infty} = \{\sin n\pi x\}_{n=1}^{\infty}$ in H^o ($\underset{\sim}{x} \sim \Sigma\, x_n \underset{\sim}{e}_n$, etc.), we get

$$[\beta^2 + (n\ )^2][\hat{u}_n(s) - x_n\hat{\phi}(s)] = [\hat{F}_n(s) + u_{0,n} - x_n s\hat{\phi}(s)]\hat{A}(s)^{-1}$$

or

$$\hat{u}_n(s) = s^{-1}[-(n\pi)^2\hat{A}(s)[\hat{u}_n(s)-x_n\hat{\phi}(s)] + \hat{F}_n(s) + u_{0,n}].$$

By comparing coefficients, then, we see that $\underset{\sim}{u}$ is the unique solution in $L^1_{loc}(\underset{\sim}{R}^+, H^0)$ of

$$\underset{\sim}{u}(t) = \int_0^t B(t-\tau)\underset{\sim}{L}^*[\underset{\sim}{u}(\tau) - \underset{\sim}{x}\phi(\tau)]d\tau + \int_0^t \underset{\sim}{F}(\tau)d\tau + \underset{\sim}{u}_o \tag{3.6}$$

($B(t) = \int_0^t A(t)dt$), a weak, integrated form of (2.1) - (2.4).

When $A'(0+) = -\infty$, instead of (3.1) we get $\alpha(s) \to \sqrt{A(0+)}$ and $e^{-\beta(s)} \to 0$ as $s \to \infty$, uniformly in any right half-plane. By (2.12), (3.2) still holds if $\{\mathcal{R}s \geq \sigma_1\}$ is free of zeros of $\Delta(s)$, and we can use (3.4) and $\alpha\beta = s$ to get (3.5) and establish existence and decay as above.

The same reasoning applies when $I > 0$ or $\varepsilon > 0$, and we get the following result.

THEOREM. Let (2.5) and (2.6) hold, and suppose

$$\Delta(s) \neq 0 \quad (\mathcal{R}s \geq \sigma_1 > -\infty).$$

When $I = 0$, $\varepsilon > 0$, and $A(0+) < \infty$, assume in addition that

$$\dot{k} < \sqrt{A(0+)}\,\frac{1-C}{1+C}\,.$$

When $\sigma_1 < 0$, assume that $\varepsilon = 0$, $\sigma_1 > s_*$, and $\Delta(s) \neq 0$ in some larger half-plane $\{\mathcal{R}s \geq \sigma_2\}$, $\sigma_2 < \sigma_1$. Then $\underline{U}(\cdot,s)$ and $U(1,s)$ are transforms of functions $\underline{u}$ and ϕ with $t \to e^{-\sigma_1 t}\underline{u}(t)$ and $t \to e^{-\sigma_1 t}\phi(t)$ in $L^2(\underline{R}^+,H^0)$ and $L^2(\underline{R})$, respectively. Moreover, $\underline{u}$ is the unique solution in $L^1_{loc}(\underline{R}^+,H^0)$ of (3.6).

Finally, we note that stronger assumptions on u_0, F and f can yield regularity estimates for u and ϕ. For an easy example, suppose F = f = 0 and, instead of (3.3), assume that $u_0 \in C_0^\infty(0,1)$. Then the integration by parts in (3.4) can be repeated indefinitely, there are no boundary terms, and the estimate is improved by $\frac{1}{\beta}$ each time. But $\frac{1}{\beta} = o(s^{-\frac{1}{2}})$ $(s \to \infty, \mathcal{R}s \geq \sigma_1)$, so (3.5) can be improved to

$$|U(x,s)| \leq \frac{Q_j}{(1 + |s|^j)} \quad (\mathcal{R}s \geq \sigma_1), \tag{3.5'}$$

j = 1,2,3,.... We can then differentiate with respect to x in (2.10) and get

$$t \to e^{-\sigma_1 t} \frac{\partial^j}{\partial x^j} u(x,t) \in L^2(\underline{R}^+,H^0)$$

for j = 0,1,2,.... When the support of u_0 includes x = 0 or 1 or u_0 has jumps, one gets stationary and propagating singularities, depending on the behaviour of A(t) near t = 0, like those discussed in [3].

References

[1] R. Datko, J. Lagnese, and M.P. Polis, An example of the effect of time delays in boundary feedback stabilization of wave equations, SIAM J. Control and Optimization 24 (1986), 152-156.

[2] K.B. Hannsgen, Y. Renardy, and R.L. Wheeler, Effectiveness and robustness with respect to time delays of boundary feedback stabilization in one-dimensional viscoelasticity, SIAM J.Control and Optimization 26 (1988).

[3] W.J. Hrusa and M. Renardy, On wave propagation in linear viscoelasticity, Quart. Appl. Math. 43 (1985), 237-254.

K.B. Hannsgen & R.L. Wheeler
Virginia Polytechnic Institute and State University
Blacksburg
Virginia 24061
USA.

M.L. HEARD & S.M. RANKIN III

Weak solutions for a class of parabolic Volterra integrodifferential equations*

ABSTRACT

We define weak solutions for a class of Volterra integrodifferential equations of the form

$$u'(t) + Au(t) = \int_0^t a(t,s)g(s,u(s))ds + f(t,u(t)), \quad t \geq 0$$

$$u(0) = 0.$$

The operator A is the negative infinestimal generator of an analytic semigroup in a Banach space X. The operator $g(t,u)$ is related to A by a special form $g(t,u) = A^{1/2}q(t,u)$, where $q(t,u)$ is an appropriate "lower order" operator. We show the existence and uniqueness of weak solutions and their continuability to infinity under suitable conditions. Using our results we study the asymptotic behaviour, as time goes to infinity, of strong solutions of a second order initial-boundary value problem.

§1. INTRODUCTION

The class of equations to be considered in this work have the form

$$u'(t) + Au(t) = \int_0^t a(t,s)g(s,u(s))ds + f(t,u(t)), \quad t \geq 0 \tag{1.1}$$

$$u(0) = u_o. \tag{1.2}$$

We shall consider (1.1), (1.2) as a Cauchy problem in a Banach space X and introduce a class of solutions which we shall call weak soltuions to (1.1), (1.2). In general, A will be a closed, linear, densely defined operator in X such that - A generates an analytic semigroup $\{T(t): t \geq 0\}$. In recent years many authors ([1], [4], [5], [6], [14], [24], [26]) have studied this problem under various assumptions on the operators g and f. When (1.1), (1.2) is applied to problems of heat flow in materials with memory ([18],

[21], [22]) it is necessary to require that g be defined on the domain D(A) of the generator A. This implies, in the applications, that g has the same spatial order as A. The operator f, meanwhile, is allowed to have "lower order" than A and is only required to be defined on an intermediate space between D(A) and X. Most often this space is taken to be the domain $D(A^{1/2})$ of the fractional power operator $A^{1/2}$. The initial data u_0 is usually required to lie in D(A) and the concept of a strong solution applied to (1.1), (1.2) means a function u(t) with values in D(A) and that both u'(t) and Au(t) are continuous in the X-norm and u(t) satisfies (1.1) for all t in some interval $0 \leq t \leq T$. Thus it is natural to consider fixed points of the integral equation

$$u(t) = T(t)u_0 + \int_0^t T(t-s)\ \{\int_0^s a(s,\tau)g(\tau,u(\tau))d\tau + f(s,u(s))\}ds. \quad (1.3)$$

In the applications of (1.1), (1.2) to typical examples of equations in heat flow, it turns out that a certain class of nonlinear operators g(t,u(t)) is considered more often than others. In the case of second order equations, these operators have a structure in divergence form:

$$g(t,u(t)) = \sum_{i=1}^{n} D_i g_i(t,x,u(t,x),\ \nabla u(t,x)) \quad (1.4)$$

where $D_i = \frac{\partial}{\partial x_i}$, i = 1,2,...,n. In this work we try to take advantage of this special form by factoring the differential operator D_i into a product

$$D_i = A^{1/2}B_i$$

where the operators B_i are bounded, linear operators on X. This allows us to represent the operator g(t,u(t)) as a special product

$$g(t,u(t)) = A^{1/2}q(t,u(t))$$

where q(t,u(t)) is associated with the first order operators $g_i(t,x,u(t,x), \nabla u(t,x))$. We can then replace the integral equation (1.3) by the equation

$$u(t) = T(t)u_0 + \int_0^t A^{1/2}T(t-s)\{\int_0^s a(s,\tau)q(\tau,u(\tau))d\tau\}ds \tag{1.5}$$

$$+ \int_0^t T(t-s)f(s,u(s))ds, \quad t \geq 0.$$

The advantage of (1.5) is that $A^{1/2}T(t)$ is a bounded linear operator of X. Thus we can regard (1.5) as a fixed point equation in an intermediate space $D(A^{1/2})$.

Roughly speaking, given $u_0 \in D(A^{1/2})$ we say that a function u(t) is a weak solution of (1.1), (1.2) if it has values in $D(A^{1/2})$ and satisfies the integral equation (1.5). We show that under suitable conditions, weak solutions of (1.1), (1.2) exist and are unique. Furthermore, they can be continued over a maximal interval of existence [0,d), $0 \leq d \leq +\infty$, and we can obtain a global solution ($d = +\infty$) if there exists an estimate on u(t) in the graph norm on $D(A^{1/2})$. In practice this amounts to obtaining an estimate on the gradient $\nabla u(t,x)$. For a different concept of weak solutions related to linear equations, we refer to [13].

An easy consequence of our definition is that every strong solution of (1.1), (1.2) is also a weak solution. Therefore our results give a partial answer to the uniqueness problem associated with (1.1), (1.2). In general terms, this problem asks under what conditions on the operators g and f will the solutions of the Volterra integrodifferential equation $u'(t) + A(t)u(t) = \int_0^t g(t,s,u(s))ds + f(t,u(t))$, $t \geq 0$, be unique. One answer was given in [14], [27] which required $f(t,\cdot)$ to be defined on D(A) and have values in $D(A^{1/2})$. Various other smoothness conditions are given in the works [19], [24]. But for the obvious case where $f(t,\cdot)$ is only defined on $D(A^{1/2})$ (which is the usual assumption when g = 0), uniqueness has only been proved for the case when X is a Hilbert space by Heard-Rankin [15]. Thus our use of weak solutions allows us to answer the question for general Banach spaces X in the case when A = A(t) is a constant operator.

The definition of weak solutions also has advantages when studying the asymptotic behaviour of solutions of (1.1), (1.2). In the last part of the paper, we consider the behaviour of solutions of a special case (1.1), (1.2). Our result extends some methods due to Engler [8], [9] in his study of the behaviour of solutions of parabolic integrodifferential equations of the type

$$u_t(t,x) - \Delta u(t,x) + \int_0^t g(t-s,u(s,x))ds = f(t,x),(t,x) \in (0,\infty) \times \Omega,$$

$$u(t,x) = 0 \text{ if } (t,x) \in (0,\infty) \times \partial\Omega,$$

$$u(0,x) = u_0(x) \text{ if } x \in \Omega.$$

Under suitable conditions Engler shows that the solution $u(t,x)$ of the above problem converges uniformly on Ω as $t \to \infty$ to a solution $u_\infty(x)$ of the integro-differential equation

$$-\Delta u_\infty + \int_0^\infty g(s,u_\infty(x))ds = f_\infty(x), \ x \in \Omega$$

$$u_\infty(x) = 0, \ x \in \partial\Omega$$

where

$$f_\infty(x) = \lim_{t\to\infty} f(t,x).$$

In [15] a similar result was obtained for (1.1), (1.2) in the case when g had the form $g(u) = Au + g_0(u)$, where $g_0(u)$ was monotone with polynomial growth. Using the idea of weak solutions, we are able to go further and discuss the asymptotic behaviour of certain initial-boundary value problems where the operator $g(t,u(t))$ has a divergence structure of the form (1.4).

We now mention other related results on asymptotic behaviour for solutions of Volterra integrodifferential equations given by Clément, MacCamy, Nohel [3], Clément and Nohel [2], Londen and Nohel [18]. In [18] the authors consider the problem

$$\frac{du}{dt} + Bu(t) + (a * Au)(t) + \frac{d}{dt}(b * u)(t) \ni f(t) \text{ a.e. } t \geq 0, \tag{1.6}$$

$$u(0) = u_0. \tag{1.7}$$

The operators A and B are assumed to be subdifferentials of proper, convex, lower-semicontinuous functionals on a Hilbert space H. The functions a and b are real-valued kernels satisfying various positivity conditions. For asymptotic behaviour the authors assume that A and B are single-valued. They show that under appropriate conditions, which are somewhat different from those we assume, each solution u(t) of (1.6), (1.7) converges in the norm of $\dot{H}$ to a limit u_∞ which satisfies the stationary equation

$$Bu_\infty + \left(\int_0^\infty a(t)dt \right) Au_\infty = f_\infty .$$

In [2] and [3] similar types of results are proven for the undifferentiated equation

$$u(t) + (b * Au)(t) \ni f_\infty .$$

The organization of the paper is as follows. In Section 2 we give the basic assumptions, definitions, and state the main theorems. In Section 3 we give the proofs to these theorems. In Section 4 we discuss the factorization problem for the differential operator D_i. In Section 5 we give an example of a second order initial-boundary value problem which has a weak solution in our sense, but no strong solution in the usual sense. In Section 6 we discuss the asymptotic behaviour of a special case of (1.1), (1.2).

§2. ASSUMPTIONS AND STATEMENTS OF MAIN RESULTS

Let X be a Banach space with norm $\|\cdot\|$ and let A be a linear operator defined on a subset of X and suppose A satisfies the following conditions:

(A1) The domain D(A) of A is a dense subset of X and $A:D(A) \to X$ is a closed operator.

(A2) The resolvent operator $R(\lambda;A) = (\lambda I - A)^{-1}$ exists for all Re $\lambda \geq 0$ and there is a constant $C > 0$ such that

$$\|R(\lambda;A)\| \leq \frac{C}{1 + |\lambda|} \text{ for all Re } \lambda \leq 0.$$

From (A1), (A2) it is well-known that $-A$ is the infinitesimal generator of an analytic semigroup $\{T(t):t \geq 0\}$ on X and there are constants $C > 0$, $\delta > 0$ such that

$$\|T(t)\| \leq Ce^{-\delta t}, \quad t \geq 0,$$

$$\|AT(t)\| \leqq \frac{Ce^{-\delta t}}{t}, \quad t > 0.$$

For details concerning these remarks and analytic semigroup theory in general, we refer to [23]. Since A has a bounded inverse on X we can define negative fractional powers of A by

$$A^{-\alpha} = \frac{1}{\Gamma(\alpha)} \int_0^\infty s^{\alpha-1} T(s)ds, \qquad 0 < \alpha < \infty,$$

where $\Gamma(\alpha)$ denotes the Gamma function. The operator $A^{-\alpha}$ is well-defined on all X and $A^{-\alpha}$ is a bounded, linear, one-one, transformation. Thus we may define positive fractional powers of A by $A^{\alpha} \equiv (A^{-\alpha})^{-1}$, $0 < \alpha < \infty$. Each mapping A^{α} is a closed linear operator with domain $D(A^{\alpha})$ dense in X and $D(A^{\alpha}) \supset D(A^{\beta})$ if $0 < \alpha < \beta < \infty$. The exponent rule

$$A^{\alpha}A^{\beta}x = A^{\beta}A^{\alpha}x = A^{\beta+\alpha}x$$

holds for all $x \in D(A^{\alpha+\beta})$. For each $t > 0$ the composition $A^{\alpha}T(t)$ is a bounded linear operator on X and there is a constant $C(\alpha) > 0$ such that

$$\|A^{\alpha}T(t)\| \leqq \frac{C(\alpha)e^{-\delta t}}{t^{\alpha}}, \quad 0 < t < \infty.$$

Furthermore, we have

$$A^{\alpha}T(t)x = T(t)A^{\alpha}x, x \in D(A^{\alpha}). \tag{2.1}$$

for all $t > 0$ and there is $C(\alpha) > 0$ such that

$$\|T(t) - I)A^{-\alpha}\| \leqq C(\alpha)t^{\alpha}, \qquad 0 < t < \infty.$$

Finally, if $\psi(t)$ is a continuous X-valued function defined for $0 \leqq t < \infty$, then

$$\int_0^t T(t-s)\psi(s)ds \in D(A^{\alpha}),$$

and

$$A^{\alpha}\int_0^t T(t-s)\psi(s)ds = \int_0^t A^{\alpha}T(t-s)\psi(s)ds. \tag{2.2}$$

For proofs of these remarks, see [23, Theorem 6.13, p. 74]. We let X_α denote $D(A^\alpha)$ equipped with the graph norm $\|x\|_\alpha = \|A^\alpha x\|$, $x \in D(A^\alpha)$. Then X_α is a Banach space with continuous dense embedding $X_\alpha \to X$. We make the following additional assumptions:

(A3) There are continuous mappings $f,q:[0,\infty) \times X_{1/2} \to X$ such that for each $t_o > 0$, $r_0 > 0$ there are constants $L = L(t_o^*,r_o^*) > 0$ and $M = M(t_o^*,r_o^*) > 0$ such that if $u,v \in X_{1/2} \leqq r_o^*$, with $\|u\|_{1/2} \leqq r_o^*$, $\|v\|_{1/2} \leqq r_o^*$ then

$$\|q(t,u) - q(t,v)\| \leqq M \|u-v\|_{1/2},$$

$$\|f(t,u) - f(t,v)\| \leqq L \|u-v\|_{1/2},$$

for all $0 \leqq t \leqq t_o^*$.

(A4) The set $W = \{(t,u) : t \geqq 0,\ u \in X_1 \text{ and } q(t,u) \in X_{1/2}\}$ is a nonempty subset of $[0,\infty) \times X_1$.

(A5) The function $a: [0,\infty) \times [0,\infty) \to C$ is continuous with continuous first partial derivative $a_1(t,s) = \frac{\partial a}{\partial t}(t,s)$, $0 \leqq t,\ s < \infty$.

On the basis of assumptions (A4) we can define a mapping $g:W \to X$ by

$$g(t,u) = A^{1/2}q(t,u). \tag{2.3}$$

We make no formal assumption on the continuity of the mapping $u \to g(t,u)$, but we observe that if X is reflexive and if $u \to q(t,u)$ maps bounded subsets of X_1 into bounded subsets of $X_{1/2}$, then $u_n \to u$ strongly in X_1 implies $g(t,u_n) \to g(t,u)$ weakly in X.

We first consider the local Cauchy problem

$$u'(t) + Au(t) = \int_{t_o}^{t} a(t,s)g(s,u(s))ds + f(t,u(t)), \quad t \geqq t_o \geqq 0 \tag{2.4}$$

$$u(t_o) = u_o. \tag{2.5}$$

DEFINITION 1. Given $u_o \in X_{1/2}$, we say that a function $u(t)$ is a weak solution of (2.4), (2.5) on $[t_o,t_o + \delta]$ if:

(i) $u \in C([t_o,t_o + \delta]; X_{1/2})$,

(ii) u satisfies the integral equation

$$u(t) = T(t-t_o)u_o + \int_{t_o}^{t} A^{1/2}T(t-s)\{\int_{t_o}^{s} a(s,\tau)q(\tau,u(\tau))d\tau\}ds \tag{2.6}$$

$$+ \int_{t_o}^{t} T(t-s)f(s,u(s))ds, \quad t_o \leqq t \leqq t_o + \delta.$$

We now recall the definition of a strong solution of (2.4), (2.5) that is most frequently used in the literature ([10, [14], [19], [24], [26], [27]). For convenience, let us suppose that $W = [0,\infty) \times X_1$ in assumption (A4).

DEFINITION 2. Given $u_o \in X_1$ we say that a function $u(t)$ is a strong solution of (2.4), (2.5) on $[t_o,t_o + \delta]$ if:

(i) $u \in C([t_o,t_o + \delta];X_1) \cap C^1([t_o,t_o + \delta];X)$

(ii) $t \to g(t,u(t))$ belongs to $L^1([t_o,t_o+\delta];X)$

(iii) $u(t_o) = u_o$

(iv) $u(t)$ satisfies (2.4) on $[t_o,t_o + \delta]$.

We are able to state our main results.

THEOREM 1. Let assumptions (A1)-(A5) hold and let $u_o \in X_{1/2}$, $t_o \geqq 0$ be given. Then there exists $\delta = \delta(t_o,u_o) > 0$ such that the Cauchy problem (2.4), (2.5) has a unique weak solution $u(t)$ on $[t_o,t_o + \delta]$.

THEOREM 2. Let assumptions (A1) - (A5) hold and let $u_o \in X_{1/2}$, $t_o \geqq 0$ be given. Then the unique weak solution $u(t)$ of (2.4), (2.5) given by Theorem 1 can be uniquely continued to the right as a weak solution of (2.4), (2.5) to a maximal domain $[t_o,d)$, where $t_o < d \leq +\infty$. If $d < +\infty$ then there must be a sequence $t_n \to d^-$ such that $\|u(t_n)\|_{1/2} \to +\infty$.

THEOREM 3. Let assumptions (A1) - (A5) hold with $W = [0,\infty) \times X_1$. Suppose $u_o \in X_1$, $t_o \geq 0$ and let u(t) be a strong solution of (2.4), (2.5) on an interval $[t_o,t_o+\delta]$. Then u(t) is also a weak solution of (2.4), (2.5) on $[t_o,t_o+\delta]$. Consequently, all strong solutions of (2.4), (2.5) are unique.

§3. PROOFS OF THE MAIN RESULTS

PROOF OF THEOREM 1. Assume (A1) - (A5) hold and let $u_o \in X_{1/2}$ and $t_o \geq 0$ be given. Fix $\delta > 0$, $r > 0$ and let $S = S(t_o,r,\delta)$ denote the set of all functions $u \in C([t_o,t_o+\delta];X_{1/2})$ such that

$$u(t_o) = u_o \text{ and } \sup_{t_o \leq t \leq t_o+\delta} \|u(t) - u_o\|_{1/2} \leq r. \tag{3.1}$$

Then S is a closed, convex subset of $C([t_o,t_o+\delta]; X_{1/2})$ and we may define an operator Φ on S by

$$(\Phi u)(t) = T(t-t_o)u_o + \int_{t_o}^{t} A^{1/2}T(t-t_o)\ \{\int_{t_o}^{s} a(s,\tau)q(\tau,u(\tau))d\tau\}ds \tag{3.2}$$
$$+ \int_{t_o}^{t} T(t-s)f(s,u(s))ds, \quad t_o \leq t \leq t_o + \delta.$$

By (2.2) it follows that

$$A^{1/2}(\Phi u)(t) = A^{1/2}T(t-t_o)u_o + A\int_{t_o}^{t} T(t-s)\ \{\int_{t_o}^{s} a(s,\tau)q(\tau,u(\tau))d\tau\}ds$$
$$+ A^{1/2}\int_{t_o}^{t} T(t-s)f(s,u(s))ds, \quad t_o \leq t \leq t_o + \delta.$$

By Lemma 2.1 of [26] we have

$$A\int_{t_o}^{t} T(t-s)\ \{\int_{t_o}^{s} a(s,\tau)q(\tau,u(\tau))d\tau\}ds = \int_{t_o}^{t} a(t,s))q(s,u(s))ds$$
$$- \int_{t_o}^{t} T(t-s)\ \{a(s,s)q(s,u(s)) + \int_{t_o}^{s} a_1(s,\tau)q(\tau,u(\tau))d\tau\}ds,$$

where $a_1(s,\tau) = \frac{\partial a}{\partial s}(s,\tau)$. So by (2.1) we have

$$A^{1/2}[\Phi u(t)-u_0] = [T(t-t_0)-I]A^{1/2}u_0 + \int_{t_0}^{t} a(t,s)q(s,u(s))ds$$
$$- \int_{t_0}^{t} T(t-s)\ \{a(s,s)q(s,u(s)) + \int_{t_0}^{s} a_1(s,\tau)q(\tau,u(\tau))d\tau\}ds$$
$$+ A^{1/2}\int_{t_0}^{t} T(t-s)f(s,u(s))ds, \quad t_0 \leqq t \leqq t_0 + \delta.$$

Let $L > 0$, $M > 0$ be the constants given in assumption (A3) with $t_0^* = t_0 + \delta$ and $r_0^* = r + \|u_0\|_{12}$. Define

$$C_q = Mr + \sup_{t_0\leqq s\leqq t_0+\delta} \|q(s,u_0)\|,$$

$$C_f = Lr + \sup_{t_0\leqq s\leqq t_0+\delta} \|f(s,u_0)\|,$$

$$a_\infty = \sup_{t_0\leqq s,t\leqq t_0+\delta} |a(t,s)|,$$

$$a'_\infty = \sup_{t_0\leqq s,t\leqq t_0+\delta} |a_1(t,s)|.$$

Then for all $t_0 \leqq t \leqq t_0 + \delta$ and all $u \in S$, we have the estimates

$$\|\int_{t_0}^{t} a(t,s)q(s,u(s))ds\| \leqq a_\infty\delta C_q,$$

$$\|\int_{t_0}^{t} T(t-s)\{a(s,s)q(s,u(s)) + \int_{t_0}^{s} a_1(s,\tau)q(\tau,u(\tau))d\tau\}ds\| \leqq C\delta(a_\infty + a'_\infty\,\delta)C_q,$$

$$\|A^{1/2}\int_{t_0}^{t} T(t-s)f(s,u(s))ds\| \leqq 2CC_f\delta^{1/2}.$$

Therefore from (3.3)

$$\|\Phi u(t)-u_0\|_{12} \leqq \|[T(t-t_0)-I]A^{1/2}u_0\| + a_\infty\delta C_q + C\delta(a_\infty + a'_\infty\delta)C_q + 2CC_f\delta^{1/2},$$

for all $t_0 \leqq t \leqq t_0 + \delta$ and $u \in S$. Since $T(t)$ is strongly continuous, it follows that there exists $\delta > 0$ sufficiently small such that

$$\sup_{t_0 \leq t \leq t_0+\delta} \|\Phi u(t) - u_0\|_{1/2} \leq r$$

for all $u \in S$. Also from (3.3) it is clear that $u \in S$ implies that $\Phi u \in C([t_0,t_0+\delta];X_{1/2})$. Consequently Φ maps S into itself. Now let $u,v \in S$, then

$$\begin{aligned} A^{1/2}[\Phi u(t) - \Phi v(t)] &= \int_{t_0}^{t} a(t,s)[q(s,u(s)) - q(s,v(s))]ds \\ &- \int_{t_0}^{t} T(t-s)a(s,s)[q(s,u(s))-q(s,v(s))]ds \\ &- \int_{t_0}^{t} T(t-s)\left\{\int_{t_0}^{s} a_1(s,\tau)[q(\tau,u(\tau))-q(\tau,v(\tau))]d\tau\right\}ds \\ &+ A^{1/2}\int_{t_0}^{t} T(t-s)[f(s,u(s))-f(s,v(s))]ds, \quad t_0 \leq t \leq t_0+\delta. \end{aligned}$$

From assumption (A3) it follows that

$$\sup_{t_0 \leq t \leq t_0+\delta} \|\{\Phi u(t)-\Phi v(t)\|_{1/2} \leq [a_\infty M\delta(1+C) + a'_\infty CM\delta^2 + 2CL\delta^{1/2}] \times \sup_{t_0 \leq t \leq t_0+\delta} \|u(t)-v(t)\|_{1/2}.$$

Thus we can choose $\delta > 0$ smaller, if necessary, so that Φ is a strict contraction on S in the norm of $C([t_0,t_0+\delta]; X_{1/2})$. Therefore Φ has a unique fixed point $u \in S$ and this fixed point is the unique weak solution of (2.4), (2.5) on $[t_0,t_0+\delta]$.

PROOF OF THEOREM 2. We wish to extend the weak solution u(t) obtained in Theorem 1 beyond the right endpoint $t_0 + \delta$. But since the memory term

$$\int_{t_0}^{t_0+\delta} a(t,s)g(s,u(s))ds$$

need not exist for weak solutions, we must continue u(t) as a solution of

(2.6). Put $t_1 = t_o + \delta$, $u_1 = u(t_1)$ and let $\delta_1 > 0$ be fixed. We say that a function $\tilde{u}(t)$ is a <u>weak continuation</u> of $u(t)$ to the interval $[t_1,t_1+\delta_1]$ if:

(a) $\tilde{u} \in C([t_1,t_1 + \delta_1];X_{1/2})$,

(b) $\tilde{u}$ satisfies the integral equation

$$\tilde{u}(t) = T(t-t_1)u_1 + \int_{t_1}^{t} A^{1/2}T(t-s)\ \{\int_{t_1}^{s} a(s,\tau)q(\tau,\tilde{u}(\tau))d\tau\}ds$$
$$+ A^{1/2}\int_{t_1}^{t} T(t-s)\ \{\int_{t_o}^{t_1} a(s,\tau)q(\tau,u(\tau))d\tau\}ds$$
$$+ \int_{t_1}^{t} T(t-s)f(s,\tilde{u}(s))ds,$$

for $t_1 \leqq t \leqq t_1 + \delta$.

The terminology "weak continuation" applied to $\tilde{u}(t)$ is justified by the observation that if we define a new function $v(t)$ on $[t_o,t_1 + \delta_1]$ by setting

$$v(t) = \begin{cases} u(t) & \text{if } t_o \leqq t \leqq t_1, \\ u(t) & \text{if } t_1 \leqq t \leqq t_1 + \delta_1, \end{cases}$$

then $v(t)$ is a weak solution of (2.4), (2.5) on $[t_o,t_1 + \delta_1]$ in the sense of Definition 1. The existence and uniqueness of the weak continuation $\tilde{u}(t)$ is demonstrated exactly as was done in Theorem 1 with only a few minor changes. We shall omit the details.

After carrying out the above process step-by-step on successive intervals, we extend our original solution to a unique weak solution of (2.4), (2.5) on some maximal interval of existence $[t_o,d)$, $t_o < d \leqq +\infty$. We shall denote this extension also by $u(t)$ and show that if $d < +\infty$ then

$$\|u(t_n)\|_{1/2} \to +\infty \text{ as } n \to -\infty \tag{3.5}$$

for some suitable sequence $t_n \to d^-$. Indeed, let us assume that (3.5) is false. Then there is a constant $C_d > 0$ such that

$$\|u(t)\|_{1/2} \leq C_d \text{ for all } t_o \leq t < d. \tag{3.6}$$

Using this estimate on u(t), we arrive at a contradiction in the following manner. Given $t_o < t_1 < d$ we consider the integral equation (3.4) for $t \geq t_1$. We show that there is a constant $\delta > 0$, independent of t_1, such that (3.4) has a unique solution $\tilde{u} \in C([t_1,t_1+\delta];X_{1/2})$. For t_1 close to d, $\tilde{u}(t)$ represents a weak continuation of u(t) beyond d and this is a contradiction.

Since this type of argument has been used before in [14] in conjuction with strong solutions, and since the details are very similar to those in the proof of Theorem 1, we shall only sketch the argument. Given $r > 0$, $\delta > 0$, we define S_1 to be the set of all $w \in C([t_1,t_1+\delta]; X_{1/2})$ such that

$$w(t_1) = u(t_1) \text{ and } \sup_{t_1 \leq t \leq t_1+\delta} \|w(t) - u(t_1)\|_{1/2} \leq r.$$

We define an operator Φ_1 on S_1 by

$$\begin{aligned}(\Phi_1 w)(t) = T(t-t_1)u(t_1) &+ \int_{t_1}^{t} A^{1/2}T(t-s)\{\int_{t_1}^{s} a(s,\tau)q(\tau,w(\tau))d\tau\}ds \\ &+ A^{1/2} \int_{t_1}^{t} T(t-s)\{\int_{t_o}^{t_1} a(s,\tau)q(\tau,u(\tau))d\tau\}ds \\ &+ \int_{t_1}^{t} T(t-s)f(s,w(s))ds,\end{aligned} \tag{3.7}$$

for $t_1 \leq t \leq t_1 + \delta$. Then for all $t_1 \leq t \leq t_1 + \delta$ and $w \in S_1$ we have

$$\begin{aligned}A^{1/2}[\Phi_1 w(t)-u(t_1)] = [T(t-t_1)-I]A^{1/2}u(t_1) &+ \int_{t_1}^{t} a(t,s)q(s,w(s))ds \quad (3.8)\\ - \int_{t_1}^{t} T(t-s)\{a(s,s)q(s,w(s)) &+ \int_{t_1}^{s} a_1(s,\tau)q(\tau,w(\tau))d\tau\}ds \\ + A \int_{t_1}^{t} T(t-s) \{\int_{0}^{t_1} &[a(s,\tau)-a(t,\tau)]q(\tau,u(\tau))d\tau\}ds \\ + [I-T(t-t_1)] \int_{0}^{t_1} &a(t,\tau)q(\tau,u(\tau))d\tau \\ + A^{1/2} \int_{t_1}^{t} &T(t-s)f(s,w(s))ds.\end{aligned}$$

From (3.8) it is clear that $\Phi_1(w) \in C([t_1,t_1+\delta];X_{1/2})$. We let $L > 0$, $M > 0$ be the constants given in (A3) with $t_0^* = d$ and $r_0^* = r + C(d)$, where $C(d)$ is given by (3.6). We define uniform constants $\bar{C}_q$, $\bar{C}_f$, $\bar{a}_\infty$, $\bar{a}'_\infty$ by

$$\bar{C}_q = Mr + 2MC(d) + \sup_{0\le s\le d} \|f(s,u_0)\|, \quad \bar{a}_\infty = \sup_{0\le s,t\le d} |a(t,s)|,$$

$$\bar{C}_f = Lr + 2LC(d) + \sup_{0\le s\le d} \|f(s,u_0)\|, \quad \bar{a}'_\infty = \sup_{0\le s,t\le d} |a_1(t,s)|.$$

These constants are independent of t_1 and as in the proof of Theorem 1 we have the estimate

$$\begin{aligned}\|\Phi w(t)-u(t_1)\|_{1/2} \le{}& \|[T(t-t_1) - I]A^{1/2}u(t_1)\| \\ &+ \left\|[I-T(t-t_1)]\int_0^{t_1} a(t,\tau)q(\tau,u(\tau))d\tau\right\| \\ &+ \bar{a}_\infty\delta C_1 + C\delta(\bar{a}_\infty + \bar{a}'_\infty\delta)\bar{C}_q \\ &+ C\bar{a}'_\infty d\bar{C}_q\delta + 2C\bar{C}_f\delta^{1/2}.\end{aligned} \tag{3.9}$$

Consider the function

$$(t,s) \to \int_0^t a(s,\tau)q(\tau,u(\tau))d\tau.$$

This is a continuous X-valued function for $0 \le t, s \le d$ and therefore has compact range in X. So by [14, Lemma 3, p. 85] we have

$$\lim_{t\to t_1} \left\|[I-T(t-t_1)]\int_0^{t_1} a(t,\tau)q(\tau,u(\tau))d\tau\right\| = 0 \tag{3.10}$$

uniformly in t_1. To estimate the term $\|[T(t-t_1)-I]A^{1/2}u(t_1)\|$, we note that since $u(t)$ is a weak solution of (2.1), (2.2) on $[0,d)$ it follows from (2.3) that

$$\begin{aligned}A^{1/2}u(t) ={}& T(t)A^{1/2}u_0 + \int_0^t a(t,s)q(s,u(s))ds \\ &+ A^{1/2}\int_0^t T(t-s)f(s,u(s))ds\end{aligned} \tag{3.11}$$

$$- \int_0^t T(t-s)\{a(s,s)q(s,u(s))$$

$$+ \int_0^s a_1(s,\tau)q(\tau,u(\tau))d\tau\}ds, \quad 0 \leq t < d.$$

We show that $A^{1/2}u(t)$ is uniformly continuous on $[0,d)$. The first two terms of (3.11) are continuous on $[0,d)$. For the third term we use [11, Lemma II.14.4] to obtain

$$\|A^{1/2}\{\int_0^{t+h} T(t+h-s)f(s,u(s))ds - \int_0^t T(t-s)f(s,u(s))ds\}\|$$

$$\leq C\bar{C}_f h^{1/2}(|\log h| + 1), \quad 0 \leq t < t + h < d.$$

So the third term is uniformly continuous on $[0,d)$. The last term is written

$$\int_0^t T(t-s)\psi(s)ds \tag{3.12}$$

where

$$\psi(s) = a(s,s)q(s,u(s)) + \int_0^s a_1(s,\tau)q(\tau,u(\tau))d\tau, \quad 0 \leq s < d.$$

Since $\psi \in L^\infty(0,d;X)$ it follows from [23, Theorem 3.1, p. 110] that (3.12) is Hölder continuous on $[0,d]$. Therefore $A^{1/2}u(t)$ is uniformly continuous on $[0,d)$ and must have precompact range in X. Thus

$$\lim_{t\to t_1} \|[T(t-t_1) - I]A^{1/2}u(t_1)\| = 0 \tag{3.13}$$

uniformly in t_1. So from (3.9), (3.10), (3.13) it follows that there exists $\delta > 0$, independent of t_1, such that

$$\sup_{t_1\leq t\leq t_1+\delta} \|\Phi_1 w(t) - u(t_1)\|_{1/2} \leq r$$

for all $w \in S_1$. A similar argument shows that if $\delta > 0$ is sufficiently small and independent of t_1, then

$$\sup_{t_1 \leq t \leq t_1+\delta} \|\Phi_1 w(t) - \Phi_1 \bar{w}(t)\|_{1/2} \leq c \sup_{t_1 \leq t \leq t_1+\delta} \|w(t)-\bar{w}(t)\|_{1/2}$$

for a suitable constant $0 < c < 1$. Thus Φ_1 maps S_1 into itself and is a strict contraction. So there is a unique fixed point $\bar{e} \in S_1$ of Φ_1 which represents the weak continuation of $u(t)$ to the interval $[t_1, t_1 + \delta]$. Since the length δ of this interval does not depend on t_1, we arrive at a contradiction when t_1 is close to d. Q.E.D.

PROOF OF THEOREM 3. If $u(t)$ is a strong solution of (2.1), (2.2) on $[t_0, t_0 + \delta]$, then the memory term

$$\int_0^t a(t,s)g(s,u(s))ds$$

is well-defined and belongs to $C^1([t_0,t_0 + \delta];X)$. Therefore $u(t)$ must satisfy the integral equation

$$u(t) = T(t-t_0)u_0 + \int_{t_0}^{t} T(t-s) \{\int_{t_0}^{s} a(s,\tau)g(\tau,u(\tau))d\tau\}ds$$
$$+ \int_{t_0}^{t} T(t-s)f(s,u(s))ds, \quad t_0 \leq t \leq t_0 + \delta.$$

By virtue of the representation (2.3) and property (2.1) it follows that $u(t)$ is a weak solution of (2.4), (2.5) on $[t_0, t_0 + \delta]$. Q.E.D.

§4. FACTORIZATION OF FIRST ORDER DIFFERENTIAL OPERATORS

In order to apply the results of Section 2 to concrete equations, it is necessary to demonstrate the factorization given by (2.3). In order to accomplish this, we must have a reasonable characterization of the domain of $A^{1/2}$ for our generator A. To obtain such characterizations is nontrival and usually requires the theory of pseudo-differential operators, singular integrals or interpolation theory. To avoid unnecessary complications, we shall restrict our discussion to the case of the Laplace operator under Dirichlet boundary conditions.

Let $1 < p < \infty$ and let $\Omega \subset \mathbf{R}^n$ be a bounded, open, connected subset with smooth boundary $\partial\Omega$. Let $X = L^p(\Omega)$ be equipped with the usual norm

$$\|u\| = \left(\int_\Omega |u(x)|^p ds\right)^{1/p}$$

and let $A_p = -\Delta$ denote the negative Laplacian in $L^p(\Omega)$ with Dirichlet boundary conditions:

$$D(A_p) = W^{2,p}(\Omega) \cap W_0^{1,p}(\Omega).$$

By [23, Theorem 3.5, p. 214] we know that A_p is the infinitesimal generator of an analytic semigroup $\{T(t): t \geq 0\}$ in $L^p(\Omega)$ and the fractional power operator $A_p^{1/2}$ is well-defined. Following the notation of Section 2 we let $X_{1/2}$ denote the domain of $A_p^{1/2}$ equipped with the graph norm. In the special case when $p = 2$, J.A. Goldstein [12] gave a simple proof that $X_{1/2} = H_0^1(\Omega)$. When $p \neq 2$, the characterization of $X_{1/2}$ is more difficult. In the case when $X = L^p(\underline{R}^n)$, D. Henry [16, p. 77] proved that $X_{1/2} = W^{1,p}(\underline{R}^n)$. In the general case when Ω is bounded, the combined works of various authors are summarized in Treibel [25] to prove that

$$D((A_p + \lambda I)^{1/2}) = W_0^{1,p}(\Omega), \ \lambda \geqq 0 \tag{4.1}$$

with equivalent norms (see especially the proof of Theorem 4.9.2, p. 335). Taking $\lambda = 0$ gives that $X_{1/2} = W_0^{1,p}(\Omega)$. We shall use this result to prove the following factorization theorem. In the statement that follows, we let $D_j = \frac{\partial}{\partial x_j}$ denote the first order partial derivative with respect to x_j in the sense of distributions. Then D_j is well-defined everywhere on all $L^p(\Omega)$ spaces.

<u>THEOREM 4</u>. Let $1 < p < \infty$ and $\Omega \subset R^n$ be a bounded, open, connected subset with smooth boundary $\partial\Omega$. Let A_p denote the negative Laplacian in $L^p(\Omega)$ with Dirichlet boundary conditions. Then there is a bounded linear operator $B_j : L^p(\Omega) \to L^p(\Omega)$ such that $B_j(W^{1,p}(\Omega)) \subseteq W_0^{1,p}(\Omega)$ and

$$D_j = A_p^{1/2} B_j \text{ on } W^{1,p}(\Omega). \tag{4.2}$$

Proof. We let $\frac{1}{p} + \frac{1}{q} = 1$ and define $\tilde{B}_j$ on $L^q(\Omega)$ by

$$\tilde{B}_j = D_j \circ A_q^{-1/2}.$$

By (4.1), $A_q^{-1/2}$ is a continuous linear mapping from $L^q(\Omega)$ onto $W_0^{1,q}(\Omega)$. So $\tilde{B}_j$ is well-defined on all of $L^q(\Omega)$ and is a bounded linear operator from $L^q(\Omega)$ into itself. Thus $\tilde{B}_j$ has a well-defined adjoint $\tilde{B}_j^*$ which is a bounded linear operator from $L^p(\Omega)$ into itself. Let $\langle\ ,\ \rangle$ denote the natural pairing between $L^p(\Omega)$ and $L^q(\Omega)$:

$$\langle f,g\rangle = \int_\Omega f(x)g(x)dx,\ f \in L^p(\Omega),\ g \in L^q(\Omega).$$

$$\langle \tilde{B}_j^* f,g\rangle = \langle f,\tilde{B}_j g\rangle \text{ for all } f \in L^p(\Omega),\ g \in L^q(\Omega). \tag{4.3}$$

Now if $f \in C_0^\infty(\Omega)$, then the definition of D_j gives that

$$\langle f,\tilde{B}_j g\rangle = -\langle (A_q^{-1/2})^* D_j f,g\rangle \text{ for all } g \in L^q(\Omega). \tag{4.4}$$

So from (4.3) and (4.4) we obtain that

$$\tilde{B}_j^* f = (A_q^{-1/2})^* D_j f \tag{4.5}$$

for all $f \in C_0^\infty(\Omega)$. Taking limits we obtain (4.5) for all f in $W_0^{1,p}(\Omega)$. Now by [23, Lemma 3.4, p. 213] we have $A_q^* = A_p$ and by [28, Theorem 2, p. 225] we have $(R(\lambda;A_q))^* = R(\lambda;A_p)$ for all $\lambda \geqq 0$. So by [23, Formula (6.4), p. 69] we have $(A_q^{-1/2})^* = A_p^{-1/2}$ in $L^p(\Omega)$ and therefore by (4.5) we have

$$\tilde{B}_j^* f = -A_p^{-1/2} D_j f \text{ for all } f \in W_0^{1,p}(\Omega).$$

Hence $\tilde{B}_j^*: W_0^{1,p}(\Omega) \to W_0^{1,p}(\Omega)$ continuously and

$$A_p^{1/2}\tilde{B}_j^* f = -D_j f \text{ for all } f \in W_0^{1,p}(\Omega).$$

We set $B_j = -\tilde{B}_j^*$ and obtain

$$B_j f = A_p^{-1/2} D_j f \text{ for all } f \in W_0^{1,p}(\Omega).$$

But the operator $f \to A_p^{-1/2} D_j f$ is well-defined and bounded from $W^{1,p}(\Omega)$ into $W_0^{1,p}(\Omega)$. Therefore B_j represents the continuous linear extension of $A_p^{-1/2} D_j$ from $W^{1,p}(\Omega)$ to all of $L^p(\Omega)$. Q.E.D.

§5. WEAK SOLUTIONS

Let $\Omega \subset R^n$ be a bounded domain with smooth boundary $\partial\Omega$. Let $D_o = I$ be the identity operator and let $D_1 = \frac{\partial}{\partial x_1}, \ldots, D_n = \frac{\partial}{\partial x_n}$. Let $\Delta = \Sigma_{i=1}^n D_i^2$ denote the Laplacian and let $\nabla = (D_1, \ldots, D_n)$ denote the gradient operator. We consider the initial-boundary value problem:

$$\frac{\partial u}{\partial t} - \Delta u(t,x) = \int_0^t a(t,s) \sum_{i=0}^{n} D_i g_i(s,x,u(s,x),\nabla u(s,x))ds \tag{5.1}$$

$$+ f(t,x,u(t,x),\nabla u(t,x)),(t,x) \in (0,\infty) \times \Omega,$$

$$u(t,x) = 0 \text{ for } (t,x) \in (0,\infty) \times \partial\Omega , \tag{5.2}$$

$$u(0,x) = u_o(x) \text{ for } x \in \Omega. \tag{5.3}$$

We shall formulate (5.1), (5.2), (5.3) as an abstract Cauchy problem in an L^p-space and demonstrate the existence of weak solutions to this problem under mild smoothness assumptions on $g_0, \ldots, g_n$ and f.

We make the following assumptions:

(i) The kernel function a(t,s) satisfies (A5).

(ii) For each $i = 0,1,\ldots,n$ the functions $g_i:[0,\infty) \times \Omega \times R \times R^n \to R$ are continuous and for every $t_o > 0$, $r_o > 0$ there exists $M_o > 0$ such that if $|u| \leq r_o$, $|v| \leq r_o$ then

$$|g_i(t,x,u,\xi)-g_i(t,x,v,\eta)| \leq M_o(|u-v| + |\xi- \eta|)$$

for all $0 \leq t \leq t_o$, $x \in \Omega$, $\eta \in R^n$, R^n, $i = 0,1,\ldots,n$.

(iii) The function $f:[0,\infty) \times \Omega \times R \times R^n \to R$ is continuous and for every $t_o > 0$, $r_o > 0$ there exists $L_o > 0$ such that if $|u| \leq r_o$, $|v| \leq r_o$ then

$$|f(t,x,u,\xi) - f(t,x,v,\eta)| \leq L(|u-v| + |\xi-\eta|)$$

for all $0 \leq t \leq t_0$, $x \in \Omega$, $\xi \in R^n$, $\eta \in R^n$.

Let $\frac{n}{2} < p < \infty$ and put $X = L^p(\Omega)$ equipped with the usual norm. Define a linear operator $A = A_p$ in X by

$$\begin{cases} A\varphi = -\Delta\varphi \\ \varphi \in D(A) = W^{2,p}(\Omega) \cap W_0^{1,p}(\Omega). \end{cases} \tag{5.4}$$

The operator A satisfies assumptions (A1), (A2) of Section 2 and -A generates an analytic semigroup$\{T(t) : t \geq 0\}$ in X. We let $X_{1/2} = D(A^{1/2}) = W_0^{1,p}(\Omega)$ then by Theorem 4 there are bounded linear operators B_j on X such that $D_j = A^{1/2}B_j$ on $X_{1/2}$ for all $0 \leq j \leq n$. (When $j = 0$ we may take $B_0 = A^{-1/2}$.) We define operators $q_j(t,\varphi)$ on $[0,\infty) \times X_{1/2}$ by

$$q_j(t,\varphi) = B_j[g_j(t,\cdot,\varphi,\nabla\varphi)] \tag{5.5}$$

and we set

$$q(t,\varphi) = \sum_{j=0}^{n} q_j(t,\varphi). \tag{5.6}$$

By the Sobolev Imbedding Theorem we have $W_0^{1,p}(\Omega) \to C(\bar{\Omega})$ with embedding constant $K > 0$. If $\varphi \in W_0^{1,p}(\Omega)$ and if $\|\varphi\|_{1,p} \leq r_0^*$ then $|\varphi(x)| \leq Kr_0^*$ for all $x \in \Omega$. So by assumption (ii) with $r_0 = Kr_0^*$, it follows that given $t_0 > 0$ there exists $M_0 > 0$ such that if $\varphi,\psi \in W_0^{1,p}(\Omega)$ with $\|\varphi\|_{1,p} \leq r_0^*$, $\|\psi\|_{1,p} \leq r_0^*$ then

$$|g_j(t,x,\varphi(x),\nabla\varphi(x)) - g_j(t,x,\psi(x),\nabla\psi(x))|$$
$$\leq M_0(|\varphi(x)-\psi(x)| + |\nabla\varphi(x) - \nabla\psi(x)|)$$

for a.e., $x \in \Omega$, $0 \leq t \leq t_0$, $j = 0,1,\ldots,n$. So there is $M > 0$ such that

$$\|g_j(t,\varphi) - q_j(t,\psi)\|_p \leq M\|\varphi - \psi\|_{1,p},$$

if $\|\varphi\|_{1,p} \leq r_0^*$, $\|\psi_{1,p} \leq r_0^*$, $0 \leq t \leq t_0$, $j = 0,1,\ldots,n$. Using this last result, it is easy to see that q_j is a continuous mapping from $[0,\infty) \times X_{1/2}$ to X. So the operator $q(t,\varphi)$ defined by (5.6) satisfies assumption (A3).

Next we define an operator $g(t,\varphi)$ on $[0,\infty) \times X_{1/2}$ by

$$g(t,\varphi) = \sum_{i=0}^{n} D_i g_i(t,x,\varphi,\nabla\varphi) \tag{5.7}$$

in the sense of distributions. Then by (5.5), (5.6) we formally have the factorization

$$g(t,\varphi) = A^{1/2}q(t,\varphi). \tag{5.8}$$

Finally, we let $f(t,\varphi)$ denote the operator defined on $[0,\infty) \times X_{1/2}$ by

$$f(t,\varphi)(x) = f(t,x,\varphi(x), \nabla\varphi(x)) \text{ a.e. } x \in \Omega.$$

Using similar arguments as given above, the function f is a continuous mapping from $[0,\infty) \times X_{1/2}$ to X and satisfies assumption (A3).

Thus for each initial function $u_0 \in W_0^{1,p}(\Omega)$, the initial-boundary value problem (5.1), (5.2), (5.3) can be formulated as an abstract Cauchy problem

$$u'(t) + Au(t) = \int_0^t a(t,s)g(s,u(s))ds + f(t,u(t)), \quad t \geq 0 \tag{5.9}$$

$$u(0) = u_0, \tag{5.10}$$

in the Banach space $L^p(\Omega)$, $n/2 < p < \infty$.

<u>THEOREM 5</u>. For the initial-boundary value problem (5.1), (5.2), (5.3) let the hypotheses (i), (ii), (iii) be satisfied. Let $u_0 \in W_0^{1,p}(\Omega)$ be given, when $n/2 < p < \infty$. Then there exists a unique weak solution u(t) of (5.1), (5.2), (5.3) on a maximum interval of existence $[0,d)$, $0 < d \leq +\infty$. In addition, suppose the functions $g_0,g_1,\ldots,g_n$, f satisfy a sublinear growth condition of the following type:

(iv) For each $t_0 > 0$ there is a constant $K(t_0) > 0$ such that

$$|g_j(t,x,u,\xi)| \leq K(t_0)(|u| + |\xi|), \quad j = 0,1,\dots,n$$

$$|f(t,x,u,\xi)| \leq K(t_0)(|u| + |\xi|)$$

for all $(t,x,u,\xi) \in [0,t_0] \times \Omega \times R \times R^n$.

Then $d = +\infty$ and $u(t)$ is a globally defined weak solution of (5.1), (5.2), (5.3).

Proof. For the Cauchy problem (5.9), (5.10) the assumptions (A1) - (A5) are satisfied and therefore we obtain a unique weak solution $u(t)$ on a maximum interval of existence $[0,d)$ by Theorem 1 and Theorem 2. If, in addition, hypothesis (iv) holds, then it follows from (3.3) (with $t_0 = 0$) that

$$\|u(t)\|_{1/2} \leq C_1 + C_2 \int_0^t \{1 + (t-s)^{-1/2}\} \|u(s)\|_{1/2} ds, \quad 0 \leq t < t_0,$$

for suitable constants $C_1 > 0$, $C_2 > 0$ depending on u_0 and t_0. Using a singular Gronwall inequality [14, Lemma 7.1.1, p. 188] we obtain an estimate of the type (3.6). This implies that $d = +\infty$. Q.E.D.

§6. ASYMPTOTIC BEHAVIOUR

In this section we consider the asymptotic behaviour of strong solutions of the initial-boundary value problem

$$\frac{\partial u}{\partial t}(t,x) - \Delta u(t,x) + cu(t,x) = \int_0^t a(t-s) \sum_{i=0}^n D_i g_i(u(s,x),\nabla u(s,x))ds + f(u(t,x)), (t,x) \in (0,\infty) \times \Omega; \tag{6.1}$$

$$u(t,x) = 0 \text{ if } (t,x) \in (0,\infty) \times \partial\Omega ; \tag{6.2}$$

$$u(0,x) = u_0(x) \text{ if } x \in \Omega. \tag{6.3}$$

Under suitable hypotheses we shall prove that $u(t,x)$ converges uniformly on Ω as $t \to \infty$ to a weak solution $u(x)$ of the quasilinear boundary value problem

$$-\Delta u + cu = a \sum_{i=0}^{n} D_i g_i(u,\nabla u) + f(u) \text{ in } \Omega, \tag{6.4}$$

$$u = 0 \text{ on } \partial\Omega \tag{6.5}$$

where $a = \int_0^\infty a(s)ds$.

Assume that the dimension $n \geqq 2$ and let $\Omega \subset R^n$ be a bounded domain with smooth boundary $\partial\Omega$. Let λ_o denote the first eigenvalue of the Laplacian in Ω under Dirichlet boundary conditions. To study the asymptotic behaviour of strong solutions of (6.1), (6.2), (6.3) we must strengthen the smoothness assumptions on the functions $a, g_o, g_1, \ldots, g_n$, f that were made in Section 5. Our basic hypotheses are as follows:

(i)' The kernel function $a \in C^2([0,\infty))$ with $a(t) \geqq 0$, $a'(t) \leqq 0$ for all $t \geqq 0$. Furthermore, $a'(t) + \lambda_0 a(t) \leqq 0$ and $a''(t) + \lambda_0 a'(t) \geqq 0$ for all $t \geqq 0$.

(ii)' For each $i = 0,1,\ldots,n$ the functions $(u,\xi) \to g_i(u,\xi)$ are continuously differentiable from $R \times R^n$ into R with first partial derivatives $\frac{\partial g_i}{\partial u}$, $\frac{\partial g_i}{\partial \xi_k}$ Lipschitz continuous on bounded subsets of $R \times R^n$.

(iii)' There is a positive C^1 convex function $H : R^{n+1} \to R$ such that $\nabla H = (g_o, g_2, \ldots, g_n)$.

(iv)' There are positive constants C_1, C_2 such that

$$|g_i(u,\xi)| \leqq C_1(|u| + |\xi|) + C_2$$

for all $(u,\xi) \in R \times R^n$, $i = 0,1,\ldots,n$.

(v)' The function $f \in C^{1+\gamma}(R)$, where $0 < \gamma \leqq 1$, and there is a constant $C_3 > 0$ such that $|f'(u)| \leqq C_3$ for all $u \in R$.

A family of kernels $a(t)$ satisfying (i) can be written in the form

$$a(t) = Ce^{-\lambda_o t} + \int_0^t e^{-\lambda_o (t-s)} h(s)ds, \quad t \geqq 0,$$

where $C > 0$ is constant and $h \in C^1([0,\infty))$ satisfies $h(t) \leqq 0$, $h'(t) \geqq 0$ for all $t \geqq 0$, and

$$\int_0^\infty e^{\lambda_0 s} |h(s)| ds \leqq C.$$

We let $n < p < \infty$ and put $X = L^p(\Omega)$ with the usual norm

$$\|\cdot\| = \|\cdot\|_p.$$

We let A be defined in X by

$$A = -\Delta + cI, \; D(A) = W^{2,p}(\Omega) \cap W_0^{1,p}(\Omega).$$

Let $X_1 = D(A)$ equipped with the graph norm. We put $X_{1/2} = D(A^{1/2}) = W_0^{1,p}(\Omega)$ with equivalent norms and define operators $g(\varphi)$, $f(\varphi)$ on X_1, X respectively, by

$$g(\varphi)(x) = \sum_{i=0}^{n} D_i g_i(\varphi(x), \nabla\varphi(x)), \text{ a.e. } x \in \Omega,$$

$$f(\varphi)(x) = f(\varphi(x)), \text{ a.e. } x \in \Omega.$$

Then the initial-boundary value problem (6.1), (6.2), (6.3) can be reformulated as an abstract Cauchy problem in $L^p(\Omega)$:

$$u'(t) + Au(t) = \int_0^t a(t-s)g(u(s))ds + f(u(t)), \quad t \geqq 0, \tag{6.6}$$

$$u(0) = u_o. \tag{6.7}$$

It is not difficult to show (using the Sobolev Imbedding Theorem) that given any $r_o > 0$ there are constants $L_o > 0$, $M_0 > 0$ such that

$$\|g(\varphi) - g(\psi)\| \leqq M_0\|\varphi - \psi\|_1$$

$$\|f(\varphi') - f(\psi')\| \leqq L_0\|\varphi' - \psi'\|,$$

for all $\varphi, \psi \in X_1$, and $\varphi', \psi' \in X$ which satisfy $\|\varphi\|_1 \leq r_0$, $\|\psi\|_1 \leq r_0$, $\|\varphi'\| \leq r_0$, $\|\psi'\| \leq r_0$. Consequently, $g : X_1 \to X$ and $f : X \to X$ are continuous mappings. Furthermore, f has a Fréchet derivative $f'(\varphi)$, defined for all $\varphi \in X$, which is locally Hölder continuous from $X_{1/2}$ to $\mathcal{L}(X)$. That is, given $r_0 > 0$ there is a constant $C_0 > 0$ such that if $\varphi, \psi \in X_{1/2}$ with $\|\varphi\|_{1/2} \leq r_0$, $\|\psi\|_{1/2} \leq r_0$ then

$$\|f'(\varphi) - f'(\psi)\|_{\mathcal{L}(x)} \leq C_0\|\varphi - \psi\|^{\gamma}_{1/2}.$$

Here $\mathcal{L}(X)$ denotes the space of all bounded linear operators on X.

As a consequence of the above remarks, we may apply the results of Heard-Rankin [15, Theorem 2 and Theorem 4] to (6.6), (6.7) and conclude that for each $u_0 \in X$, there exists a <u>strong solution</u> u(t) on a maximal interval of existence $[0,d)$, $0 < d \leq +\infty$. Furthermore, this solution is twice strongly continuously differentiable on each subinterval $(0,T]$, $0 < T < d$, in the X-norm and satisfies the differentiated integrodifferential equation

$$u''(t) + Au'(t) = \int_0^t a'(t-s)g(s(s))ds + a(0)g(u(t)) \qquad (6.8)$$
$$+ f'(u(t))u'(t).$$

To demonstrate that u(t) is also a weak solution of (6.6), (6.7), we shall assume that $c \geqq 0$ in (6.1) and use Theorem 4 to write $D_i = A^{1/2}B_i$ where $B_i : X \to X$ is a bounded linear operator for $i = 0,1,\ldots,n$ (Theorem 4 was proved for $-\Delta$ but it also holds for the perturbed operator $-\Delta + cI$). Now define

$$q(\varphi) = \sum_{i=0}^{n} G_i[g_i(\varphi,\nabla\varphi)], \quad \varphi \in X_{1/2}. \qquad (6.9)$$

It is easy to see that $q(\varphi)$ satisfies assumption (A3) and

$$g(\varphi) = A^{1/2}q(\varphi) \text{ for all } \varphi \in X_1. \qquad (6.10)$$

So by Theorem 3 it follows that u(t) is also a weak solution of (6.6), (6.7). We summarize the above results in the following theorem.

THEOREM 6. Let hypotheses (i)' - (v)' be satisfied and let $u_0 \in D(A)$. Then there exists a unique globally defined strong solution u(t) of (6.6), (6.7) on $[0,\infty)$. Furthermore, u(t) is twice strongly continuously differentiable on $[0,\infty)$ in the X-norm and satisfies the differential integrodifferential equation (6.8) on $(0,\infty)$.

Proof. The uniqueness part follows from Theorem 1 since u(t) is a weak solution. The fact that u(t) is globally defined on $[0,\infty)$ follows from Theorem 5 since hypotheses (iv)' and (v)' imply that $g_i(u,\xi)$ and f(u) satisfy sublinear growth conditions. Q.E.D.

A central part of the study of asymptotic behaviour of solutions of (6.1), (6.2), (6.3) in Engler's energy inequality which he proved in [8, p. 17] for the case of a single function. A generalized version of Engler's result was proved in [15, Lemma 3] for the case of n+1 functions. We quote this result below.

LEMMA 1. Let $a,b \in C^1([0,\infty))$ and let continuous real-valued functions $g_0,g_1,\ldots,g_n \in C(\underline{R}^{n+1})$ be given such that:

(1) there exists a C^1 convex function $H : \underline{R}^{n+1} \to \underline{R}$ such that $\nabla H = (g_0,g_1,\ldots,g_n)$,

(2) $a(t) \geqq 0$, $a'(t) \leqq 0$ for all $t \geqq 0$;

(3) $b(t) \geqq 0$ for all $t \geqq 0$;

(4) for each $\tau \geqq 0$ the map $t \to b(t+\tau)a'(t)$ is nondecreasing.

Then for every set $v_0,v_1,\ldots,v_n \in C^1([0,\infty))$ we have

$$\sum_{i=0}^{n} \int_0^t b(s)v_i'(s) \{\frac{d}{ds} \int_0^s a(s-\tau)g_i(v_0(\tau),\ldots,v_n(\tau))d\tau\}ds$$

$$\geqq a(t)b(t)H(v_0(t),\ldots,v_n(t))-a(0)b(0)H(v_0(0),\ldots,v_n(0))$$

$$- \int_0^t [\frac{d}{ds}(a(s)b(s))]H(v_0(s),\ldots,v_n(s))ds.$$

Our first result gives the basic L^2-estimate for solutions (6.1), (6.2), (6.3).

LEMMA 2. Let hypotheses (i)' - (v)' be satisfied and let $u_o \in D(A)$. Let $u(t,x)$ be the corresponding global strong solution of (6.1), (6.2), (6.3) with maximal domain $[0,\infty) \times \Omega$. Let $b(t) = e^{\lambda_o t}$ for $t \geq 0$. Then for $c > 0$ sufficiently large there is a constant $C(u_o) > 0$, depending only on u_o, such that

$$\int_\Omega b(t)|u_t(t,x)|^2 dx \leq C(u_o), \tag{6.11}$$

$$\int_0^t b(s) \int_\Omega |\nabla u_t(s,x)|^2 dxds \leq C(u_o), \tag{6.12}$$

for all $0 \leq t < +\infty$.

Proof. We differentiate (6.1) with respect to t, multiply by $b(t)u_t(t,x)$ and integrate over $[0,T] \times \Omega$ (we use the summation convention for repeated indices):

$$\begin{aligned}&\int_{[0,T]\times\Omega} b(t)u_t(t,x)u_{tt}(t,x)dxdt - \int_{[0,T]\times\Omega} b(t)u_t(t,x)\Delta u_t(t,x)dsdt \\ &+ c\int_{[0,T]\times\Omega} b(t)u_t^2(t,x)dxdt \\ &= \int_{[0,T]\times\Omega} b(t)u_t(t,x)\ (\frac{\partial}{\partial t}\int_0^t a(t-s)D_i g_i(u(s,x),\nabla u(s,x))ds)dxdt \\ &+ \int_{[0,T]\times\Omega} b(t)f'(u(t,x))u_t^2(t,x)dxdt.\end{aligned} \tag{6.13}$$

By Lemma 1 with $v_o = u$, $v_i = D_i u$ we obtain

$$\begin{aligned}\int_0^T b(t)D_i u_t(t,x)(\frac{\partial}{\partial t}\int_0^t a(t-s)g_i(u(s,x), \nabla u(s,x))ds)dt \\ \geq -a(0)H(u_o(x), \nabla u_o(x)),\ x \in \Omega.\end{aligned} \tag{6.14}$$

By Poincaré's inequality, we have

$$\begin{aligned}\int_0^T b(t)\int_\Omega |\nabla u_t(t,x)|^2 dxdt &\leq \lambda_0 \int_0^T b(t)\int_\Omega |u_t(t,x)|^2 dxdt \\ &= \int_0^T b'(t)\int_\Omega |u_t(t,x)|^2 dxdt.\end{aligned} \tag{6.15}$$

By assumption (v)', we have

$$\int_0^T b(t) \int_\Omega f'(u(t,x))u_t^2(t,x)dxdt \leq C_3 \int_0^T b(t) \int_\Omega |u_t^2(t,x)|dxdt. \quad (6.16)$$

Now choose $c > C_3$, then after integrating by parts in (6.13) and using (6.14), (6.15), (6.16) we have

$$\frac{1}{2}\int_\Omega b(T)|u_t(T,x)|^2dx + \frac{1}{2}\int_0^T b'(t) \int_\Omega |u_t(t,x)|^2dxdt \quad (6.17)$$

$$\leq a(0) \int_\Omega H(u_o(x), \nabla u_o(x))ds + \frac{1}{2}\int_\Omega b(0)|u_t(0,x)|^2 dx.$$

From the equation (6.1), we have

$$\int_\Omega |u_t(0,x)|^2dx = \int_\Omega |\Delta u_o(x)-cu_o(x) + f(u_o(x))|^2dx. \quad (6.18)$$

So it follows from (6.17), (6.18) that there is a constant $C(u_o) > 0$ such that

$$\frac{1}{2}\int_\Omega b(T)(u_t(T,x)|^2dx \leq C(u_o), \quad (6.19)$$

$$\frac{\lambda_o}{2} \int_0^T b(t) \int |u_t(t,x)|^2dxdt \leq C(u_o). \quad (6.20)$$

Using (6.13) and integrating by parts again, we have

$$\int_0^T b(t) \int_\Omega |\nabla u_t(t,x)|^2dxdt \leq \frac{1}{2}\int_0^T |u_t(0,x)|^2dx \quad (6.21)$$

$$+ \frac{1}{2}\int_0^T b'(t) \int_\Omega |u_t(t,x)|^2dxdt + a(0) \int_\Omega H(u_o(x), \nabla u_o(x))dx.$$

From (6.18), (6.20) we obtain

$$\int_0^T b(t) \int_\Omega |\nabla u_t(t,x)|^2dxdt \leq C(u_o).$$

This proves the lemma.

Q.E.D.

THEOREM 7. Let hypotheses (i)' - (v)' be satisfied and let $u_o \in D(A)$. Let $u(t,x)$ be the corresponding global strong solution of (6.1), (6.2), (6.3). Then $u(t,x)$ converges uniformly on Ω as $t \to \infty$ to a weak solution $u(x)$ of the quasilinear boundary value problem (6.4), (6.5).

Proof. Let $\|\cdot\|_2$ denote the L^2-norm on Ω and assume that $c > C_3$. Then by (6.11) we have

$$\int_0^t \|u_t(s)\|_2 ds \leqq \frac{2}{\lambda_o} \sqrt{C(u_o)}, \quad 0 \leqq t < \infty.$$

Likewise, by (6.12) and the Cauchy-Schwarz inequality, we have

$$\int_0^t \|\nabla u_t(s)\|_2 ds \leqq \left(\int_0^t \frac{1}{b(s)} ds\right)^{1/2} \left(\int_0^t b(s) \|\nabla u_t(s)\|_2^2 ds\right)^{1/2}$$

$$\leqq \left(\frac{C(u_o)}{\lambda_o}\right)^{1/2}, \qquad 0 \leqq t < \infty.$$

Combining these we obtain

$$\sup_{0 \leqq t < \infty} \|u(t)\|_{1,2} \leqq K(u_o) \tag{6.22}$$

for some constant $K(u_o) > 0$ depending only on u_o.

Next we obtain an L^p-estimate on $A^{1/2}u(t)$. Let $T(t)$ denote the analytic semigroup with infinitesimal generator $-A$. Then there is a constant $C_4 > 0$ such that

$$\|T(t)\| \leqq C_4 e^{-ct}, \quad t \geqq 0, \tag{6.23}$$

$$\|A^{1/2}T(t)\| \leqq C_r \frac{e^{-ct}}{\sqrt{t}}, \quad t > 0. \tag{6.24}$$

From hypotheses (iv)' and (v)' there are constants $C_1 > 0$, $C_2 > 0$ such that

$$\|q(u(t))\|_p \leqq C_1 \|A^{1/2}u(t)\|_p + C_2, \tag{6.25}$$

$$\|f(u(t))\|_p \leqq C_1 \|A^{1/2}u(t)\|_p + C_2 \tag{6.26}$$

for all $0 \leq t < \infty$. From equation (3.11) (with $a(t,s) = a(t-s)$) we have

$$\|A^{1/2}u(t)\|_p \leq C_4 e^{-ct} \|A^{1/2}u_o\|_p + \int_0^t a(t-s) \|q(u(s))\|_p \, ds \tag{6.27}$$
$$+ C_4 \int_0^t (t-s)^{-1/2} e^{-c(t-s)} \| f(u(s))\|_p ds$$
$$+ a(0)C_4 \int_0^t e^{-c(t-s)} \|g(u(s))\|_p ds$$
$$+ \left\|\int_0^t T(t-s) \int_0^s a'(s-\tau)q(u(\tau))d\tau ds\right\|_p, \quad 0 \leq t < \infty.$$

By Fubini's Theorem

$$\int_0^t T(t-s) \int_0^s a'(s-\tau)q(u(\tau))d\tau ds = \int_0^t \hat{T}(t-\tau)q(u(\tau)d\tau$$

where

$$\hat{T}(t) = \int_0^t a'(t-\sigma)T(\sigma)d\sigma.$$

By (6.23) - (6.27) we have therefore

$$\|A^{1/2}u(t)\|_p \leq C_4 e^{-ct} \|A^{1/2}u_o\|_p + C_2 \int_0^t a(s)ds \tag{6.28}$$
$$+ C_2 \int_s^t s^{-1/2} e^{-cs} ds + a(0)C_2C_4 \int_0^t e^{-cs} ds$$
$$+ C_2 \int_0^t \|\hat{T}(s)\| \, ds + \int_0^t \{c_1 a(t-s)$$
$$+ C_1 \frac{e^{-c(t-s)}}{\sqrt{t - s}} + a(0)C_1C_4 e^{-c(t-s)}$$
$$+ C_1 \|\hat{T}(t-s)\| \} \|A^{1/2}u(s)\|_p ds.$$

We let

$$M = \max \{C_1, C_2, C_4, a(0)C_4C_1, a(0)C_4C_2\},$$

$$\beta(t) = e^{ct} \|A^{1/2}u_o\|_p + \int_0^t a(s)ds + \int_0^t s^{-1/2}s^{-cs}ds + \int_0^t e^{-cs}ds + $$
$$+ \int_0^t \|\hat{T}(s)\| ds,$$
$$\alpha(t) = a(t) + t^{-1/2}e^{-ct} + e^{-ct} + \|\hat{T}(t)\|,$$

for $0 \leqq t < \infty$. Then by (6.28) we have

$$\|A^{1/2}u(t)\|_p \leqq M \{\beta(t) + \int_0^t \alpha(t-s) \|A^{1/2}u(s)\|_p ds\}, \quad 0 \leqq t < \infty. \tag{6.29}$$

From hypothesis (i)' we have $0 \leqq a(t) \leqq a(0)e^{-\lambda_o t}$ and $0 \leqq |a'(t)| \leqq |a'(0)|e^{-\lambda_o t}$ for all $t \geqq 0$. By Young's inequality it follows that

$$\int_0^\infty \|\hat{T}(t)\| dt \leqq \int_0^\infty |a'(\sigma)\| d\sigma < \infty.$$

Therefore $\beta \in L^\infty([0,\infty))$ and $\alpha \in L^1([0,\infty))$. Put

$$\alpha_M(t) = M\alpha(t), \; \beta_M(t) = M\beta(t),$$

and define

$$\psi(t) = \|A^{1/2}u(t)\|_p - \int_0^t \alpha_M(t-s) \|A^{1/2}u(s)\|_p ds,$$

for $0 \leqq t < \infty$. Then

$$\|A^{1/2}u(t)\|_p = \psi(t) + \int_0^t \alpha_M(t-s) \|A^{1/2}u(s)\|_p ds \tag{6.30}$$

and by (6.29) we have

$$\psi(t) \leqq \beta_M(t) \tag{6.31}$$

for all $0 \leqq t < \infty$. Let $r_M(t)$ be the resolvent operator corresponding to $\alpha_M(t)$. Then from (6.30) we have

$$\|A^{1/2}u(t)\|_p = \psi(t) - \int_0^t r_M(t-s)\psi(s)ds$$
$$= \psi(t) + \int_0^t |r_M(t-s)|\psi(s)ds, \quad 0 \leq t < \infty,$$

since $\alpha_M \geq 0$ implies $r_M \leq 0$ by construction ([20, Theorem 2.3, p. 197]). So from (6.31) we have

$$\|A^{1/2}u(t)\|_p \leq \beta_M(t) + \int_0^t |r_M(t-s)|\beta_M(s)ds, \quad 0 \leq t < \infty. \tag{6.32}$$

From [20, Corollary 4.3, p. 407] we have that $r_M \in L^1([0,\infty))$ if and only if

$$\int_0 \alpha_M(t)e^{-iyt}dt \neq 1 \text{ for all } y \in \underline{R}. \tag{6.33}$$

Now (6.33) is equivalent to

$$\int_0^\infty \alpha(t)e^{-iyt}dt \neq \frac{1}{M} \text{ for all } y \in \underline{R}. \tag{6.34}$$

To achieve this condition, we observe that M can be arbitrarily large without changing (6.29). So we only need to verify (6.34) for a sufficiently large M. If this were not possible, then there would exist a constant $M_0 > 0$ such that if $M > M_0$ then

$$\int_0^\infty \alpha(t)e^{-iyt}dt = \frac{1}{M} \text{ for some } y \in \underline{R}. \tag{6.35}$$

So if we define a function F by

$$F(y) = \int_0^\infty \alpha(t)e^{-iyt}dt, \quad y \in \underline{R}.$$

then (6.35) implies that the open interval $(0,1/M_0)$ belongs to the range of F. By continuity, this implies that

$$\text{Im } F(y) = 0 \text{ for all } a < y < b, \tag{6.36}$$

for suitable constants a and b. We now show that (6.36) implies that $\text{Im } F(y) = 0$ for all $y \in \underline{R}$. Indeed, consider the function

$$\Phi(z) = \int_0^\infty \alpha(t) \sin zt \, dt, \quad z = y + ir.$$

From the definition of $\alpha(t)$ it is easy to see that there exist constants $\eta > 0$, $K > 0$ such that $|\alpha(t)| \leq Ke^{-\eta t}$ for all $t \geq 0$. It follows that $\Phi(z)$ is analytic in the open strip $S = \{z = y + ir : |r| < \eta\}$. Since $\Phi(y) = 0$ for all $a < y < b$ by (6.36), we must have $\Phi(z) \equiv 0$ in S.

Next, we define a function G by

$$G(z) = \int_0^\infty \alpha(t) e^{-izt} dt, \quad z = y + ir.$$

Then G(z) is well-defined and analytic in the half plane $\Pi = \{z = y + ir : -\infty < \tau < \eta\}$. We also have the estimate $|G(z)| \leq K/\eta$ for all $r \leq 0$. Hence G(z) is bounded in the lower half plane $\{z = y + ir : \alpha \leq 0\}$ and is real on the real axis. By the Schwarz Reflection Principle we can extend G to an entire function defined everywhere which is also bounded. Hence G(z) must be constant and this contradicts (6.35). So (6.34) must hold for a sufficiently large constant M.

Consequently, we have $r_M \in L^1([0,\infty))$ and by (6.32) it follows that

$$u \in L^\infty([0,\infty); W_0^{1,p}(\Omega)). \tag{6.37}$$

By the Nirenberg-Gagliardo inequalities [9, Theorem I.10.1] we have

$$\|u\|_\infty \leq C \|u\|_{1,p}^{a} \|u\|_{r}^{1-a} \text{ for all } u \in W^{1,p}(\Omega) \cap L^r(\Omega)$$

where $n < p < \infty$, $1 < r < \infty$ and $a = \frac{np}{pr-nr+np}$. Now suppose that $2 < r < \frac{2n}{n-2}$, then $W^{1,2}(\Omega) \to L^r(\Omega)$ by the Sobolev Imbedding Theorem and we have

$$\|u\|_\infty \leq C \|u\|_{1,p}^{a} \|u\|_{1,2}^{1-a} \quad \text{for all } u \in W^{1,p}(\Omega).$$

Applying the estimate (6.37) it follows that

$$\|u(t)-u(s)\|_\infty \leq C \|u(t)-u(s)\|_{1,2}^{1-a} \text{ for all } 0 \leq s, t < \infty. \tag{6.38}$$

By (6.11), (6.12) we have

$$\|u(t) - u(s)\|_{1,2} \leq \int_s^t (\|u_t(\tau)\|_2 + \|\nabla u_t(\tau)\|_2)d\tau$$

$$\leq \sqrt{C(u_o)} \{\int_s^t e^{-\lambda_o r/2} d + (\int_s^t e^{-\lambda_o r} d\tau)^{1/2}\}$$

$$\to 0 \text{ as } s,t \to +\infty.$$

It follows from (6.38) that there is a function $u \in L^\infty(\Omega) \cap W_0^{1,2}(\Omega)$ such that

$$\lim_{t\to\infty} \|u(t) - u\|_\infty = 0, \quad \lim_{t\to\infty} \|u(t)-u\|_{1,2} = 0. \tag{6.39}$$

We multiply (6.1) by a test function $\varphi \in C_0^\infty(\Omega)$ and integrate by parts. We obtain:

$$\frac{d}{dt}\int_\Omega u(t,x)\varphi(x)dx + \int_\Omega \nabla u(t,x) \cdot \nabla\varphi(x)dx + c\int_\Omega u(t,x)\varphi(x)dx$$

$$= -\int_0^t a(t-s) \sum_{i=0}^n \int_\Omega g_i(u(s,x),\nabla u(s,x))D_i\varphi(x)dx + \int_\Omega f(u(t,x))\varphi(x)dx.$$

From (6.39) we have

$$\lim_{t\to\infty}\int_\Omega u(t,x)\varphi(x)dx = \int_\Omega u(x)\varphi(x)dx,$$

$$\lim_{t\to\infty}\int_\Omega \nabla u_t(t,x) \cdot \nabla\varphi(x)dx = \int_\Omega \nabla u(x) \cdot \nabla\varphi(x)dx,$$

$$\lim_{t\to\infty}\int_\Omega f(u(t,x))\varphi(x)dx = \int_\Omega f(u(x))\varphi(x)dx.$$

Because of the growth conditions in hypothesis (iv)' the operators

$$(\varphi_o,\varphi_1,\ldots,\varphi_n) \to g_i(\varphi_o,\varphi_1,\ldots,\varphi_n)$$

are bounded and continuous from the product space $\Pi_{i=0}^n L^2(\Omega)$ into $L^2(\Omega)$ ([17, Chapter 1, Theorem 2.1). Therefore

$$\lim_{t\to\infty} \|g_i(u(t), \nabla u(t)) - g_i(u,\nabla u)\|_2 = 0, \quad 0 \leq i \leq n.$$

Thus letting $t \to \infty$ in (6.40) gives

$$\int_\Omega \nabla u(x) \cdot \nabla\phi(x)dx + c\int_\Omega u(x)\varphi(x)dx$$
$$= -\int_0^\infty a(s)ds \sum_{i=0}^n \int_\Omega g_i(u(x),\nabla u(x))D_i\varphi(x)dx$$
$$+ \int_\Omega f(u(x))\varphi(x)dx.$$

Hence u is a weak solution of the quasilinear boundary value problem (6.4), (6.5). Q.E.D.

References

[1] V. Barbu and M.A. Malik. Semilinear equations in Hilbert spaces. J. Math. Anal. Appl. 67 (1979), 452-475.

[2] Ph. Clément and J.A. Nohel. Asymptotic Behavior of Solutions of Nonlinear Volterra Equations with completely positive kernels. SIAM J. Math. Anal. 12, no. 4, (1981), 514-535.

[3] Ph. Clément, R.C. MacCamy and J.A. Nohel. Asympotitic Properties of Solutions of nonlinear abstract Volterra equations. J. Integral Eqs. 3, (1981), 185-216.

[4] M.G. Crandall, S.-O. Londen and J.A. Nohel. An abstract nonlinear Volterra integrodifferential equation. J. Math. Anal. Appl. 64, (1978), 701-735.

[5] G. Da Prato and M. Iannelli. Linear integro-differential equations in Banach spaces. Rend. Sem. Math. Univ. Padova. 62, (1980), 207-219.

[6] G. Da Prato and M. Iannelli. Existence and regularity for a class of integrodifferential equations of parabolic type. Universita Degli Studi di Trento. Departimento di Matematica U.T.M. 138 - Marzo 1984.

[7] G. Da Prato, M. Iannelli and E. Sinestrari. Regularity of solutions of a class of linear integrodifferential equations in Banach spaces. J. Integral Eqs. 8 (1985), 27-40.

[8] H. Engler. Stabilization of solutions for a class of parabolic integro-differential equations. MRC Technical Report, 2555, Madison, Wisconsin 1983.

[9] H. Engler. On Some parabolic integrodifferential equations: existence and asymptotics of solutions. Lecture Notes in Mathematics. 1017, Springer-Verlag, New York, (1983), 161-167.

[10] W.E. Fitzgibbon. Semilinear integrodifferential equations in Banach space. Nonlinear Analysis, 4, (1980), 745-760.

[11] A. Friedman. Partial Differential Equations. Holt Rinehart and Winston, New York, 1969.

[12] J.A. Goldstein. Semigroups and Second-Order Differential Equations. J. Functional Anal. 4, (1969), 50-60.

[13] R. Grimmer and W. Schappacher. Weak solutions of integrodifferential equations and resolvent operators. J. Integral Eqs. 6, no. 3, (1984), 205-229.

[14] M.L. Heard. An abstract parabolic Volterra integrodifferential equations. SIAM J. Math. Anal. 13, (1982), 81-105.

[15] M.L. Heard and S.M. Rankin, III. A semilinear parabolic Volterra integrodifferential Equation. Submitted to Journal of Differential Equations.

[16] D. Henry. Geometric Theory of Semilinear Parabolic Equations. Lecture Notes in Mathematics. 840, Springer-Verlag, New York, 1981.

[17] M.A. Krasnosel'ski. Topological methods in the theory of nonlinear integral equations. International Series of Monographs of Pure and Applied Mathematics. 45, MacMillan Co., New York, 1964.

[18] S.O. Londen and J.A. Nohel. Nonlinear Volterra Integrodifferential equation occurring in heat flow. J. Integral Eqs. 6, no. 1, (1984), 11-50.

[19] A. Lunardi and E. Sinestrari. Fully nonlinear integrodifferential equations in general Banach spaces. Math. Zeitshift, 190, (1985), 225-248.

[20] R.K. Miller. Nonlinear Volterra Integral Equations. Benjamin Press, New York, 1971.

[21] J. Milota and H. Retzeltová. An existence theorem for semilinear functional parabolic equations. Casopis pro pestovani matematiky. 110, (1985), no. 3, 274-288.

[22] J.A. Nohel. Nonlinear Volterra equations for heat flow in materials with memory. Integral and Functional Differential Equations. Edited by Herdman, Rankin and Stech. Lecture Notes in Pure and Applied Mathematics. 67, Marcel Dekker, Inc., New York (1981), 3-82.

[23] A. Pazy. Semigroups of linear operators and applications to partial differential equations. Applied Mathematical Sciences. 44, Springer-Verlag, New York, 1983.

[24] E. Sinestrari. Continuous interpolation spaces and spatial regularity in nonlinear Volterra integrodifferential equations. J. Integral Eqs. 45 (1983), 287-308.

[25] H. Tribel. Interpolation theory, Function Spaces, Differential Operators. North-Holland Publishing Company, New York, 1978.

[26] G.F. Webb. An abstract second order semilinear Volterra integro-differential equation. Proc. Amer. Math. Soc. 69, (1978), 225-260.

[27] G.F. Webb. Abstract Volterra integrodifferential equations and a class of reaction-diffusion equations. Lecture Notes in Mathematics. 737, Springer-Verlag, New York, (1979), 295-303.

[28] K. Yosida. Functional Analysis. (6th edition), Springer-Verlag, New York, (1980).

Melvin L. Heard
Department of Mathematics,
Statistics, and Computer Science
University of Illinois at Chicago
Chicago
Illinois 60680
USA.

Samuel M. Rankin, III
Department of Mathematics
West Virginia University
Morgantown
West Virginia 26506
USA.

W.J. HRUSA & M. RENARDY

A model problem in one-dimensional viscoelasticity with a singular kernel

ABSTRACT

We discuss the existence of solutions to initial value problems for a class of model equations in one-dimensional viscoelasticity. The proofs are based on the energy method. The results allow the memory function to have a singularity.

1. INTRODUCTION

The equation of motion for a deforming continuum reads

$$\rho u_{tt} = \operatorname{div} \underline{S} + f, \tag{1}$$

where u denotes the displacement vector, ρ the reference density, $\underline{S}$ the Piola-Kirchhoff stress tensor and f is a given body force. The properties of the material composing the continuum are embodied in a constitutive equation, which relates $\underline{S}$ to the motion. For viscoelastic materials, $\underline{S}$ depends on the history of the deformation gradient; for a popular class of models (see [1], [8]) $\underline{S}$ is given by an integral expression of the form

$$\underline{S}(x,t) = \int_{-\infty}^{t} m(t-\tau)\underline{H}(\nabla u(x,t),\nabla u(x,\tau))d\tau. \tag{2}$$

In this paper, we shall be concerned with one-dimensional motions, in which the equation of motion corresponding to the constitutive relation (2) reduces to

$$\rho u_{tt}(x,t) = \int_{-\infty}^{t} m(t-\tau)h(u_x(x,t),u_x(x,\tau))_x \, d\tau + f(x,t). \tag{3}$$

This equation is to be solved together with appropriate boundary and initial conditions. In this paper, we shall limit our attention to Dirichlet boundary conditions; the problem is posed for $x \in [0,1]$ and we require that

$$u(0,t) = u(1,t) = 0, \quad t > 0. \tag{4}$$

The initial conditions are

$$u(x,t) = v(x,t),\ x \in [0,1],\ t < 0;\ u(x,0+) = u_0(x),$$
$$u_t(x,0^+) = u_1(x),\ x \in [0,1], \tag{5}$$

where u_0 and u_1 are not necessarily compatible with v.

It is usual to assume that $m > 0$, $m' < 0$ and that $h_{,1}$ (the derivative of h with respect to the first argument) is positive. Under these assumptions, plus appropriate smoothness and compatibility assumptions for the data, local existence of a smooth solution was proved by Dafermos and Nohel [2] for a special case of (3); the arguments can easily be extended to the general case and also to three dimensions (see [11]). (Under further assumptions, it is also possible to prove global existence in time for small data; we shall not pursue this issue here). The local existence argument in [2] is based on the iteration

$$\rho u_{tt}^{n+1}(x,t) = u_{xx}^{n+1}(x,t) \cdot \int_{-\infty}^{t} m(t-\tau)h_{,1}(u_x^n(x,t),u_x^n(x,\tau))d\tau$$
$$+ \int_{-\infty}^{t} m(t-\tau)h_{,2}(u_x^n(x,t),u_x^n(x,\tau))u_{xx}^n(x,\tau)d\tau + f(x,t). \tag{6}$$

At each step of this iteration, one must thus solve a linear wave equation with variable coefficients.

This method requires that m must be sufficiently regular. There are some indications, however, both theoretical and experimental, that many visco-elastic materials can be described by memory functions which are singular at 0. This case was recently investigated by the authors [5], [7]. In what follows, we give a summary of the results and briefly sketch the idea of the proofs. For details, we refer to [5] and [7].

The existence theorems stated below are local in time; in [5], we have also proved a global existence result for small data. What happens for large data remains an interesting open question. If m is smooth, then it is well known that singularities can develop. Such results were proved by a number of authors; the most general results are in [3] and [11]. Singular kernels have been shown to lead to smoothing of discontinuities in linear viscoelasticity [6], [9], [10]. On the other hand, Greenberg's [4] results on the existence of steady shock waves still apply when m has an integrable singularity [11]. This latter result makes it appear possible that despite the smoothing of singularities in linear viscoelasticity, one might still have development of shocks in the nonlinear situation.

2. STATEMENT OF RESULTS

We shall state our results for a special case; a generalization will be discussed later. Without loss of generality, we shall henceforth assume that $\rho = 1$. We consider an equation of motion of the form

$$u_{tt}(x,t) = \phi(u_x(x,t))_x - \int_{-\infty}^{t} m(t-\tau)\psi(u_x(x,\tau))_x \, d\tau + f(x,t). \tag{7}$$

With the notations

$$\chi = \phi - \psi \cdot \int_0^\infty m(s)\, ds, \quad a(t) = \int_t^\infty m(s) ds \tag{8}$$

we can write (7) in the following alternative form

$$u_{tt} = \chi(u_x)_x + \int_{-\infty}^{t} a(t-\tau)\psi(u_x(x,\tau))_{x\tau} \, d\tau + f. \tag{9}$$

We are interested in the case when $m(0) = \infty$. The first theorem given below covers the case when the singularity in m is integrable. In the form (9), the equation of motion is also meaningful when only a is integrable, but $m = -a'$ is not. This case is addressed in the second theorem.

Let us consider equation (7) with the Dirichlet boundary conditions (4) and initial conditions (5). We make the following assumptions:

(s1) $\phi, \psi \in C^3(\mathbb{R})$.

(s2) For some $T > 0$, $f, f_x, f_t \in L^\infty((-\infty,T);L^2(0,1)) \cap L^2((-\infty,T);L^2(0,1))$, $f_{tt} \in L^2((-\infty,T);\ L^2(0,1))$.

(s3) v and all its derivatives up to third order belong to $L^\infty((-\infty,0);\ L^2(0,1)) \cap L^2((-\infty,0);\ L^2(0,1))$. Moreover, $v_{xxtt} \in L^2((-\infty,0);\ L^2(0,1))$.

(m) $a, m \in L^1(0,\infty)$, $a \geq 0$, $m \geq 0$, $m' \leq 0$ (in the sense of distributions), and the measure m' has a nontrivial absolutely continuous component.

(c) The history v satisfies the differential equation and boundary conditions, and $u_0(x) = v(x,0-)$, $u_1(x) = v_t(x,0-)$ for $x \in [0,1]$.

(e) $\psi' > 0$, $\phi' > 0$.

THEOREM 1. Assume that (s1)-(s3), (m), (c) and (e) hold. Then there is a solution u of equation (7), with boundary conditions (4) and initial conditions (5), which exists on a maximal interval $(-\infty, T_0)$, $0 < T_0 \leq T$. The solution u and all its derivatives up to third order lie in $L^\infty_{loc}((-\infty,T_0); L^2(0,1))$.

REMARKS.

1. Although it is nowhere assumed that m is singular, this is the case of interest. If m is smooth, some of the assumptions are unnecessary and the proof of the theorem is much easier than the one outlined below.
2. The assumption that v satisfies the differential equation can always be satisfied by modifying the forcing term. However, smoothness assumptions on f then translate into compatibility conditions across $t = 0$.
3. If the initial conditions are not compatible with the history, then the singularity in m generates singular behaviour of the solution as $t \to 0+$. The assumed compatibility avoids this.
4. The result can be generalized to equations of the form (3), see the remarks in the next section.

We now consider equation (9) with a kernel a such that a' need not be integrable at 0. We split up the integral in (10) into a part from $-\infty$ to 0, which is absorbed into the forcing term, and a part from 0 to t. In this fashion, we obtain the initial value problem

$$u_{tt} = \chi(u_x)_x + \int_0^t a(t-\tau)\psi(u_x(x,\tau))_{x\tau}\, d\tau + f,\ 0 \leq x \leq 1,\ t > 0,$$

$$u(0,t) = u(1,t) = 0,\ t \geq 0, \tag{10}$$

$$u(x,0) = u_0(x),\ u_t(x,0) = u_1(x).$$

The assumptions formulated below are what is required to solve (10). In order to reduce (9) to (10) we must assume that u_0 is compatible with the history. Also we have to require that ψ has four continuous derivatives and we need more smoothness of the history than is obtained below for the solution. We make the following assumptions:

(s1*) $\chi, \psi \in C^3(\mathbf{R})$.

(s2*) For some $T > 0$, $f, f_x, f_{xx} \in L^2((0,T); L^2(0,1))$.

(s3*) $u_0 \in H^3(0,1)$, $u_1 \in H^2(0,1)$.

(m*) $a \in L^1(0,\infty)$, $a \geqq 0$, $a' \leqq 0$, $a'' \geqq 0$, $a''' \leqq 0$.

(e*) $\psi' > 0$, $\chi' + a(0)\psi' > 0$.

(c1*) $u_0(0) = u_0(1) = u_1(0) = u_1(1) = 0$.

(c2*) $u_0''(0) = u_0''(1) = 0$ and $f(0,t) = f(1,t) = 0$ for a.e. $t \in (0,T)$.

THEOREM 2. Assume that (s1*)-(s3*), (m*), (e*) and (c1*)-(c2*) hold. Then there is a solution u of (10), defined on a maximal interval $[0,T_0)$. The solution satisfies $u, u_x, u_t, u_{xx}, u_{xt}, u_{xxx}, u_{xxt} \in L^\infty_{loc}([0,T_0);L^2(0,1))$ and $u_{tt}, u_{xtt} \in L^2_{loc}([0,T_0);L^2(0,1))$.

REMARKS.

1. Note that we have made no assumptions on the temporal smoothness of f. This is because the splitting of the integral in (9) and reduction to the initial value problem (10) will lead to a forcing term which has a singularity in its derivative, even if the history is smooth.

2. In contrast to Theorem 1 above, the smoothness of the solution guaranteed by Theorem 2 is not enough to guarantee that the equation (10) is satisfied in the classical sense. It is, however, satisfied almost everywhere and the proof we give below also guarantees uniqueness.

3. The compatibility condition (c2*) is not physically natural but cannot be avoided unless other assumptions are introduced. It guarantees that u can be continued as an odd periodic function of x without loss of regularity.

4. Note that the second part of condition (e*) is vacuous if $a(0) = +\infty$. In this case no sign condition on χ is necessary.

3. IDEA OF PROOFS

The proofs of both theorems are based on the energy method. The strategy of this method for solving nonlinear problems can be summarized as follows:

1. Define an appropriate iteration leading to a sequence of linear problems.
2. For the linear problem, find approximate problems which are known to have a solution.
3. Find uniform estimates for the solutions of the approximate problems.
4. Pass to the limit to obtain a solution of the linear problem.
5. Use a contraction argument to show that the iteration converges. Typically, the contraction is in a norm weaker than the one associated with the a priori bounds found in step 3.

The main part of the proof is usually the derivation of uniform estimates (step 3), and we shall focus on this aspect of the proof here. See [5] and [7] for details of a full proof.

In the case of Theorem 1, we use the iteration

$$u_{tt}^{n+1} = \phi'(u_x^n)u_{xx}^{n+1} - \int_{-\infty}^{t} m(t-\tau)\psi'(u_x^n(x,\tau))u_{xx}^{n+1}(x,\tau)d\tau + f. \tag{11}$$

This leads to a linear problem of the form

$$u_{tt}(x,t) = \alpha(x,t)u_{xx}(x,t) - \int_{-\infty}^{t} m(t-\tau)\beta(x,\tau)u_{xx}(x,\tau)d\tau + f(x,t). \tag{12}$$

To find approximate problems known to have solutions, one typically uses finite differences or a Galerkin approach. In the present case, however, it is simpler to take advantage of the fact that an existence theorem for problems with smooth kernels is already known. With ρ_δ denoting a standard mollifier, we set

$$m_\delta(s) := \int_{-\delta}^{\delta} \rho_\delta(\tau)m(s + \delta - \tau)d\tau \tag{13}$$

and define an approximate problem in which m is replaced by m_δ. In order to maintain the validity of the assumption that the history satisfies the equation, we must modify the forcing term. Thus one obtains the approximate problem

$$u^\delta_{tt} = \alpha u^\delta_{xx} - \int_{-\infty}^{t} m_\delta(t-\tau)\beta(x,\tau)u^\delta_{xx}(x,\tau)d\tau + f^\delta(x,t), \tag{14}$$

with the same initial and boundary conditions as the original equation (7).

We introduce the following notations:

$$\Delta_h v(x,t) := v(x,t) - v(x,t-h), \tag{15}$$

$$\gamma^\delta(x,t) := \alpha(x,t) - a_\delta(0)\beta(x,t), \tag{16}$$

$$Q(v,t,a_\delta) := \int_{-\infty}^{t}\int_0^1 v(x,s) \int_{-\infty}^{s} a_\delta(s-\tau)v(x,\tau)d\tau \, dx \, ds. \tag{17}$$

Here a_δ is related to m_δ in the same way as a is to m. We apply Δ_h to (7), multiply by $\Delta_h(\beta u_{xx})_t$ and integrate over $[0,1] \times (-\infty,t]$. After a number of integrations by parts one finds an "energy equation" of the form

$$\frac{1}{2}\int_0^1 \{\beta\gamma^\delta(u^\delta_{xxt})^2 + \beta(u^\delta_{xtt})^2\}(x,t)\,dx + \lim_{h\downarrow 0}\frac{1}{h^2} Q(\Delta_h(\beta u^\delta_{xx})_t,t,a_\delta) = \ldots, \tag{18}$$

where the dots indicate terms which can be controlled. In [5], we establish that for every $\varepsilon > 0$, we have an inequality of the form

$$\lim_{h\downarrow 0}\frac{1}{h^2} Q(\Delta_h(\beta u^\delta_{xx})_t,t,a_\delta) \geqq (\tfrac{1}{2}\,a_\delta(0) - \varepsilon)\,\|(\beta u^\delta_{xx})_t(\cdot,t)\|^2_{L^2} - C(\varepsilon)\int_{-\infty}^{t}\|(\beta u^\delta_{xx})_t(\cdot,\tau)\|^2_{L^2}\,d\tau. \tag{19}$$

Because of this inequality, the left hand side of (18) provides an upper bound for the L^2-norms of $u^\delta_{xxt}(\cdot,t)$ and $u^\delta_{xtt}(\cdot,t)$. Bounds for u^δ_{ttt} and u^δ_{xxx} are obtained by differentiating (14) with respect to t and x, respectively.

For an equation of the general form (3), the iteration analogous to (11) leads to linear problems of the form

$$u_{tt}(x,t) = \alpha(x,t)u_{xx}(x,t) - \int_{-\infty}^{t} m(t-\tau)\beta(x,t,\tau)u_{xx}(x,\tau)d\tau + f(x,t). \tag{20}$$

For this problem, we split β into two terms as follows:

$$\beta(x,t,\tau) = \beta(x,\tau,\tau) + [\beta(x,\tau,\tau) - \beta(x,t,\tau)]. \tag{21}$$

The first contribution can be treated as above and the second can be treated as a perturbation.

For Theorem 2, we use the iteration

$$u_{tt}^{n+1} = \chi'(u_x^n)u_{xx}^{n+1} + \int_0^t a(t-\tau)[\psi'(u_x^n(x,\tau))u_{xx}^{n+1}(x,\tau)]_\tau \, d\tau + f(x,t), \tag{22}$$

which leads to linear problems of the form

$$u_{tt}(x,t) = \gamma(x,t)u_{xx}(x,t) + \int_0^t a(t-\tau)[\beta u_{xx}]_\tau(x,\tau)d\tau + f(x,t). \tag{23}$$

We approximate a by a_δ and we approximate f by some f^δ which has some extra smoothness. Instead of using a temporal difference operator, we use a spatial one:

$$D_h v(x,t) := v(x,t) - v(x - h,t), \tag{24}$$

and we now define

$$Q(v,t,a_\delta) := \int_0^t \int_0^1 v(x,s) \int_0^s a_\delta(s-\tau)v(x,\tau)d\tau \; dx \; ds. \tag{25}$$

We note that our compatibility assumptions are such that we can think of u as being extended as an odd periodic function of x. Near x = 0, D_h is defined accordingly.

We now apply D_h to (the approximated version of) (23), multiply by $D_h(\beta u_{xx}^\delta)_t$ and integrate over $[0,1] \times [0,t]$. From this, we find an energy estimate of the form

$$\frac{1}{2}\int_0^1 \beta(u_{xxt}^\delta)^2 + \beta\gamma(u_{xxx}^\delta)^2(x,t)dx + \lim_{h\downarrow 0} \frac{1}{h^2} Q(D_h(\beta u_{xx}^\delta)_t,t,a_\delta) = \ldots \tag{26}$$

From the left hand side, we can obtain bounds for the L^2-norms of u_{xxt}^δ and u_{xxx}^δ in an analogous fashion as before. However, to control the terms on the right hand side, we also need a bound for u_{xtt}^δ. Such a bound is obtained by differentiating (23) with respect to x and using the following inequality of Staffans [12] in the integral term:

$$\int_0^t \| \int_0^s a_\delta(s-\tau)\phi(\tau)\, d\tau \|^2 \, ds \leq 5 \, \|a_\delta\|_{L_1} \, Q(\phi,t,a_\delta). \tag{27}$$

In this fashion, one obtains a bound for

$$\int_0^t \|u^\delta_{xtt}(\cdot,s)\|^2_{L^2} \, ds. \tag{28}$$

Acknowledgement

This research was supported by the National Science Foundation under Grant No. DMS-8796241 and by the Air Force Office of Scientific Research under Grant No. AF05R-85-0307.

References

[1] B. Bernstein, E.A. Kearsley and L.J. Zapas, A study of stress relaxation with finite strain, Trans. Soc. Rheology 7 (1963), 391-410.

[2] C.M. Dafermos and J.A. Nohel, A nonlinear hyperbolic Volterra equation in viscoelasticity, Amer. J. Math. Supplement (1981), 87-116.

[3] C.M. Dafermos, Development of singularities in the motion of materials with fading memory, Arch. Rat. Mech. Anal. 91 (1986), 193-205.

[4] J.M. Greenberg, The existence of steady shock waves in nonlinear materials with memory, Arch. Rat. Mech. Anal. 24 (1967), 1-21.

[5] W.J. Hrusa and M. Renardy, On a class of quasilinear partial integro-differential equations with singular kernels, J. Diff. Eq. 64 (1986), 195-220.

[6] W.J. Hrusa and M. Renardy, On wave propagation in linear viscoelasticity, Quart. Appl. Math. 43 (1985), 237-254.

[7] W.J. Hrusa and M. Renardy, A model equation for viscoelasticity with a nonintegrable memory function, SIAM J. Math. Anal. 19 (1988), 257-269.

[8] A. Kaye, Non-Newtonian flow in incompressible fluids, College of Aeronautics, Cranfield, England, Tech. Note 134 (1962).

[9] J. Prüss, Positivity and regularity of hyperbolic Volterra equations in Banach spaces, Math. Ann. 279 (1987), 317-344.

[10] M. Renardy, Some remarks on the propagation and non-propagation of discontinuities in linearly viscoelastic liquids, Rheol. Acta 21 (1982), 251-254.

[11] M. Renardy, W.J. Hrusa and J.A. Nohel, Mathematical Problems in Viscoelasticity, Longman Scientific and Technical, Essex (England), and John Wiley and Sons, New York, 1987.

[12] O.J. Staffans, An inequality for positive definite Volterra kernels, Proc. Amer. Math. Soc. 58 (1976), 205-210.

William J. Hrusa
Department of Mathematics
Carnegie-Mellon University
Pittsburgh
PA 15213
USA.

Michael Renardy
Department of Mathematics
Virginia Polytechnic Institute
and State University
Blacksburg
VA 24061
USA.

U. JIN CHOI & R.C. MACCAMY

Fractional order Volterra equations

This work was supported by the National Science Foundation under Grant DMS 8400986.

1. INTRODUCTION

Volterra integrodifferential equations have been much studied as models for physical systems with memory effects. In this connection there has been some very interesting recent work by Renardy and Hrusa, [3] and [7] dealing with viscoelastic models with integrals involving singular kernels. This is closely related to the idea of fractional calculus which has been introduced by engineers.

The present paper has two goals. First we describe classes of models involving fractional order Volterra operators which interpolate between various well known equations as the order varies. The second idea we want to describe is the effect on error estimates for Galerkin approximations of the order of singularity.

The error estimates, in turn, are influenced by two classical ideas. The first is contained in the basic paper of Douglas and DuPont, [1] on parabolic equations. They showed that the best results on convergence rates for Galerkin approximation involved fractional order spaces in the time variable. For our problems, which have fractional derivatives in time, such estimates are completely natural.

The second idea is contained in the work of Wheeler [8] and involves the idea of elliptic projections. For parabolic equations she showed that one obtained optimal error estimates by comparing with solutions of elliptic problems, that is static problems. We show that there is a corresponding idea for our general models. We term this the quasi elliptic projection since in the physical contexts it corresponds to the well known quasi-static approximations.

In section two we present a summary of our fractional order models and indicate the way in which they can be used for interpolation. In section three we discuss the Galerkin estimates for a particular case. The quasi-elliptic projection idea is presented for the same case in section three. A complete discussion of the other cases, with details of proof will appear elsewhere.

Our results are related to but different from those of Renardy and Hrusa. Their model is described in section two. What they showed is that making the models more singular increases the spatial regularity of solutions. Our concern is with Galerkin approximations so that the spatial smoothness is fixed by the choice of spaces. What we observe is an increase in regularity in time, for the same spatial norms, as the singularity increases.

2. DESCRIPTION OF MODELS

We will describe first a special case of our models. At the end of the paper we indicate the extensions.

We begin by defining an operator M on $H = L_2(0,1)$ by:

$$\mathcal{D}_M = H_2(0,1) \cap \overset{0}{H}_1(0,1), \quad Mu = -u'' \tag{2.1}$$

Associated with this operator we mention three physical problems:

(P) $\ddot{u} + MU = f$ (Parabolic heat flow)

(E) $\ddot{u} + Mu = f$ (Linear elasticity)

(K) $\ddot{u} + M\dot{u} = Mu = f$ (Kelvin solid)

These problems are all one-dimensional and well studied.

We next introduce a linear hereditary operation L_{a_0} by

$$L_{a_0}[\varphi](t) = \int_0^t a_0(t-\tau)\varphi(\tau)d\tau, \quad a_0(t) = e^{-t} \tag{2.2}$$

We can compose the operators M and L_{a_0} to form two new models:

$(GP)^*$ $\dot{u} + L_{a_0}[Mu] = f.$ (Gurtin-Pipkin heat flow)

$(V)^*$ $\ddot{u} + 2Mu - L_{a_0}[Mum] = f$ (Viscoelasticity).

(See [2] and [5]).

There are some qualitative properties which models of the above type may possess. First we mention the idea of propagation speed. We say any of the models above has propagation speed c if given initial data ($u(x,\cdot)$ or $u(x,\cdot)$ and $u_t(x,\cdot)$) with support in $[\alpha,\beta]$. $0 < \alpha < \beta < 1$ the solution remains identically zero for

$$0 \leqq t \leqq \frac{x - \beta}{c} \text{ for } \beta < x < 1, \; 0 \leqq t \leqq \frac{\alpha - x}{c} \text{ for } 0 < \alpha < 1.$$

We include the case of infinite propagation speed $c = \infty$. The second idea is asymptotic stability which means that for any initial data $u(\cdot,t) \to 0$, in some norm, as $t \to \infty$. In contrast to this is conservation of energy which means that some norm $\|u(\cdot,t)\|$ is independent of time. We can place the models above relative to these properties by the following schematic diagram.

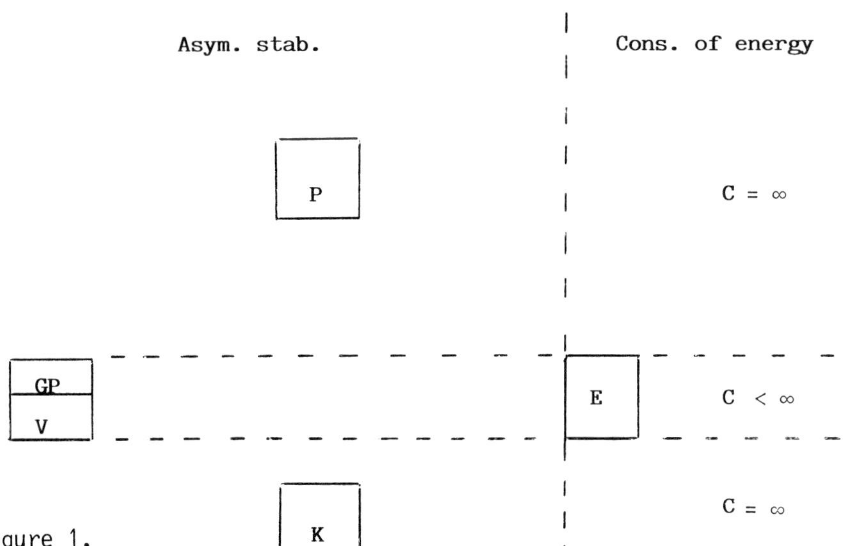

Figure 1.

(*) In these the initial history of u is zero up to time $t = 0$.

We now describe our fractional order models. We first introduce the fractional order derivatives D^α: $0 \leqq \alpha < 1$:

$$D\varphi(t) = \dot{\varphi}(t);\ D^\alpha\varphi(t) = \Gamma(1-\alpha)^{-1}\ D\int_0^t (t-\tau)^{-\alpha}\varphi(\tau(d\tau), \tag{2.3}$$

where $\Gamma(z)$ is the gamma function. We also introduce the singular kernels.

$$\tilde{a}_\alpha(t) = t^{-\alpha}e^{-t},\ a_\alpha(t) = \Gamma(1-\alpha)^{-1}\tilde{a}_\alpha(t). \tag{2.4}$$

We have the following limit relations, in obvious notation:

$$D^\alpha\varphi \begin{array}{l}\nearrow D\varphi \\ \searrow \varphi\end{array} : L_{a_\alpha}[\varphi] \begin{array}{l}\nearrow \varphi \\ \searrow L_{a_0}[\varphi]\end{array} ;\ DL_{a_\alpha}[\varphi) \begin{array}{l}\nearrow D\varphi \\ \searrow \varphi - L_{a_0}[\varphi]\end{array} \text{ as } \alpha \begin{array}{l}\nearrow 1 \\ \searrow 0\end{array} \tag{2.5}$$

We can now describe our models. They are:

(GP-P) $Du + L_{a_\alpha}[Mu] = f$

(E-P) $Du + D^\alpha Mu = f$

(V-E) $D^2u + 2Mu - L_{a_\alpha}[Mu] = f$

(V-K) $D^2u + Mu + DL_{a_\alpha}[Mu] = f$

(E-K) $D^2u + (\frac{1}{2}\alpha + \frac{1}{2})(Mu + D^\alpha Mu) = f$

The notation $M_0 - M_1$ indicates that the equation formally tends to M_0 (M_1) as α tends to 0 (1) as indicated schematically in Figure two below.

We want to describe the Rendardy-Hrusa model in our notation. A special case of it can be expressed in the form,

(RH) $D^2u + (1 + \Gamma(1-\alpha))Mu - L_{\tilde{a}_\alpha}[Mu] = f$.

They have an extension of their model for the range $1 \leqq \alpha < 2$, defined as follows. Put

$$\tilde{L}_{\tilde{a}_\alpha}[\varphi](t) = \int_{-\infty}^{t} \tilde{a}_\alpha(t-\tau)(\varphi(t) - \varphi(\tau))d\tau$$

Then their model extends to

$$(\text{RH}') \quad D^2u + Mu + \tilde{L}_{\tilde{a}_\alpha}[Mu] = f \qquad 1 \leqq \alpha < 2.$$

For $u(t) = 0$ on $t < 0$, (RH) is the same as (RH)' for $0 \leqq \alpha < 1$ but (RH) does not have meaning for $1 \leqq \alpha < 2$.

We now indicate the interpolations. The regions indicated by dashed lines have the same meaning as in Figure 1. The arrows indicate the direction for α increasing from 0 to 1 except for (RH)' where α increases from 1 to 2.

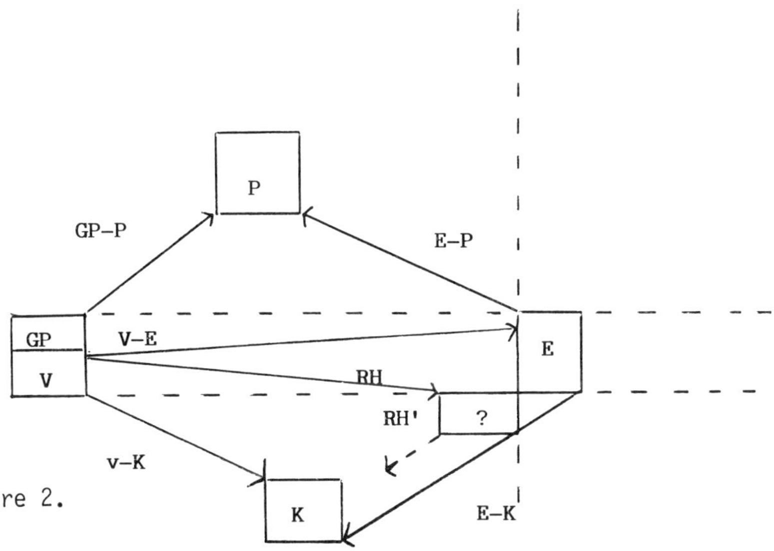

Figure 2.

REMARK 1. The arrow E - P is to be interpreted as the statement that if one integrates the limit of E - K as $\alpha \to 1$ with respect to t one obtains (P) with f replaced by $\int_0^t f(x,\tau)d\tau + \dot{u}(x,0)$.

REMARK 2. The limit of RH as $\alpha \to 1$ is a model which is hard to describe. It does have asymptotic stability and infinite propagation speed hence the designation with the box with question mark. The limit of RH' as $\alpha \to 2$ does not seem to exist.

REMARK 3. There is an essential difference between V - E and RH. For $0 \leq \alpha < 1$ V - E has propagation speed $\sqrt{2}$ while for (E) c = 1. On the other hand the propagation speed for RH is $c^2 = 1 + \Gamma(1 - \alpha)$ which varies from $\sqrt{2}$ at $\alpha = 0$ to ∞ as $\alpha \to 1$.

REMARK 4. The model E - K is essentially the "spring-pot" idea of Koeller [4], an interpolation between a spring and a spring-dash-pot combination.

3. GALERKIN METHOD FOR GP-P

As a model problem we study GP-P with zero initial value, that is

$$\dot{u} + L_{a_\alpha}[Mu] = f, \ t > 0; \ u(0) = 0 \tag{3.1}$$

We give a variational formulation in a standard way. Put $V = H_1^0(0,1)$ and $H = L_2(0,1)$ and let $A(u,v)$ denote the bilinear form on $V \times V$ defined by

$$A(u,v) = \int_0^1 u'(x)v'(x)dx. \tag{3.2}$$

We consider the variational problem: Find $u : [0,\infty) \to V$ such that

$$(\dot{u},v)_H + L_{a_\alpha}[A(u,v)] = (f,v)_H \quad \forall v \in V, \ u(0) = 0 \tag{VP}$$

We also consider approximate problems. Suppose V^h is a family of finite dimensional subspaces, $V^h = \langle \phi_1^h, \dots, \phi_{N_h}^h \rangle$, of V. Then the approximate variational problems are: Find $u^h : [0,\infty) \to V^h$ such that

$$(\dot{u}^h,v^h)_H + L_{a_\alpha}[A(u^R,v^h)] = (f,v^h) \quad \forall v^h \in V^h, \ u^h(0) = 0 \tag{AVP}$$

We want to Laplace transform the above problems. We let $\tilde{V}$ and H be the complexifications of V and H. We extend A to a form $\tilde{A}$ on $\tilde{V} \times \tilde{V}$ by $\tilde{A}(\hat{u},\hat{v}) = \int_0^1 \hat{u}'(x)\bar{\hat{v}}'(x)dx$. We see from (2.4) that the transform of a_α is $\hat{a}_\alpha(s) = (s + 1)^{-1+\alpha}$. Then if u and u^h are solutions of (VP) and (AVP) the transforms $\hat{u}$ and $\hat{u}^h$ satisfy:

$$s(\hat{u},\hat{v})_H + \hat{a}_\alpha(s)\hat{A}(\hat{u},\hat{v}) = (\hat{f},\hat{v})_H \ \forall\ \hat{v} \in \tilde{V} \qquad (\widehat{VP})$$

$$s(\hat{u}^h,\hat{v}^h)_{\tilde{H}} + \hat{a}_\alpha(s)\tilde{A}(\hat{u},\hat{v}) = (\hat{f},\hat{v})_{\tilde{H}} \ \forall\ \hat{v}^h \in \tilde{V}^h \qquad (AVP)$$

We begin our analysis by observing that A is coercive on V which means that $\tilde{A}$ is coercive on $\tilde{V} \times \tilde{V}$, that is,

$$\tilde{A}(\hat{u},\hat{u}) \geq a \|u\|_V^2, \quad \underline{a} > 0. \qquad (3.3)$$

Next we observe that

$$\hat{a}(s) = s^{-1+\alpha} + (\alpha - 1)s^{-2+\alpha} + O(s^{-3+\alpha}) \text{ as } s \to \infty \qquad (3.4)$$

It follows from (3.3) and (3.4) that,

$$\hat{a}_\alpha(s)\tilde{A}(\hat{u},\hat{v}) = s^{-1+\alpha}A(\hat{u},\hat{v}) + o(s^{-1+\alpha}). \qquad (I)$$

Next we observe that the real part of $\hat{a}_\alpha(i\eta)$ is bounded below by a positive constant for η in any compact set $|\eta| \leq N$. It follows that

$$\forall\ N > 0\ \exists\, \delta(N) > 0 \ \text{ s.t. } \ \mathrm{Re}\ \hat{a}_\alpha(i\eta)\tilde{A}(\hat{u},\hat{u}) \geq \delta(N) \|\hat{u}\|_{\tilde{V}}^2 \ \forall\ \hat{u} \in \hat{V} \qquad (II)$$

and $|\eta| \leq N$.

We impose the condition,

$$f \in L_1(0, \infty : H) \qquad (F)$$

It follows that $\hat{f}$ exists for Re $s \leq 0$ and is analytic in Re $s > 0$. We obtain then the following result.

<u>THEOREM 3.1</u>. The problems $\widehat{VP}$ and $A\hat{V}P$ have unique solutions $\hat{u}$ and $\hat{u}^h$ in Re $s \geq 0$. Moreover $\hat{u}(s)$ and $\hat{u}^h(s)$ are analytic in Re $s > 0$.

<u>Proof</u>. Consider the form Re $\hat{a}_\alpha(s)\tilde{A}(\hat{u},\hat{u})|\ \|\hat{u}\|_V^2$. For a fixed $\hat{u}$ this is a harmonic function in Re $s > 0$. By (II) it is positive on Re $s = 0$ and by (I)

it tends to zero as $s \to \infty$. Hence by the maximum principle it is positive for any s in R $s \geq 0$. The existence and uniqueness of $\hat{u}$ and $\hat{u}^h$ follows from the Lax-Milgram lemma. The analyticity follows by differentiation of $(\widehat{VP})$ and $(A\hat{V}P)$ with respect to $\bar{s}$.

REMARK. It is possible to show that if one defines functions u(t) and $u^h(t)$ by the complex inversion formulas

$$u(t) = (2\pi)^{-1} \int_{-\infty}^{+\infty} d^{i\eta t}\hat{u}(i\eta)d\eta, \quad u^h(t) = (2\pi)^{-1} \int_{-\infty}^{+\infty} e^{i\eta t}\hat{u}^h(i\eta)d\eta,$$

then one will have solutions of (VP) and (AVP).

We want to estimate $\hat{u} - \hat{u}^h$. We put $\hat{v} = \hat{v}^h$ in $\widehat{VP}$ and subtract from $A\hat{V}P$. Let $\hat{w}^h$ be an arbitrary element of $\hat{V}^h$ and put $\hat{\varepsilon} = \hat{u}^h - \hat{w}^h$ and $\hat{e} = u - \hat{w}^h$. Then one obtains,

$$i\eta(\hat{\varepsilon}(i\eta),\hat{v}^h)_{\tilde{H}} + \hat{a}_\alpha(i\eta)\hat{A}(|(i\eta),\hat{v}^h)$$

$$= i\eta(\hat{e}(i\eta),\hat{v}^h)_{\tilde{H}} + \hat{a}_\alpha(i\eta)\tilde{A}(\hat{e}(i\eta),\hat{v}^h) \ \forall\ \hat{v}^h \in V^h.$$

We put $\hat{v}^h = \hat{\varepsilon}$ and find that,

$$i\eta\|\hat{\varepsilon}(i\eta)\|^2_{\tilde{H}} + \hat{a}_\alpha(i\eta)\hat{A}(\hat{\varepsilon}(i\eta),\hat{\varepsilon}(i\eta))$$

$$= i\eta(\hat{e}(i\eta),\varepsilon(i\eta))_{\tilde{H}} + a_\alpha(i\eta)A >> (\hat{e}(i\eta),\varepsilon(i\eta)). \tag{3.5}$$

We first consider η such that $|\eta| \leq N$. The right side of (3.5) is bounded above by a constant times $\|\hat{e}(i\eta)\|_{\tilde{V}} \|\varepsilon(i\eta)\|_{\tilde{V}}$. Then if we take the real part of (3.5) and use (I) to bound the left side below we obtain an estimate of the form

$$\|\hat{\varepsilon}(i\eta)\|_H \leq \|\hat{\varepsilon}(i\eta)\|_V \leq \delta'(N)\ \|\hat{e}(i\eta)\|_V \text{ for } |\eta| \leq N \tag{3.6}$$

We can also obtain an estimate for large η but this is more subtle. Let R and L denote the right and left sides of (3.5) and we let c denote a generic constant. From (I) we can bound R above:

$$|R| \leq |\eta| \; \|\hat{e}(i\eta)\|_{\tilde{H}} \|\hat{\varepsilon}(i\eta)\|_{\tilde{H}} + c|\eta|^{-1+\alpha}\|\hat{e}(i\eta)\|_{\tilde{V}} \|\hat{\varepsilon}(i\eta)\|_{\tilde{V}}. \tag{3.7}$$

for η large. Next we observe that for $\eta > 0$

$$(i\eta)^{-1+\alpha} = |\eta|^{-1+\alpha}(\gamma_\alpha - i\tau_\alpha), \; \gamma_\alpha = \cos(1-\alpha)\,\frac{\pi}{2}, \; \tau_\alpha = \sin(1-\alpha)\,\frac{\pi}{2} \tag{3.8}$$

It follows then from (I) that for large positive η.

$$Re\; L \geqq c\gamma_\alpha \eta^{-1+\alpha}\|\hat{\varepsilon}\|_V^2,$$

hence if we take the real part of (3.5) we obtain, for large positive η,

$$\eta^{-1+\alpha}\|\hat{\varepsilon}\|_{\tilde{H}}^2 \leq |R| + c\eta^{-1+\alpha} \;\; \|\hat{\varepsilon}\|_V^2 \tag{3.9}$$

If we take the imaginary part of (3.5) and use (I) again we obtain, for large positive η,

$$\eta\|\hat{\varepsilon}\|_H^2 \leq |R| + c\eta^{-1+\alpha} \;\; \|\hat{\varepsilon}\|_V^2 \tag{3.10}$$

Put $\alpha = \eta^{1/2}\,\|\hat{\varepsilon}\|_{\tilde{H}}$, $a = \eta^{1/2}\,\|\hat{e}\|_{\tilde{H}}$, $\beta = \eta^{-\frac{1+\alpha}{2}}\,\|\hat{\varepsilon}\|_{\tilde{V}}$, $b = \eta^{-\frac{1+\alpha}{2}}\,\|e\|_{\tilde{V}}$.

Then if we combine (3.7), (3.9) and (3.10) we obtain the estimates

$$\beta^2 \leq c_1\alpha^{-1}(a\alpha + b\beta), \; \alpha^2 \leqq c_2(a\alpha + b\beta + \beta^2).$$

From this one obtains the estimate

$$\alpha^2 + \beta^2 \leqq c\gamma^{-1}(a^2 + b^2) \tag{3.11}$$

for large positive η and by reflection for large $|\eta|$.

Given a function φ: $(0,\infty) \to V$ let us define norms $|||\varphi|||_\alpha$ by

$$|||\varphi|||_\alpha^2 = (1 + |\eta|^2)^{1/2}\,\|\hat{\varphi}(i\eta)\|_{\tilde{H}}^2 + (1+|\eta|^2)^{-\frac{1-\alpha}{2}}\,\|\hat{\varphi}(i\eta)\|_{\tilde{V}}^2 \tag{3.12}$$

Then (3.6) and (3.11) together yield,

$$|||\hat{\varepsilon}|||_{\alpha} \leq c\gamma_{\gamma}^{-1} \; |||\hat{e}|||_{\alpha} \, . \tag{3.13}$$

The estimate (3.13) is an optimality result in the fractional order Sobolev norms $|||\cdot|||_{\alpha}$. It states that the u^h approximate the exact solution u as well as it is possible to approximate u in that norm, with functions taking values in V^h. Note the symmetry: the same norm appears on both sides. When $\alpha = 1$ so tha GP - P formally becomes (P) the $|||\cdot|||_{\alpha}^2$ norm is $\|\varphi\|^2_{H^{1/2}(0,\infty):H)} + \|\varphi\|_{L_2(0,\infty:V)}$ and our result coincides with the one given by Douglas and Dupont for the problem (P).

REMARK. A typical result which can be derived from (3.13) is the following. Suppose one chooses piecewise linear finite element spaces for V^h with h the mesh length. Then interpolation results say that if u belongs to $H^2(0,1)$ there is a $w^h \in V^h$ such that $\|u-w^h\|_V \leq ch \|u\|_{H_2(0,1)}$ and $\|u-w^h\|_H \leq ch^2 \|u\|_{L_2(0,1)}$. It follows then from (3.13) that if $u \in H^{1/2}(0,\infty:H_2)$ we will have $|||u-u^h|||_{\alpha} = O(h)$. The idea of the next section is to improve on this estimate by showing that if one considers only the $H^{1/2}(0,\infty:H)$ norm of $u-u^h$ then one obtains $O(h^2)$ convergence.

We see from (3.8) that $\gamma_{\alpha} \to 0$ as $\alpha \to 0$ hence the estimate (3.13) degenerates as $\alpha \to 0$ that is GP - P tends to GP. In order to study the case of small α different techniques are required and will be discussed elsewhere.

4. THE QUASI-ELLIPTIC PROJECTION

Suppose u is a solution of (VP). We define the quasi-elliptic projection of u as that function $U^h : (0,\infty) \to V$ such that,

$$L_{a_{\alpha}}(A(U^h,v^h)) = L_{a_{\alpha}}(A(u,v^h)) \forall \; v^h \in V^h. \tag{4.1}$$

One sees easily that if u^h is the solution of (AVP) and U^h the solution of (4.1) then,

$$(\dot{u}^h-\dot{U}^h,v^h)_H + L_{a_{\alpha}}(A(u^h-U^h,v^h)) = (\dot{u}-\dot{U}^h,v^h)_H \; \forall \; v^h \in v^h \tag{4.2}$$

We transform and obtain

$$s(\hat{u}-\hat{U}^h,\hat{v}^h)_{\tilde{H}} + \hat{a}_\alpha \tilde{A}(\hat{u}^h-\hat{U}^h,\hat{v}^h) = s(\hat{u}-\hat{U}^h,v^h)_{\tilde{H}} \forall \hat{v}^h \in \tilde{V}^h \ . \tag{4.3}$$

Equation (4.3) is similar to (3.5). We put $s = i\eta$, $\lambda(i\eta) = \hat{u}^h(i\eta) - \hat{U}^h(i\eta)$, $\Lambda(i\eta) = \hat{u}(i\eta) - U(i\eta)$. Then for $v^h = \lambda(i\eta)$, (4.3) yields

$$i\eta \,\|\lambda(i\eta)\|_H^2 + a_\alpha(i\eta)\tilde{A}(\lambda(i\eta),\lambda(i\eta)) = i\eta(\lambda(i\eta),\Lambda(i\eta))_{\tilde{H}} \tag{4.4}$$

Again let R and L denote the right and left sides. Instead of (3.7) we have now,

$$|R| \leq |\eta| \;\|\lambda(i\eta)\|_{\tilde{H}} \|\Lambda(i\eta)\|_{\tilde{H}} \tag{4.5}$$

The real part of L can be bounded below just as before and we obtain as an analog of (3.9),

$$\eta^{-1+\alpha}\|\lambda\|_{\tilde{V}}^2 \leq c\gamma_\alpha^{-1}|\eta| \;\|\lambda\|_{\tilde{H}} \;\|\Lambda\|_{\tilde{H}} \tag{4.6}$$

Similarly if we take the imaginary part we obtain an analog of (3.10)

$$\eta\|\lambda\|_{\tilde{H}}^2 \leq \{|\eta| \;\|\lambda_{\tilde{H}} \;\|\Lambda\|_{\tilde{H}} + |\eta|^{-1+\alpha} \;\|\lambda\|_{\tilde{V}}\} \tag{4.7}$$

From (4.6) and (4.7) we obtain the estimate

$$\|\lambda\|_{\tilde{H}} \leq c\gamma_\alpha^{-1}\|\Lambda\|_{\tilde{H}}$$

Finally we have $\|\hat{u} - \hat{u}^h\|_{\tilde{H}} \leq \|\hat{u} - \hat{U}^h\|_{\tilde{H}} + \|\lambda\|_{\tilde{H}}$ hence,

$$\|\hat{u} - \hat{u}^h\|_{\tilde{H}} \leq c\gamma_\alpha^{-1} \;\|\hat{u} - \hat{U}^h\|_{\tilde{H}} \tag{4.8}$$

Equation (4.8) is to be contrasted with (3.13) in that it involves only the $\tilde{H}$ norms. It can be used with Nitsche's trick to provide the improvement indicated in the Remark at the end of the last section. The calculation proceeds as follows.

We transform (4.1) and obtain $\tilde{A}(\hat{U}^h,\hat{v}^h) = \hat{A}(\hat{u},\hat{v}^h)$ for all $\hat{v}^h \in \hat{V}^h$ or

$$\tilde{A}(\hat{u} - \hat{U}^h, \hat{v}^h) = 0 \; \forall \hat{v}^h \in \tilde{V}^h \tag{4.9}$$

We now consider the problem: Find $\hat{w} \in \hat{V}$ such that

$$\tilde{A}(\hat{v}, \hat{w}) = (\hat{u} - \hat{U}^h, \hat{v})_H \; \forall \; \hat{v} \in \hat{V} \tag{4.10}$$

If we recall the meaning of A in (3.2) we see that (4.10) is a simple elliptic problem for $\hat{w}$. It will have a unique solution $\hat{w} \in \tilde{H}_2(0,1)$ and furthermore elliptic regularity will yield,

$$\|\hat{w}\|_{\tilde{V}} \leq c \, \|\hat{u} - \hat{U}^h\|_{\tilde{H}} \; ; \; \|\hat{w}\|_{\tilde{H}_2(0,1)} \leq c \, \|\hat{u} - \hat{U}^h\|_{\tilde{H}} \tag{4.11}$$

We put $\hat{v} = \hat{u} - \hat{U}^h$ in (4.10) and obtain,

$$\tilde{A}(\hat{u} - \hat{U}^h, \hat{w}) = \|\hat{u} - \hat{U}^h\|^2_{\tilde{H}}$$

Then we subtract (4.9) to obtain,

$$\tilde{A}(\hat{u} - \hat{U}^h, \hat{w} - \hat{v}^h) = \|\hat{u} - \hat{U}^h\|^2_{\tilde{H}} \forall \; \hat{v}^h \in \hat{V}^h \tag{4.12}$$

Thus

$$\|\hat{u} - \hat{U}^h\|^2_{\tilde{H}} \leq c \, \|\hat{u} - \hat{U}^h\|_{\tilde{V}} \, \|\hat{w} - \hat{v}^h\|_{\tilde{V}} \; \forall \; \hat{v}^h \in \hat{V}^h \tag{4.13}$$

We recall that for piecewise linear elements we can find $\hat{v}^h \in \tilde{V}^h$ such that $\|\hat{w} - \hat{v}^h\|_{\tilde{V}} \leq ch \, \|\hat{w}\|_{\tilde{H}_2(0,1)}$ hence (4.13) and (4.11) yield

$$\|\hat{u} - \hat{U}^h\|_{\tilde{H}} \leq ch \, \|\hat{w}\|_{\tilde{H}_2(0,1)} \, \|\hat{u} - \hat{U}^h\|_{\tilde{V}} \tag{4.14}$$

Now return to the defining equation for $\hat{U}^h$ and write, $\tilde{A}(\hat{U}^h - \hat{w}^h, \hat{v}^h) = \tilde{A}(\hat{u} - \hat{w}^h, \hat{v}^h)$, $\forall \; \hat{w}^h, \; \hat{v}^h \in \tilde{V}$. This yields $\|\hat{U}^h - \hat{w}^h\|_{\tilde{V}} \leq c \, \|\hat{u} - \hat{w}^h\|_{\tilde{V}} \; \forall \; \hat{w}^h \in \tilde{V}^h$. Hence

$$\|\hat{u} - \hat{U}^h\|_{\tilde{V}} \leq \|\hat{u} - \hat{w}^h\|_{\tilde{V}} + \|\hat{w}^h - \hat{U}^h\|_{\tilde{V}} \leq c \, \|\hat{u} - \hat{w}^h\|_{\tilde{V}} \; \forall \; \hat{w}^h \in \tilde{V}^h \; .$$

But the approximation property for piecewise linear elements says that one can choose $\hat{w}^h \in \hat{V}^h$ such that $\|\hat{u} - \hat{w}^h\|_{\tilde{V}} \leq ch \|\hat{u}\|_{\tilde{H}_2(0,1)}$. Finally then (4.14) yields

$$\|\hat{u} - \hat{U}^h\|_{\tilde{H}} \leq ch^2 \|\hat{u}\|_{\tilde{H}_2(0,1)},$$

or on integration with respect to η,

$$\|u - U^h\|_{L_2(0,\infty:H)} \leq ch^2 \|\hat{u}\|_{L_2(0,\infty:H_2(0,1))} \tag{4.15}$$

Thus if piecewise linear elements are used, one obtains order h^2 convergence in the $L_2(0,\infty:H)$ norm.

5. GENERALIZATIONS

The above theory admits of considerable generalization. First we could replace the operator M by an elliptic operator of order $2m$ in a region Ω with general boundary conditions. The requirement on M is that there exist a subspace V of $H_m(\Omega)$ on which the bilinear form associated with M is coercive.

The next extension is that we can take more general kernels $a(t)$. They need to be chosen, with a coercive M so that conditions (I) and (II) of section three are satisfied. a should belong to $L^1(0,\infty)$ so that its transform $\hat{a}(s)$ exists for $\mathrm{Re}\ s \geq 0$. Then condition (I) will be satisfied if,

$$\forall N \ \exists\ \delta(N) > 0 \quad \text{s.t.} \quad \mathrm{Re}\ \hat{a}(i\eta) \geq \delta(N) \text{ for } |\eta| \leq N \tag{5.1}$$

(5.1) is a well-known condition in the study of Volterra integral equations, see [6]. Condition II will be satisfied if

$$\hat{a}(s) = as^{-1+\alpha} + O(s^{-1+\alpha}) \text{ as } s \to \infty \tag{5.2}$$

Condition (5.2) indicates that the function $a(t)$ has a singularity of the form $t^{-\alpha}$ at $t = 0$.

One can make a much greater extension. The models mentioned so far were all compositions of a fixed differential operator with a scalar heriditary operator. Such compositions are not to be expected in two or three dimensional models in continuum mechanics. The theory can be extended to the following more general situation.

Let V and H be Hilbert spaces. Identity H with a part of V' by the formula $\langle h,v\rangle = (h,v)_H$ for $h \in H$, $v \in V$. Then one has, with identification, $V \subset H \subset V'$. Let A be a map from $(0,\infty)$ into bounded linear mappings from V to V'. Then one can replace the operator $L_a[M]$ by L_A.

$$L_A[M] = \int_0^t A(\tau-\tau)u(\tau)d\tau$$

A should have the following properties:

1. $\langle A(t)u,v\rangle = \langle A(t)v,\dot{u}\rangle \;\; \forall t \in [0,\infty)$ and $u,v \in V$
2. $A \in L_1(0,\infty : BL(V,V\Sigma))$
3. $\forall N \; \exists \; \delta(N) \in Re\langle\hat{A}(i\eta)\hat{u},\hat{u}\rangle \leq \delta(N) \; \|\hat{u}\|^2_V$ for $|\eta| \leq N$ and any $\hat{u} \in \tilde{V}$
4. $\tilde{A}(s) = As^{-1+\alpha} + O(s^{-1+\alpha})\langle A\hat{u},\hat{u}\rangle \geqq \underline{a} \; \|\hat{u}\|^2_V$ as $|s| \to \infty$.

References

[1] Douglas, J. and DuPont, T., Galerkin methods for parabolic equations, SIAM Journ. of Num. Anal. Vol. 7 (1970) pp. 575-526.

[2] Gurtin, M.E. and Pipkin, A.C., A general theory of heat conduction with finite waves speeds, Arch. for Rat. Mech. and Anal., Vol. 31 (1968) pp. 113-126.

[3] Hrusa, W.J. and Renardy, M., On wave propagation in linear visco-elasticity, Quart. of Appl. Math., Vol. 23 (1985) pp. 237-253.

[4] Koeller, R.C., Applications of fractional calculus to the theory of viscoelasticity, Journ. of App. Mech., Vol., 51 (1984) pp. 299-307.

[5] MacCamy, R.C., A model for one-dimensional nonlinear viscoelasticity, Quart. of Appl. Math., Vol. 35 (1977) pp. 21-33.

[6] MacCamy, R.C. and Wong, J.S.W., Stability theories for functional equations, Trans. Amer. Math. Soc., Vol. 164, pp. 1-37.

[7] Renardy, M., Some remarks on the propagation and non-propagation of singularities in linearly viscoelastic fluids, Rheaol. Acta., Vol. 21 (1982) pp. 251-254.

[8] Wheeler, M., A priori L_2-error estimates for Galerkin approximations to parabolic differential equations, SIAM Journ. of Numer. Anal., Vol. 10 (1973) pp. 723-759.

U. Jin Choi & R.C. MacCamy
Department of Mathematics
Carnegie-Mellon University
Pittsburgh
PA 15213
USA.

V. LAKSHMIKANTHAM

Stability results for integrodifferential equations

INTRODUCTION

In investigating the qualitative behaviour of solutions of nonlinear integrodifferential equations of Volterra type by various means, one is naturally lead to the study of scalar integrodifferential inequalities. Since finding the behaviour of solutions of even linear integrodifferential equations is itself very difficult, it is desirable to reduce the study of integrodifferential inequalities to that of differential inequalities [3]. Furthermore, to extend the linearization techniques to nonlinear integrodifferential systems we have to deal with variational systems that are linear integrodifferential systems, and to bring out the inherent rich behaviour of linear integrodifferential systems, compared to that of linear differential systems, is more difficult. In this paper, we shall first develop comparison results which reduce the study of integrodifferential inequalities to that of differential inequalities. We then obtain nonlinear variation of parameters formula for perturbed integrodifferential systems. Employing the idea of finding an equivalent linear differential system given a linear integrodifferential system and also using the nonlinear variation of parameters formula, we discuss the stability properties of nonlinear integrodifferential systems by utilizing the theory of integrodifferential inequalities developed.

1. COMPARISON RESULTS

We consider the integrodifferential inequality

$$D_-m(t) \leq f(t,m(t)) + \int_{t_0}^{t} H(t,s,m(s))ds, \quad t \geq t_0, \tag{1.1}$$

where $m \in C[R_+,R_+]$, $f \in C(R_+^2,R]$, $H \in C[R_+,R]$, $H(t,s,u)$ is nondecreasing in u for each (t,s) and $D_-m(t) = \liminf_{h \to 0^-} \frac{1}{h}[m(t+h) - m(t)]$. In order to reduce the study of (1.1) to that of differential inequality, we need the following comparison result.

THEOREM 1.1. Let $m \in C[R_+,R_+]$, $f \in C[R_+^2,R]$, $K \in C[R_+^3,R]$, $g_0 \in C[R_+^2,R]$ and

$$D_-m(t) \leqq f(t,m(t)) + \int_{t_0}^{t} K(t,s,m(t))ds, \ t \in I_0, \tag{1.2}$$

where $I_0 = [t \geqq t_0;\ m(s) \leqq \eta(s,t,m(t)),\ t_0 \leqq s \leqq t]$, $\eta(t,T,v_0)$ being the left maximal solution of

$$u' = g_0(t,u),\ u(T) = v_0 \geqq 0, \tag{1.3}$$

existing on $[t_0,T]$. Assume that

$$g_0(t,u) \leqq F(t,u;t_0) \tag{1.4}$$

where

$$F(t,u;t_0) = f(t,u) + \int_{t_0}^{t} K(t,s,u)ds \tag{1.5}$$

and $r(t)$ is the right maximal solution of

$$u' = F(t,u;t_0),\ u(t_0) = u_0 \geqq 0, \tag{1.6}$$

existing on $[t_0,\infty)$. Then $m(t_0) \leqq u_0$ implies

$$m(t) \leqq r(t),\ t \geqq t_0.$$

Proof. Since it is known [2] that $\lim_{\varepsilon\to 0} u(t,\varepsilon) = r(t)$, where $u(t,\varepsilon)$ is any solution of

$$u' = F(t,u;t_0) + \varepsilon,\ u(t_0) = u_0 + \varepsilon,$$

for $\varepsilon > 0$ sufficiently small, on any compact set $[t_0,T]$ of $[t_0,\infty)$, it is enough to prove that

$$m(t) < u(t,\varepsilon),\ t_0 \leqq t \leqq T.$$

If this is not true, there exists a $t_1 \in (t_0,T]$ such that

$$m(s) < u(s,\varepsilon),\ t_0 \leqq s < t_1 \text{ and } m(t_1) = u(t_1,\varepsilon).$$

This implies that

$$D_-m(t_1) \geqq u'(t_1,\varepsilon) = F(t_1,u(t_1,\varepsilon);t_0) + \varepsilon. \tag{1.7}$$

Consider now the left maximal solution $\eta(s,t_1,m(t_1))$, $t_0 \leqq s \leqq t_1$, of $u' = g_0(t,u)$, $u(t_1) = m(t_1)$. By Lemma 2.1 in [5], we have $r(s) = r(s,t_0,u_0) \leqq \eta(s,t_1,m(t_1))$, $t_0 \leqq s \leqq t_1$. Since $r(t_1,y_0,u_0) = \lim_{\varepsilon\to 0} u(t_1,\varepsilon) = m(t_1) = \eta(t_1,t_1,m(t_1))$ and $m(s) \leqq u(s,\varepsilon)$. $t_0 \leqq s \leqq t_1$, it follows that $m(s) \leqq r(s,t_0,u_0) \leqq \eta(s,t_1,m(t_1))$, $t_0 \leqq s \leqq t_1$. This inequality implies that $t_1 \in I_0$ and as a result, inequality (1.2) yields $D_-m(t_1) \leqq F(t_1,m(t_1);t_0)$ which contradicts (1.7). Thus $m(t) \leqq r(t)$ for $t \geqq t_0$ is true and the proof is complete.

We shall now utilize Theorem 1.1 to study the integrodifferential inequality (1.1). The advantage of this approach will be clear when we consider a special case.

THEOREM 1.2.

Let the inequality (1.1) hold with m, f and H satisfying the assumptions listed above. Suppose further that g_0 and F are such that (1.4) is satisfied with

$$F(t,u;t_0) = f(t,u) + \int_{t_0}^{t} K(t,s,u)ds,$$

where

$$K(t,s,u) = H(t,s;\eta(s,t,u)),$$

$\eta(t,T,v_0)$ being the left maximal solution of (1.3) existing on $[t_0,T]$. Then $m(t) \leqq r(t)$, $t \geqq t_0$, provided $m(t_0) \leqq u_0$, where $r(t) = r(t,t_0,u_0)$ is the right maximal solution of (1.6) existing on $[t_0,\infty)$.

Proof. We set $I_0 = [t \geqq t_0 : m(s) \leqq \eta(s,t,m(t)),\ t_0 \leqq s \leqq t]$. Since $H(t,s,u)$ is nondecreasing in u, we get using (1.1) the inequality

$$D_-m(t) \leq f(t,m(t)) + \int_{t_0}^{t} H(t,s,\eta(s,t,m(t)))ds, \ t \in I_0$$

Hence by Theorem 1.1, we have the stated result.

An interesting special case of Theorem 1.2 is the following Corollary.

COROLLARY 1.1. Consider the special case with $f \geq 0$ and $H \geq 0$ in Theorem 1.2. Then $m(t) \leq r(t)$, $t \geq t_0$, where $r(t)$ is the right maximal solution of the differential equation

$$u'(t) = f(t,u(t)) + \int_{t_0}^{t} H(t,s,u(t))ds, \ u(t_0) = u_0,$$

existing on $[t_0,\infty)$, provided $m(t_0) \leq u_0$. If, in particular, $f(t,u) = a(t)u$, $H(t,s,u) = H(t,s)u$ with $a(t) \geq 0$ and $H(t,s) \geq 0$, then $m(t) \leq m(t_0)$ $\exp\left[\int_{t_0}^{t} A(s,t_0)ds\right]$, $t \geq t_0$, where $A(t,t_0) = a(t) + \int_{t_0}^{t} H(t,s)ds$.

Proof. We choose $g_0(t,u) \equiv 0$ so that $I_0 = [t \geq t_0 : m(s) \leq m(t)]$.

2. NONLINEAR VARIATION OF PARAMETERS FORMULA

Let us now consider the IVP for nonlinear integrodifferential equation of the form

$$x'(t) = f(t,x(t)) + \int_{t_0}^{t} g(t,s,x(s))ds, \ x(t_0) = x_0. \tag{2.1}$$

We need the following result relative to the differentiability of solutions of (2.1). See [1].

THEOREM 2.1. Assume that $f \in C[R_+ x R^n, R^n]$, $g \in C[R_+ x R_+ x R^n, R^n]$ and f_x, g_x exist and are continuous on $R_+ x R^n$, $R_+ x R_+ x R^n$ respectively. Let $x(t,t_0,x_0)$ be the unique solution of (2.1) existing on some interval $t_0 \leq t < a < \infty$ and set $J = [t_0, t_0 + T]$, $t_0 + T < a$. Define $H(t,t_0,x_0) = f_x(t,x(t,t_0,x_0))$ and $G(t,x,t_0,x_0) = g_x(t,s,x(s,t_0,x_0))$. Then

(i) $\Phi(t,t_0,x_0) = \frac{\partial x}{\partial x_0}(t,t_0,x_0)$ exists and is the solution of

$$y'(t) = H(t,t_0,x_0)y(t) + \int_{t_0}^{t} G(t,s;t_0,x_0)y(s)ds, \tag{2.2}$$

such that $\Phi(t_0,t_0,x_0) = I$;

(ii) $\psi(t,t_0,x_0) = \frac{\partial x}{\partial t_0}(t,t_0,x_0)$ exists and is the solution of

$$z'(t) = H(t,t_0,x_0)z(t) + \int_{t_0}^{t} G(t,s;t_0,x_0)z(s)ds - g(t,t_0,x_0), \qquad (2.3)$$

such that $\psi(t_0,t_0,x_0) = -f(t_0,x_0)$;

(iii) the functions $\Phi(t,t_0,x_0)$, $\psi(t,t_0,x_0)$ satisfy the relation

$$\psi(t,t_0,x_0) + \Phi(t,t_0,x_0)f(t_0,x_0) + \int_{t_0}^{t} R(t,\sigma;t_0,x_0)g(\sigma,t_0,x_0)d\sigma = 0, \qquad (2.4)$$

where $R(t,s;t_0,x_0)$ is the solution of the IVP

$$\frac{\partial R}{\partial s}(t,s;t_0,x_0) + R(t,s;t_0,x_0)H(s,t_0,x_0) + $$
$$+ \int_{s}^{t} F(t,\sigma;t_0,x_0)G(\sigma,s;t_0,x_0)d\sigma = 0, \qquad (2.5)$$

$R(t,t;t_0,x_0) = I$ on the interval $t_0 \leqq s \leqq t$ and $R(t,t_0;t_0,x_0) = \Phi(t,t_0,x_0)$.

We are now in a position to prove the nonlinear variation of parameters formula for solutions $y(t,t_0,x_0)$ of perturbed IVP for integro-differential equation

$$y'(t) = f(t,y(t)) + \int_{t_0}^{t} g(t,s,y(s))ds + F(t,y(t),(Sy)(t)), y(t_0) = x_0. \qquad (2.6)$$

<u>THEOREM 2.2</u>. Suppose that the hypotheses of Theorem 2.1 hold. Assume further that $F \in C[R_+ \times R^n \times R^n, R^n]$ and $(Sy)(t) = \int_{t_0}^{t} K(t,s,y(s))ds$ with $K \in C[R_+ \times R_+ \times R^n, R^n]$. If $x(t,t_0,x_0)$ is the solution of (2.1) existing on J, any solution $y(t,t_0,x_0)$ of (2.6) existing on J satisfies the integral equation

$$y(t,t_0,x_0) = x(t,t_0,x_0) + \int_{t_0}^{t} \Phi(t,s,y(s))F(s,y(s),(Sy)(s))ds$$
$$+ \int_{t_0}^{t}\int_{s}^{t} [\Phi(t,\sigma,y(\sigma)) - R(t,\sigma;s,y(s)]g(\sigma,s,y(s))d\sigma ds \quad (2.7)$$

for $t_0 \geqq t_0$, where $y(t) = y(t,t_0,x_0)$ and $R(t,s;t_0,x_0)$ is the solution of the IVP (2.5).

Proof. Setting $p(s) = x(t,s,y(s))$ where $y(s) = y(s,t_0,x_0)$, we have

$$p'(s) = \psi(t,s,y(s)) + \Phi(t,s,y(s))y'(s).$$

Substituting for $y'(s)$, integrating from t to t_0 and using Fubini's theorem we get

$$p(t) - p(t_0) - \int_{t_0}^{t} \Phi(t,s,y(s))F(s,y(s),(Sy)(s))ds = \int_{t_0}^{t} [\psi(t,s,y(s))$$
$$+ \Phi(t,s,y(s))f(s,y(s)) + \int_{s}^{t} \Phi(t,\sigma,y(\sigma))g(\sigma,s,y(s))d\sigma]ds.$$

Now the relation (2.4) together with the fact $x(t,t,y(t)) = y(t,t_0,x_0)$ yields (2.7) completing the proof.

REMARKS. (i) Let $f(t,x) = A(t)x$ and $g(t,s,x) = B(t,s)x$ in (2.1) where A,B are $n \times n$ continuous matrices. In this special case $x(t,t_0,x_0) = R(t,t_0)x_0$ and $\Phi(t,t_0,x_0) = \frac{\partial x}{\partial x_0}(t,t_0,x_0) = R(t,t_0)$, where $R(t,s)$ is the solution of (2.5) such that $R(t,t) = I$. Consequently, it is easy to see that the relation (2.7) reduces to the linear variation of parameters formula

$$y(t,t_0,x_0) = R(t,t_0)x_0 + \int_{t_0}^{t} R(t,s)F(s,y(s),(Sy)s)ds, \; t \geqq t_0,$$

since $\Phi(t,\sigma,y(\sigma)) = R(t,\sigma;s,y(s))$ being independent of y, and s.

(ii) If $g(t,s,x) \equiv 0$, the relation (2.7) reduces to the usual Alekseev's formula [1]. However, when $g \not\equiv 0$ or is not linear in x, $\Phi(t,t_0,x_0) \neq R(t,s;t_0,x_0)$ and hence the relation (2.7) can not have a simpler form that corresponds to Alekseev's formula as one would like to expect.

3. STABILITY RESULTS

Let us begin by recalling the following known result [4] concerning linear integrodifferential equations.

THEOREM 3.1. Assume that

(a) there exists an $n \times n$ continuous matrix function $L(t,s)$ on $R_+ \times R_+$ such that $L_s(t,s)$ exists, continuous and satisfies

$$K(t,s) + L_s(t,s) + L(t,s)A(s) + \int_s^t L(t,u)K(u,s)du = 0 \tag{3.1}$$

where $A(t)$, $K(t,s)$ are continuous $n \times n$ matrices on R_+ and $R_+ \times R_+$ respectively. Then the IVP for linear integrodifferential system.

$$u'(t) = A(t)u(t) + \int_{t_0}^t K(t,s)u(s)ds + F(t),\ u(t_0) = x_0, \tag{3.2}$$

where $F \in C[R_+,R^n]$, is equivalent to the IVP for linear differential system

$$v'(t) = B(t)v(t) + L(t,t_0)x_0 + H(t), v(t_0) = x_0, \tag{3.3}$$

where $B(t) = A(t) - L(t,t)$ and $H(t) = F(t) + \int_{t_0}^t L(t,s)F(s)ds.$

We shall first employ Theorem 3.1 to discuss the stability properties of integrodifferential system (2.1). For this purpose, we assume that $f \in C[R_+xR^n,R^n]$, $g \in C[R_+^2xR^n,R^n]$, f_x, g_x exist and are continuous, and $f(t,0) \equiv g(t,s,0) \equiv 0$. Setting

$$x'(t) = A(t)x(t) + \int_{t_0}^t K(t,s)x(s)ds + F(t,x(t)) + \int_{t_0}^t G(t,s,x(s))ds$$

with $x(t_0) = x_0$, where

$$F(t,x) = \int_0^1 [f_x(t,x\theta) - f_x(t,0)]d\theta x,$$

and

$$G(t,s,x) = \int_0^1 [g_x(t,s,x\theta) - g_x(t,x,0)d\theta x.$$

We shall now prove the following result.

THEOREM 3.2. Assume that

(i) the trivial solution of the linear integrodifferential equation

$$x'(t) = A(t)x(t) + \int_{t_0}^{t} K(t,s)x(s)ds, \; x(t_0) = x_0$$

is exponentially asymptotically stable;

(ii) $\|F(t,x)\| \leq w_1(t, \|x\|)$, $\|G(t,s,x)\| \leq w_2(t,s, \|x\|)$, where $w_1 \in C[R_+^2, R_+]$, $w_2 \in C[R_+^3, R_+]$ and $w_1(t,u)$, $w_2(t,s,u)$ are nondecreasing in u.

Then, the stability properties of the trivial solution of the differential system

$$u'(t) = -\alpha u(t) + M[w_1(t,u(t)) + \int_{t_0}^{t} w(t,s,u(t))ds, \; u(t_0) = u_0, \tag{3.7}$$

where $\alpha > 0$, $M > 0$ and

$$w(t,s,u(s)) = w_2(t,s,u(s)) + \int_{t_0}^{t} \|L(t,s)\| \, w_1(s,u(s)) \; + \\ + \int_{t_0}^{s} \|L(s,\sigma)\| \, w_2(s,\sigma,u(\sigma))d\sigma, \tag{3.8}$$

$L(t,s)$ being any solution of (3.1), imply the corresponding stability properties of the system (2.1).

Proof. By Theorem 3.1, system (3.5) is equivalent to

$$x'(t) = B(t)x(t) + L(t,t_0)x_0 + F(t,x(t)) + \int_{t_0}^{t} L(t,s)F(s,x(s))ds \\ + \int_{t_0}^{t} G(t,s,x(s))ds + \int_{t_0}^{t}\int_{t_0}^{s} L(s,\sigma)G(s,\sigma,x(\sigma))d\sigma ds, \tag{3.9}$$

with $x(t_0) = x_0$, where $B(t) = A(t) - L(t,t)$ and $L(t,s)$ is any solution of (3.1). Also, it is clear that (3.6) is equivalent to

$$x'(t) = B(t)x(t) + L(t,t_0)x_0, x(t_0) = x_0 \tag{3.10}$$

and hence assumption (i) shows that the trivial solution of (3.10) is exponentially asymptotically stable. Consequently, (see [1]) there exists a Lyapunov function V such that

(a) $V \in C[R_+ x s(\rho), R_+]$, $V(t,x)$ is Lipschitzian x for a constant $M > 0$ and $\|x\| \leq V(t,x) \leq K\|x\|$, $(t,x) \in R_+ x s(\rho)$;

(b) $D^+V(t,x) \leq -\alpha V(t,x)$, $(t,x) \in R_+ x s(\rho)$,

where $s(\rho) = [x \in R^n: \|x\| < \rho]$.

Let $x(t) = x(t,t_0,x_0)$ be any solution of (2.1) existing on some interval $t_0 \leq t < t_1$. Then, using (a) and (b) and setting $m(t) = V(t,x(t))$ we get the integrodifferential inequality

$$D^+m(t) \leq -\alpha m(t) + Mw_1(t,m(t)) + M\int_{t_0}^{t} w(t,s,m(s))ds, \quad t_0 \leq t < t_1,$$

where $w(t,s,u)$ is given by (3.8). Setting $v(t) = m(t)\, e^{\alpha(t-t_0)}$, we obtain

$$D^+v(t) \leq M\, e^{\alpha(t-t_0)}[w_1(t,v(t)\, e^{-\alpha(t-t_0)}) + \int_{t_0}^{t} w(t,s,v(s)\, e^{-\alpha(s-t_0)})ds]$$

for $t_0 \leq s < t_1$, which implies by Corollary 1.1 that

$$v(t) \leq r(t,t_0,m(t_0)), \quad t_0 \leq t < t_1,$$

where $r(t,t_0,u_0)$ is the maximal solution of the differential equation

$$u'(t) = M\, e^{\alpha(t-t_0)}[w_1(t,u(t)\, e^{-\alpha(t-t_0)}) + \int_{t_0}^{t} w(t,s,u(t)e^{-\alpha(s-t_0)})ds],$$

with $u(t_0) = u_0$, existing on $t_0 \leq t < t_1$. As a result, we have

$$V(t,x(t)) \leq r(t,t_0,V(t_0,x_0))\, e^{-\alpha(t-t_0)}, \quad t_0 \leq t < t_1,$$

and it is easy to see that $R(t,t_0,u_0) = r(t,t_0,u_0)\, e^{-\alpha(t-t_0)}$ is the maximal solution of the differential equation (3.7). Hence the desired stability properties of the trivial solution of (2.1) follow from the corresponding stability properties of the trivial solution of (3.7) in view of the properties of $V(t,x)$. The proof is therefore complete.

We shall next discuss, as an application of nonlinear variation of parameters formula, the stability properties of the perturbed system (2.6). For this purpose, we require the following result which is interesting in itself.

THEOREM 3.3. Let the assumptions of Theorem 2.1 hold. Suppose that $x(t,t_0,x_0)$ and $x(t,t_0,y_0)$ are the two solutions of (2.1) existing on $[t_0,\infty)$. Then we have for $t \geq t_0$,

$$x(t,t_0,x_0) = x(t,t_0,y_0) = \int_0^1 \Phi(t,t_0,sx_0 + (1-s)y_0)ds(x_0-y_0), \tag{3.11}$$

where $\Phi(t,t_0,x_0) = \frac{\partial x}{\partial x_0}(t,t_0,x_0)$.

Proof. We set $p(s) = x(t,t_0,x_0 s + (1-s)y_0)$ for $0 \leq s \leq 1$ so that

$$p'(s) = \frac{\partial x}{\partial x_0}(t,t_0,x_0 s + (1-s)y_0)(x_0-y_0).$$

Integrating this expression from 0 to 1 and noting that $p(1) = x(t,t_0,x_0)$, $p(0) = x(t,t_0,y_0)$, we obtain the stated result (3.11).

THEOREM 3.4. Suppose that the assumptions of Theorem 2.1 hold and $f(t,0) \equiv g(t,s,0) \equiv 0$. Assume further that there exist positive numbers α, M and ρ such that if $\|x_0\| \leq \rho$, we have

$$\|R(t,s;t_0,x_0)\| \leq M e^{-\alpha(t-s)}, \quad \|\Phi(t,t_0,x_0)\| \leq M^{-\alpha(t-t_0)} \tag{3.12}$$

for $t_0 \leq s \leq t$. Suppose also that

$$\|g(t,s,x)\| \leq K \|x\| e^{-2\alpha t} \tag{3.13}$$

whenever $\|x\| \leq \rho$ and

$$\|F(t,x,(sy)(t))\| \leq \lambda(t) \|x\| + \int_{t_0}^t \beta(t,s) \|y(s)\| ds \tag{3.14}$$

where $\lambda \in C[R_+,R_+]$, $\beta \in C[R_+^2,R_+]$ and $\|x\|, \|y(s)\| \leq \rho$.

Then, any solution $y(t,t_0,x_0)$ of (2.6), with $\|x_0\|$ small enough, satisfies the estimate

$$\|y(t,t_0,x_0)\| \leqq M \|x_0\| e^{-\alpha(t-t_0)} + \int_{t_0}^{t} \sigma(s,t_0)ds, \ t \geqq t_0, \tag{3.15}$$

where $\sigma(t,t_0) = M\lambda(t) + \frac{2MK}{\alpha} e^{-2\alpha t} + m \int_{t_0}^{t} \beta(t,s) e^{\alpha(t-s)} ds$ and $\sigma \in L[t_0,\infty)$.

<u>Proof</u>. Let $x(t) = x(t,t_0,x_0)$ and $y(t) = y(t,t_0,x_0)$ be the solutions of (2.1) and (2.6) respectively existing on $[t_0,\infty)$. Then it follows from (3.11) and (3.12) that

$$\|x(t,t_0,x_0)\| \leqq M \|x_0\| e^{-\alpha(t-t_0)}, \ t \geqq t_0.$$

For those values of $t \geqq t_0$ for which $\|y(t)\| \leqq \rho$, relations (2.7), (2.12) (3.13) and (3.14) yield

$$\begin{aligned}\|y(t)\| \leqq\ & M \|x_0\| e^{-\alpha(t-t_0)} + M \int_{t_0}^{t} \lambda(s) e^{-\alpha(t-s)} \|y(s)\| ds \\ & + M \int_{t_0}^{t} e^{-\alpha(t-s)} \int_{t_0}^{s} \beta(s,\sigma) \|y(\sigma)\| d\sigma\, ds \\ & + \int_{t_0}^{t} \int_{s}^{t} 2MK e^{-\alpha(t-\sigma)} e^{-2\alpha\sigma} \|y(s)\| d\sigma ds.\end{aligned}$$

Using Fubini's theorem and setting $v(t) = \|y(t)\| e^{\alpha t}$, we get

$$v(t) \leqq M \|x_0\| e^{\alpha t_0} + \int_{t_0}^{t} \lambda_0(s)v(s)ds + \int_{t_0}^{t} (\int_{s}^{t} M e^{\alpha(\alpha-s)} \beta(\sigma,s)d\sigma)v(s)ds$$

where $\lambda_0(t) = M\lambda(t) + \frac{2MK}{\alpha} e^{-2\alpha t}$. Setting the right hand side of (3.16) by $p(t)$ and noting that $p(t)$ is increasing we get

$$p'(t) \leq \sigma(t,t_0)p(t), \ P(t_0) = M \|x_0\| e^{\alpha t_0},$$

which yields

$$\|y(t)\| \leqq M \|x_0\| \exp[-\alpha(t-t_0) + \int_{t_0}^{t} \sigma(s,t_0)ds] \leqq \rho, \ t \geqq t_0,$$

provided $\|x_0\| \leq \frac{\rho}{M} e^{-N(t_0)}$ where $\int_{t_0}^{\infty} \sigma(s,t_0)ds \leq N(t_0)$. Thus (3.15) holds and the proof is complete.

References

[1] Hu, Shouchuan, Lakshmikantham, V. and Rama Mohana Rao, M., Nonlinear variation of parameters formula for integro-differential equations of Volterra type. To appear in J.M.A.A.

[2] Lakshmikantham, V. and Leela, S., Differential and integral inequalities, Vol. I, Academic Press, New York, 1969.

[3] Lakshmikantham, V., Leela, S., and Rama Mohana Rao, M., Integral and integro-differential inequalities. Applicable Analysis, 24 (1987), 157-164.

[4] Rama Mohana Rao, M., and Srinivas, P., Asymptotic behavior of solutions of Volterra integro-differential equations. Proc. Amer. Math. Soc. 94 (1985), 55-60.

[5] Shendge, G.R., A new approach to the stability theory of functional differential systems. J.M.A.A. 95 (1983), 319-334.

V. Lakshmikantham
The University of Texas at Arlington
Arlington
Texas 76019
U.S.A.

G. LEUGERING

Boundary controllability of a viscoelastic string

ABSTRACT

A vibrating linear viscoelastic string of finite length is considered to be controlled from the boundary by a $L_2(0,T)$-function entering the Dirichlet boundary conditions. It is shown that materials with regular stress/strain relaxation kernels reveal the same controltheoretic behaviour as the purely elastic string, while for weakly singular kernels (spring/pot material) exact controllability does not hold. In this case we show approximate controllability in finite time T.

1. INTRODUCTION

Let u(x,t) be the deviation from the equilibrium state at the material point x at time t of a viscoelastic one dimensional, isotrotopic and incompressible material, which is confined between x = 0 and x = ℓ. It is well known, see for instance [2], [6], that the equation governing the motion can be stated as follows:

$$u_{tt}(x,t) = b\, u_{xx}(x,t) + \int_{-\infty}^{t} m(t-\tau)(u_{xx}(x,t) - u_{xx}(x,\tau))d\tau. \tag{1}$$

Here the subscripts denote partial differentiation w.r.t. the indicated variable, b > 0 and $\int_0^\infty m(\tau)d\tau$, if the integral exists (which is assume throughout this paper), are final and initial moduli of stress, while m(t) is the stress/strain relaxation kernel, which may be regular (see [8], [9], [10]) in some cases and weakly singular at t = 0 else. The kernel is supposed to adequately model the long run memory of the material w.r.t. the deformation history. It is plain that the value m(0) at t = 0 has a great impact on the propagation properties of discontinuities coming in, say, from the boundary. How this works out mathematically has been investigated in a series of papers, which we would like to refer to, [6], [8], [9], [10].

We distinguish between regular kernels, m, satisfying

$$\int_0^\infty m(t)dt < \infty, \ m(0) < \infty \tag{2}$$

and weakly singular ones, in sense of

$$\int_0^\infty m(t)dt < \infty, \ m(0) = \infty. \tag{3}$$

In either case, m is a positive monotone decreasing function. The common examples are the standard-linear solid (Poynting-Thomson-model) i.e. material of the Boltzmann-type, with $m(t) = \gamma \exp(-dt)$ for the regular case (2), and the fractional-derivative model (spring-pot-model) (see [1], [7]) with $m(t) = \gamma/\Gamma(1-\alpha)\cdot t^{-\alpha} \exp(-dt)$, $0 < \alpha < 1$, $d > 0$, or $m(t) = \sum_{n=1}^{\infty} \exp(-n^{\alpha}t)$, $\alpha > 1$, see [10], for the weakly singular case.

To be more specific, we consider a virgin-type material, that is a material being at rest before $t = 0$. Thus, initial Cauchy data have to be imposed

$$u(x,0) = u_0(x), \quad u_t(x,0) = v_0(x), \quad x \in (0,\ell). \tag{4}$$

Since we are going to control the process (1), (4) from the boundary, say at $x = 0$, we have the boundary conditions

$$u(0,t) = \varphi(t), \ u(\ell,t) = 0 \quad t > 0. \tag{5}$$

The control problem is the following:

(CP) Given any pair (u_0,v_0), $(u_1,v_1) \in L_2(0,\ell) \times H^{-1}(0,\ell)$ and $T > 0$. Does there exist a function $\varphi \in L_2(0,T)$ such that the solution $u(x,t,u_0,v_0,\varphi)$ of (1), (4), (5) satisfies $u(x,T,u_0,v_0,\varphi) = u_1(x)$, $u_t(x,T,u_0,v_0,\varphi) = v_1(x)$, $x \in (0,\ell)$?

If the answer to this question is affirmative we can ask whether this job can be done by a bounded control, φ, s.t. $\|\varphi\|_{L_2} \leq M$, $M > 0$ given in advance. If this, in turn, is possible we ask for controls doing all this in the least time possible. Of course, since the process is dissipative, if (CP) holds it does so for bounded controls and possibley large T. The existence of time optimal controls then follows by standard techniques.

As a result, since in the regular case (2) the speed of wave-propagation is given by $c = (b + \int_0^\infty m(t)dt)^{1/2} < \infty$ the control signals have to travel along the string with finite speed, thus exact controllability cannot be expected for times smaller than $2\ell c^{-1}$. Further, since in this case there is no smoothing, exact controllability does indeed occur for $T \geqq 2\ell c^{-1}$. On the other hand, in case of weakly singular kernels, (3), as was worked out by Renardy and others, there is a finite speed of propagation, too, the remarkable difference, however, is that we have a certain degree of smoothing (C^∞-smoothing of wave fronts, finite-gain else). This amounts to saying that exact controllability never holds, while approximative controllability is likely to occur, and is indeed shown, for times $T \geqq 2\ell c^{-1}$. From the control theoretic point of view this appears to be a new type of controllability result, since, somehow, it interpolates between controllability results for hyperbolic and parabolic equations.

2. REPRESENTATION OF SOLUTIONS

Almost everything given in this section has already been done in [6], [8], [10]. What has, so far, perhaps not been considered is the inhomogeneous boundary value problem with $L_2(0,T)$-data in a weak energy-space setting. However, supplied with the previous work mentioned above this is rather easily done. We need the Laplace transform, $\bar{u}(x,s) = \int_0^\infty \exp(-st)u(x,t)dt$, of u w.r.t. the time, when applied to (1), (4), (5):

$$\left.\begin{aligned} s^2\bar{u} &= (b + \bar{m}(0) - \bar{m}(s))\bar{u}_{xx} + su(0) + u_t(0) \\ \bar{u}(0,s) &= \bar{\varphi}(s),\ \bar{u}(\ell,s) = 0 \end{aligned}\right\} \tag{6}$$

which is treated separately

$$\text{P1:}\ \begin{cases} \bar{u}_{xx} = s^2/(b + \bar{m}(0) - \bar{m}(s))\bar{u} + 1/(b+\bar{m}(0)-\bar{m}(s))\cdot(su_0 + v_0) \\ \bar{u}(0,s) = \bar{u}(\ell,s) = 0 \end{cases}$$

$$\text{P2:}\ \begin{cases} \bar{u}_{xx} = s^2/(b + \bar{m}(0) - \bar{m}(s))\bar{u} \\ \bar{u}(0,s) = \bar{\varphi}(s),\ \bar{u}(\ell,s) = 0 \end{cases}$$

Problem P1 is the transformed system with homogeneous boundary conditions and non vanishing initial conditions, while P2 is the transform of the control-to-state part of the problem, which is the one of main interest to us. Defining $\eta^2(s) := s^2/(b + \bar{m}(0) - \bar{m}(s))$ the solution of P2 is given by

$$P_2' \qquad \bar{u}(s,x) = \bar{\varphi}(s) \frac{\sinh((\ell-x)\eta(s))}{\sinh(\ell\eta(s))}$$

$$= \bar{\varphi}(s)\{e^{-x\eta(s)} + [e^{-(x+2\ell)\eta(s)} -$$

$$- e^{-(2\ell-x)\eta(s)}] + [\ldots] + \ldots\}.$$

Obviously, in this representation, $e^{-x\eta(s)}$ plays a fundamental role. Fortunately, this is the key expression in the very well studied half-space-case, and is, therefore, very well understood. With the usual branch cut along the negative axis, we first consider this half space problem, see [6], [8], [10]

$$\tilde{u}(x,t) := \frac{1}{2\pi i} \int_{\gamma-i\infty}^{\gamma+i\infty} e^{st} \frac{1}{s} e^{-xs/(b+\bar{m}(0)-\bar{m}(s))^{1/2}} ds$$

$$= \frac{1}{2\pi i} \int_{\gamma-i\infty}^{\gamma+i\infty} \frac{1}{s} e^{s(t-\alpha x)} e^{-m(0)x\alpha^3/2\cdot(1+0(s^{-\theta}))} ds$$

where we have assumed in addition to (2):

$$m \in (W^{1,1}[0,\infty),\ W^{1,2}[0,\infty))_{\theta,q},\ 0 \leqq \theta \leqq 1,\ 1 < q < \infty \tag{7}$$

and defined $\alpha = c^{-1}$ $(c = (b + \bar{m}(0))^{1/2})$, $\gamma > 0$. It can be shown along the lines of [8] that $\tilde{u}(x,t) \equiv 0$ for $x > ct$, that $\tilde{u}(x,\alpha x) = \exp(-m(0)x\alpha^3/2)$ is the discontinuity at the front $x = ct$ and that $u(x,t)$ is as regular as m for $x < ct$. As for the singular case (3) we recall the results of Renardy [10] or [6]: Again $\tilde{u}(x,t) \equiv 0$ for $x > ct$. However, on the line $x = ct$ the solution is C^∞-smooth, while for $x < ct$ $u(x,t)$ is even real analytic. In any case, defining $\bar{\gamma} = -m(0)/(b + \bar{m}(0))$, we can represent the solution $u(x,t)$ of the subproblem P2:

$$u(x,t) = \sum_{\substack{k=0\\ k\ \text{even}}}^{N} e^{\alpha(\ell\cdot k+x)\bar{\gamma}/2}\,\varphi(t-\alpha\ell k-\alpha x)$$

$$-\sum_{\substack{k=1\\ k\ \text{odd}}}^{N} e^{\alpha(\ell(k+1)-x)\bar{\gamma}/2}\,\varphi(t-\alpha(k+1)\ell+\alpha x)$$

$$+\sum_{\substack{k=0\\ \text{even}}}^{N} \int_{\alpha(x+k\ell)}^{t} \varphi(t-s)f_s(s,x+k\ell)ds \tag{8}$$

$$-\sum_{\substack{k=1\\ \text{odd}}}^{N} \int_{\alpha((k+1)\ell-x)}^{t} \varphi(t-s)f_s(s,\ell(k+1)-x)ds$$

$kN\alpha \leqq t < (\alpha+1)\ kN.$

Note that this formula is understood in the singular case, (3), too, since then $\bar{\gamma} = -\infty$ and the first two series vanish.

As for the solution, u, of the subproblem P1 we have:

$$\text{P1}'\quad \bar{u}(x,s) = \frac{1}{(b+\bar{m}(0)-\bar{m}(s))}\cdot\frac{1}{\sinh(\ell\cdot\eta(s))}$$

$$\cdot\ \{\sinh(\eta(s)(\ell-x))\int_0^x \sinh(\eta(s)\xi)(su_0(\xi)+v_0(\xi))d\xi$$

$$+\sinh(\eta(s)x)\int_x^{\ell} \sinh(\eta(s)(\ell-\xi))(su_0(\xi)+v_0(\xi))d\xi\}.$$

Again, in P_1' the hyperbolic sine can be developed into an infinite series of $\exp(-ys/(b+\bar{m}(0)-\bar{m}(s))^{1/2})$-terms. One can argue as in [3] that the inverse Laplace-transformation procedure can be carried out term by term. It is then easy to obtain the following theorem:

THEOREM 1. Let $u_0 \in L_2(0,\ell)$, $v_0 \in H^{-1}(0,\ell)$, $\varphi \in L_2(0,T)$, $T > 0$ be given in advance. Then there exists a unique solution, $u(x,t,u_0,v_0,\varphi)$, of (1), (4), (5), which is explicitly given by superposition of the solutions of P1 and P1. Furthermore, as regards the regularity, we have for

i) regular kernels, m, with (2), (7):

$$u \in C(0,T,L_2(0,\ell)) \cap C^1(0,T,H^{-1}(0,\ell));$$

ii) weakly singular kernels, m, with

$$m(t) = \frac{\gamma}{\Gamma(1-\alpha)} \frac{e^{-\delta t}}{t^\alpha},\ 0 < \alpha < 1,\ \delta > 0 \text{ or}$$

$$m(t) = \sum_{n=1}^{\infty} e^{-n^\alpha t},\ \alpha > \frac{1}{2}$$

$$u \in C(0,T,H^1(0,\ell)) \cap C^1(0,T,L_2(0,\ell)).$$ □

REMARK 1.

i) As regards the weakly singular kernels, much more can be said, using the recent paper by Prüss [9]. In particular if $a_1(t) = \int_t^\infty m(\tau)d\tau$ is positive and logarithmic convex, then the solution of (1) is shown to reveal the same properties as in the cases above, with m given explicitly. The method of proof, however, is much more involved, and cannot be given by simple evaluations of contour integrals.

ii) As indicated above, much more can be said concerning stationary singularities, decay of the amplitude of the wave fronts, etc. The interested reader is referred to [6], [8], [9], [10]. □

3. CONTROLLABILITY

We recall the definitions of controllability:

DEFINITION 1. Given (u_0,v_0), $(u_1,v_1) \in L_2(0,\ell) \times H^{-1}(0,\ell)$ and $T > 0$, we say that (1), (4), (5) is exactly controllable if there exists a $\varphi \in L_2(0,T)$ such that the solution, $u(x,t,u_0,v_0,\varphi)$ of (1), (4), (5) satisfies $u(x,T,u_0,v_0,\varphi) = u_1(x)$, $u_t(x,T,u_0,v_0,\varphi) = v_1(x)$. The same system is said to be approximatively controllable in case for each $\varepsilon > 0$ there is a $\varphi \in L_2(0,T)$ s.t. $\|u(\cdot,T,u_0,v_0,\varphi) - u_1(\cdot)\|_{L_2} < \varepsilon$ and $\|u_t(\cdot,T,u_0,v_0,\varphi) - v_1(\cdot)\|_{H^{-1}} < \varepsilon$.

One can restate the controllability properties in terms of the solution operators:

DEFINITION 2. Let $u(x,t,0,0,\varphi)$ be the solution of P2, in fact (1), (4), (5) with $u_0 = v_0 = 0$. Then the continuous linear map $L(t) : L_2(0,T) \to L_2(0,\ell) \times H^{-1}(0,\ell)$ given by $L(t)\varphi := (u(\cdot,t,0,0,\varphi), u_t(\cdot,t,0,0,\varphi))$ is called the control-to-state map. □

It is plain, by Theorem 1, that (1) is exactly controllable if and only if the control-to-state map is surjective at $t = T$, while (1) is approximatively controllable if and only if $L(T)$ has dense range.

We set $t = T = 2\alpha\ell$. In order to give a precise meaning, at least in the $L_2(0,\ell)$ sense, to:

$$\begin{aligned} u(x,T,u_0,v_0,\varphi) &= u_1(x) \\ u_t(x,T,u_0,v_0,\varphi) &= v_1(x). \end{aligned} \tag{9}$$

We integrate the second equation w.r.t. x, that is we are looking for an identity in terms of the representers of the H^{-1}-functions. Then (9) reads as follows (look at (8)):

$$\begin{aligned} &e^{\alpha x\bar{\gamma}/2}\varphi(2\alpha\ell-\alpha x) - e^{\alpha(2\ell-x)\bar{\gamma}/2}\varphi(\alpha x) \\ &+ \int_{\alpha x}^{2\alpha\ell} \varphi(2\alpha\ell-s)f_s(s,x)ds - \int_{\alpha(2\ell-x)}^{2\alpha\ell} \varphi(2\alpha\ell-s)f_s(s,2\ell-x)ds = u_1(x) \end{aligned} \tag{10$_1$}$$

and the second identity of (9):

$$\begin{aligned} &-\frac{1}{2}\, e^{\alpha x\bar{\gamma}/2}\,\varphi(2\alpha\ell-\alpha x) - \frac{1}{\alpha}\, e^{\alpha(2\ell-x)\bar{\gamma}/2}\,\varphi(\alpha x) + \\ &+ \int_0^x \int_{\alpha\tau}^{2\alpha\ell} \varphi(2\alpha\ell-s)f_s(s,\tau)ds\, d\tau - \\ &- \int_0^x \int_{\alpha(2\ell-\tau)}^{2\alpha\ell} \varphi(2\alpha\ell-s)f_s(s,2\ell-\tau)ds\, d\tau = \int_0^x v_1(s)ds. \end{aligned} \tag{10$_2$}$$

Since $x \in [0,\ell]$ it is reasonable to view $\varphi(2\alpha\ell-\alpha x)$ and $\varphi(\alpha x)$ as different unknown functions in $L_2(0,\ell)$:

$$\psi(x) := \varphi(\alpha x)$$

$$\theta(x) := \psi(2\ell - x) = \varphi(\alpha x), \quad x \in [0,\ell].$$

It is then a matter of straightforward, but lengthly and cumbersome calculus to transform the system (10) into an identity, which is much more appealing, i.e. we can find a matrix valued square integrable kernel, $K_{ij}(x,t)$, i,j = 1,2 such that ψ,θ, satisfy the itnegral equation in $L_2(0,\ell)^2$

$$\begin{bmatrix} 0 & e^{\alpha x\bar{\gamma}/2} \\ -e^{\alpha x\bar{\gamma}/2} & e^{\alpha(2\ell-x)\bar{\gamma}/2} \end{bmatrix} \begin{bmatrix} \psi \\ \theta \end{bmatrix}(x) + \tag{11}$$

$$+ \int_0^\ell \begin{bmatrix} K_{11} & K_{12} \\ K_{21} & K_{22} \end{bmatrix} (x,s) \begin{bmatrix} \psi \\ \theta \end{bmatrix} (s)\, ds = \begin{bmatrix} F_1 \\ F_2 \end{bmatrix} (x)$$

with $F_1(x) = \frac{1}{2}\,(u_1(x) + \alpha\int_0^x v_1(s)ds)$, $F_2(x) = u_1(x)$.

Note that without the decaying memory, that is in the purely elastic case (perfect memory) (11) is just the equation one meets in the control problem for the string, which can, in turn, evidently be solved explicitly. However, with long memory, (11) is a Fredholm-integral-equation of the second kind, if the relaxation kernel is regular, (3), and of the first kind, if it is weakly singular (4).

As a partial result, in the regular case, the control-to-state map has closed range and one might expect an exact controllability result, while in the weakly singular one definitely cannot expect such a property. Of course, in order to get more information in this direction, the adjoint control-to-state map has to be investigated as regards its kernel.

$$L(2\alpha\ell) : L_2(0,2\alpha\ell) \to L_2(0,\ell) \times H^{-1}(0,\ell)$$

$$L(2\alpha\ell)^*: L_2(0,\ell) \times H_0^1(0,\ell) \to L_2(0,2\alpha\ell)$$

with quite a bit of patience one calculates

$$L(2\alpha\ell)^*(h^1,h^2)(t) = - ch^1(ct)e^{(2\alpha\ell-t)\bar{\gamma}/2}$$

$$- c^2 \dot{h}^2(ct)e^{(2\alpha\ell-t)\bar{\gamma}/2}$$

$$- \int_{ct}^{\ell} h^1(x)f_t(2\alpha\ell-t, \ell-x)dx$$

$$+ ch^2(ct) \frac{\bar{\gamma}}{2} e^{(2\alpha\ell-t)\bar{\gamma}/2} + ch^2(ct)f_t(t,2\ell-ct)$$

$$+ \int_{ct}^{\ell} h^2(\tau)f_{tt}(2\alpha\ell-t, 2\ell-c\tau)d\tau \text{ for } t \in [0,2\ell]$$

and

$$L(2\alpha\ell)^*(h^1,h^2)(t) = ch^1(2\ell-ct)e^{(2\alpha\ell-t)\bar{\gamma}/2} - c^2 e^{(2\alpha\ell-t)\bar{\gamma}/2} h^2(2\ell-ct) +$$

$$+ \int_0^{2\ell-ct} h^1(x)f_t(2\alpha\ell-t,x)dx - ch^2(2\ell-ct) \frac{\bar{\gamma}}{2} e^{(2\alpha\ell-t)\frac{\bar{\gamma}}{2}}$$

$$-c h^2(2\ell-ct)f_t(2\alpha\ell-t, 2\ell-ct) -$$

$$- \int_0^{2 -ct} h^2(\tau)f_{tt}(2\alpha\ell-t,\tau)d\tau \text{ for } t \in [\alpha\ell,2\alpha\ell].$$

The experience leads one to suggest that $L(2\alpha\ell)^*(h^1,h^2)$ (1) is intimately related to a homogeneous backwards running adjoint equation, or more conveniently, that after reflection, $L^*(2\alpha\ell)(h^1,h^2)(2\alpha\ell-t)$, is connected with the original problem. Moreover, since looking for the kernel $\ker(L^*(2\alpha\ell)(\cdot))$ amounts to assuming $L^*(2\alpha\ell)(h^1,h^2) \equiv 0$ for some $(h^1,h^2) \in L_2(0,\ell) \times H_0^1(0,\ell)$, we expect an additional boundary condition to hold, as is usual in boundary control problems.

Because of the limited space available, we dispense here with presenting the details, even though it is quite interesting to see how everything works out explicitly, due to the very explicit presentation of solutions. Suffice it to state, that one precisely confirms the suggestions above, in that $L(2\alpha\ell)^*(h^1,h^2)(2\alpha\ell-t)$ is given by (we confine ourselves to $[0,\alpha\ell]$)

$$L(2\alpha\ell)(h^1,h^2)^*(2\alpha\ell-t) = \frac{\partial}{\partial x}\Big|_{x=0^+} (bw(x,t) + \int_{-\infty}^{t} m(t-s)(w(x,t)-w(x,s))ds) \tag{12}$$

where $w(x,t) = u(x,t,0,-h^2,h^1)$, u being a mild solution of (1) with homogeneous boundary conditions and initial condition $u(x,0) = -h^2 \in H_0^1(0,\ell)$ and $u_t(x,0) = h^1 \in L_2(0,\ell)$, that is a solution satisfying $(u,u_t) \in H_0^1(0,\ell) \times L_2(0,\ell)$ or equivalently having finite "energy". It is important to observe that (12) leads, in case $(h^1,h^2) \in \ker L^*(2\alpha\ell)(\cdot)$, to an additional boundary condition:

$$(h^1,h^2) \in \ker L^*(2\alpha\ell) \iff \frac{\partial}{\partial x}\Big|_{x=0^+} (bw(x,t) + \int_{-\infty}^{t} m(t-s)(w(x,t)-w(x,s))ds) = 0 \text{ on } (0,2\alpha\ell) \tag{13}$$

that is the stress is zero at $x = 0$, too.

We proceed to show that this cannot happen without h^1,h^2 being zero. To begin with, we observe an Abelian type result: If $F(t)$ has a Laplace transform $f(s)$ having a half plane of uniform convergence, then

$$F(t) = 0 \text{ on } (0,T) \Longrightarrow f(s) = O(\exp(-\mathrm{Re}\,s)T) \quad s \to \infty \tag{14}$$

Now, since m is supposed to be integrable, $L^*(2\alpha\ell)(h^1,h^2)$ is zero on $(0,T)$ iff $\frac{\partial}{\partial x}\Big|_{x=0^+} w(x,t) = 0$ on $(0,2\alpha\ell)$. Hence, by (14) even for real t: We put $u_o = -h^2$, $v_0 = h^1$):

$$\frac{\partial}{\partial x}\Big|_{x=0^+} \bar{w}(x,t) = \frac{\eta^2(t)}{t} \int_0^{\ell} \frac{\sinh(\eta(t)(\ell-\xi))}{\sinh(\xi(t,\ell)} (u_0(\xi) + \frac{1}{t} v_0(\xi))d\xi$$
$$= O(\exp(-t2\alpha\ell)) \quad t \to \infty. \tag{15}$$

By our positiveness assumption on $m(t)$:

$$\frac{\sinh \eta(t)(\ell-\xi)}{\sinh \eta(t)\ \ell} = \frac{\sinh \alpha t(\ell-\xi)}{\sinh \alpha t\ \ell} (1 + O(\tfrac{1}{t})) \quad t \to \infty, \tag{16}$$

reflecting once again the fact that at the very beginning the viscoelastic string behaves almost elastic. Expressing the hyperbolic sine in terms of exponentials, we are, by (15), (16), left with:

$$e^{2\alpha\ell t} \frac{\partial}{\partial x}\bigg|_{x=0^+} \bar{w}(x,t) = \frac{\eta^2(t)}{t} \int_0^{2\alpha\ell} e^{t\xi}(U_0(\xi) + \frac{1}{t} V_0(\xi))d\xi$$
$$\cdot (1 + o\,\frac{1}{t}) = o(\frac{1}{t}), \quad t \to \infty. \tag{17}$$

With

$$U_0(\xi) = \begin{cases} u_0(\xi) & \xi \in [0,\alpha\ell] \\ -u_0(2\alpha\ell-\xi) & \xi \in (\alpha\ell,2\alpha\ell] \end{cases}$$

and $V_0(\xi)$ in the same way. Now, (17) implies

$$\int_0^{2\alpha\ell} e^{t\xi}(U_0(\xi) + \frac{1}{t} V_0(\xi))d\xi = 0 \quad \forall t > 0 \tag{18}$$

which appears to be equivalent to the moment problem:

$$\int_0^{2\alpha\ell} \xi^n\, U_0(\xi)d\xi = \frac{1}{n+1} \int_0^{2\alpha\ell} \xi^{n+1}\, V_0(\xi)d\xi$$
$$\int_0^{2\alpha\ell} V_0(\xi)d\xi = 0 \tag{19}$$

which in turn implies

$$U_0'(\xi) = V_0(\xi) \text{ on } [0,2\alpha\ell] \tag{20}$$

which is a contradiction to the definition of U_0 and V_0. We conclude that (15) implies $h^1 = v_0 = h^2 = -u_0 = 0$, i.e., that the adjoint $L^*(2\alpha\ell)$ has a trivial kernel.

THEOREM 2. Let $T \geq 2\alpha\ell$ be given. Then the system (1), (4), (5) is:

i) exactly controllable if m is regular;

ii) approximately controllable if m is weakly singular, in the sense of Definition 1.

REMARK 2.

While in the case of a differential equation the controllability defined in Definition 1 automatically implies that one can stay at the rest configuration, by switching off the control when arriving there, this property is not shared by system (1), simply because "zero" is not an equilibrium. That is to say, if one is interested not only to "hit a target" state (u_1,v_1), but to reach the zero state, say, and stay there one would have to solve an exact tracking problem, the reference path being just the history of the process. To this author, it seems very unlikely that this can be done. What, definitely, can be done is minimizing the viscoelastic energy at the final point $T = 2\alpha\ell$, when one has reached zero, in other words one can minimize the effect of the history on the energy w.r.t. all controls steering the initial data to zero. Since by Dafermos [2], the process can be shown to be asymptotically stable, we can then switch off the control at $t = 2\alpha\ell$ and be sure that there will be no blow up of energy afterwards. A different device would be to add a feedback control as in [4], [5].

References

[1] R.L. Bagley and P.J. Torvik, Fractional calculus - a different approach to the analysis of viscoelastic damped structures, AIAA Journal 21 (1983), 741-748.

[2] C.M. Dafermos, An abstract Volterra equation with application to linear viscoelasticity, J. Diff. Equations 7 (1970), 554-569.

[3] G. Doetsch, Handbuch der Laplace-Transformation Bd. I, Verlag Birkhäuser, Basel (1950).

[4] K.B. Hannsgen, A boundary control problem for viscoelastic materials, this volume.

[5] K.B. Hannsgen and R.L. Wheeler, Time delays and boundary feedback stabilization in one-dimensional viscoelasticity, Proc. Conf. on Control of Distr. Parameter Systems, Vorau (1986), to appear.

[6] W.J. Hrusa and M. Renardy, On wave propagation in linear viscoelasticity, Quart. Appl. Math. 43 (1985), 237-253.

[7] R.C. Koeller, Applications of fractional calculus to the theory of viscoelasticity, J. of Appl. Mech. 51 (1984), 299-307.

[8] A. Nerrain and D.D. Joseph, Linearized dynamics for step jumps of velocity and displacement of shearing flows of a simple fouid, Rheol. Acta 21 (1982), 228-250.

[9] J. Prüss, Positivity and regularity of hyperbolic Volterra Equations in Banach spaces, Math. Annalen, to appear.

[10] M. Renardy, Some remarks on the propagation and non-propagation of discontinuities in linearly viscoelastic liquids, Rheol. Acta 21 (1982), 251-254.

Günter Leugering
Fachbereich Mathematik
Technische Hochschule Darmstadt
Schlossgartenstrasse 7,
6100 Darmstadt
West Germany.

A. LORENZI & E. SINESTRARI

Stability results for a partial integrodifferential inverse problem

Both authors are partially supported by the Ministero della Pubblica Istruzione of Italy.

0. INTRODUCTION

This paper is devoted to the existence of maximal and global solutions and to the investigation of stability with respect to the data of the solution to an inverse problem related to the linear theory of materials with memory.

Such a problem was studied in [2] from the point of view of existence and uniqueness in the small. It consists of determining two functions $u: [0,T] \times \bar{\Omega} \to R$ and $K: [0,T] \to R$ such that

$$u_t(t,x) = Au(x,t) + \int_0^t (t-s)^{\alpha-1}K(t-s)Au(s,x)ds + f(t,x) \qquad (t,x) \in [0,T] \times \bar{\Omega} \tag{0.1}$$

$$u(0,x) = u_0(x) \qquad x \in \bar{\Omega} \tag{0.2}$$

$$u(t,x) = 0 \qquad (t,x) \in [0,T] \times \Gamma \tag{0.3}$$

$$\langle\phi,u(t,\cdot)\rangle = g(t) \qquad t \in [0,T] \tag{0.4}$$

Here $\alpha \in]0,1[$ is an assigned constant and Ω is a bounded open connected set in R^n with a boundary Γ of class $C^{2+\delta}(\delta \in]0,1[)$.

The linear differential operator A is defined by the equation

$$(Au)(x) = \sum_{h,k}^{1,n} a_{h,k}(x)u_{x_hx_k}(x) + \sum_{h}^{1,n} a_h(x)u_{x_h}(x)+a(x)u(x) \quad x \in \bar{\Omega} \tag{0.5}$$

the coefficients $a_{h,k}$, a_h, $a: \bar{\Omega} \to R$ $(h,k = 1,\dots,n)$ being continuous functions. Moreover we assume that A is uniformly elliptic in $\bar{\Omega}$, i.e.

$$\sum_{h,k}^{1,n} a_{h,k}(x)\xi_h\xi_k > 0 \quad x \in \bar{\Omega},\ \xi \in R^n\setminus\{0\} \tag{0.6}$$

Then we observe that f, u_0 and g are prescribed functions, while ϕ is a given finite Borel measure on Ω. In particular ϕ may be chosen to be a Dirac measure concentrated at some point $x_1 \in \Omega$ (i.e. $\langle\phi, u(\cdot,t)\rangle = u(t,x_1)$).

Finally we observe that, similarly to what we did in [2], we shall study here an abstract generalization of problem (0.1)-(0.4) by the analytic semigroup theory.

Our paper is organized as follows: in Section 1 we give the exact assumptions and the statement of our main result (Theorem 1.1); Section 2 contains the proofs of some auxiliary results; in Section 3 we give the proof of Theorem 1.1; Section 4 is devoted to the existence of maximal and global solutions to our problem.

1. ASSUMPTIONS AND RESULTS

In this section we list assumptions and we state the main result of this paper.

We denote by X a Banach space with norm $\|\cdot\|$ and consider $\alpha, \beta \in]0,1[$ such that $\beta < \min(\alpha, 1-\alpha)$.

We suppose that

(H.1) $A: D_A\ X \to X$ is the generator of an analytic semigroup $t \to e^{At}$ such that

$$\|t^h A^h e^{At}\|_{L(X)} \leq M_h \qquad h = 0,1,2,\ldots .$$

Moreover we can suppose without any loss of generality that A is invertible and $A^{-1} \in L(X)$ (see Remark 2.1. in [2]).

(H.2) $B: D_B \subset X \to X$ is a closed linear operator such that $D_B \supset D_A$;

(H.3) $f \in C^{1+\beta}([0,T];X)$;

If $Y = D_A$ is endowed with the graph norm, we denote by K a given closed subspace of $L(X) \cap L(Y)$. We shall also assume that

(H.4) $\Phi \in L(X;K)$;

(H.5) $G \in C^1([0,T];\mathcal{K}) \cap C^2([0,T];\mathcal{K})$ and $t \to t^{1-\alpha}G''(t)$ belongs to $C^\beta([0,T];\mathcal{K})$;

(H.6) $u_0 \in Y$, $Bu_0 \in Y$, $v_0 = Au_0 + f(0) \in Y$ and $Av_0 + f'(0) \in D_A(\beta,+\infty)$, where $D_A(\beta,+\infty)$ is the interpolation space defined by

$D_A(\beta,+\infty) = \{x \in X: [u]_\beta = \sup_{t>0} \|t^{1-\beta}Ae^{At}u\| < +\infty\}$

with norm $\|u\|_{D_A(\beta,+\infty)} = \|u\| + [u]_\beta$

(for more details and properties of these spaces see Section 1 of [3]);

(H.7) $G(0) = \Phi u_0$;

(H.8) $G'(0) = \Phi(Au_0 + f(0))$;

(H.9) for any $H \in \mathcal{K}$ there exists a unique solution $K = \Psi(H) \in \mathcal{K}$ of the equation $\Phi(K(Bu_0)) = H$ and Ψ belongs to $L(\mathcal{K})$.

REMARK 1.1. The motivation for the introduction of the subspace $\mathcal{K}$ may be found in [2, Section 5].

REMARK 1.2. A necessary condition for Ψ to exist is obviously $Bu_0 \neq 0$.

Under these assumptions we proved in [2] that for sufficiently small $T > 0$ there exists a unique pair (u,K) of functions

$$u \in C^1([0,T];Y) \cap C^2(]0,T];X), \quad K \in C^\beta([0,T];\mathcal{K}) \tag{1.1}$$

solution to the problem

$$u'(t) = Au(t) + \int_0^t (t-s)^{\alpha-1}K(t-s)Bu(s)ds+f(t) \quad t \in [0,T] \tag{1.2}$$

$$u(0) = u_0 \tag{1.3}$$

$$\Phi(u(t)) = G(t) \quad t \in [0,T]. \tag{1.4}$$

In Section 3 we shall prove the following stability result.

THEOREM 1.1. Let f_i, u_{oi} and G_i $(i = 1,2)$ verify the same assumptions as f, u_o and G and let (u_i, K_i) be the corresponding solution of (1.1)-(1.4) in a given interval $[0,T]$. Then we have:

$$\begin{aligned}
&\|u_2-u_1\|_{C^1([0,T];Y)} + \|K_2-K_1\|_{C^\beta([0,T];K)} \leq C(T,\tilde{M},\bar{M}) \times \\
&\times \{\|f_2-f_1\|_{C^{1+\beta}([0,T];X)} + \|u_{o2}-u_{o1}\|_Y + \|Bu_{o2}-Bu_{o1}\|_Y + \\
&+ \|A^2u_{o2}-A^2u_{o1} + Af_2(0)-Af_1(0)+f_2'(0)-f_1'(0)\|_{D_A(\beta,+\infty)} + \\
&+ \|t^{1-\alpha}G_2''-t^{1-\alpha}G_1''\|_{G^\beta([0,T];K)}\}.
\end{aligned} \tag{1.5}$$

Here $c(\cdot,\tilde{M},\bar{M})$ is an increasing function from R_+ into itself, $\tilde{M}$ is such that

$$\max\{\|Bu_{o1}\|_Y, \|Bu_{o2}\|_Y\} \leq \tilde{M} \tag{1.6}$$

and $\bar{M}$ verifies

$$\max\{\|K_1\|_{C^\beta([0,T];K)}, \|K_2\|_{C^\beta([0,T];K)}\} \leq \bar{M} \tag{1.7}$$

C depends also on α,β, M_o, M_1, M_2, $\|A^{-1}\|_{L(X)}$, $\|BA^{-1}\|_{L(X)}$, $\|\Psi\|_{L(K)}$ and $\|\Phi\|_{L(X,K)}$.

2. A PRIORI ESTIMATES

Suppose that (u,K) is a solution to (1.2)-(1.4) verifying (1.1). Then, taking advantage of the results in [2], we get that the pair (w,K), where

$$w(t) = Au'(t) \qquad t \in [0,T] \tag{2.1}$$

belongs to $C([0,T];X) \times C^\beta([0,T];K)$ and is a solution to the following system

$$w = w_o + BA^{-1}[u_1(K,w) + u_2(K,u_o)] \tag{2.2}$$

$$K = \Psi K_o - \Psi\Phi[v_1(K,w) + v_2(K,u_o) + v_3(K,w)]. \tag{2.3}$$

The functions w_o, K_o, u_i, v_i $(i = 1,2,3)$ appearing in (2.2), (2.3) are so defined in $[0,T]$:

$$w_o(t) = B(e^{At}v_o + \int_0^t e^{A(t-s)}f'(s)ds) \tag{2.4}$$

$$K_o(t) = t^{1-\alpha}G''(t) - t^{1-\alpha}\Phi A(e^{At}v_o + \int_0^t e^{A(t-s)}f'(s)ds) - t^{1-\alpha}\Phi f'(t) \tag{2.5}$$

$$u_1(K,w)(t) = A\int_0^t e^{A(t-s)}u_3(K,w)(s)ds \tag{2.6}$$

$$u_2(K,u_o)(t) = \int_0^t e^{A(t-s)}s^{\alpha-1}AK(s)Bu_o ds \tag{2.7}$$

$$u_3(K,w)(t) = \int_0^t (t-s)^{\alpha-1}K(t-s)w(s)ds \tag{2.8}$$

$$v_1(K,w)(t) = t^{1-\alpha}u_1(K,w)(t) \tag{2.9}$$

$$v_2(K,u_o)(t) = t^{1-\alpha}u_2(K,u_o)(t) \tag{2.10}$$

$$v_3(K,w)(t) = t^{1-\alpha}u_3(K,w)(t) \tag{2.11}$$

<u>LEMMA 2.1</u>. Setting for each $t \in]0,T]$

$$\lambda(w)(t) = \|w\|_{C([0,t];X)} \tag{2.12}$$

$$\mu(K)(t) = \|K\|_{C^\beta([0,T];K)} \tag{2.13}$$

and $u_i(\cdot,\cdot)(t) = u_i(t)$, $v_i(\cdot,\cdot)(t) = v_i(t)$ $(i = 1,2,3)$, we have for $0 < t \leq T$:

$$\|u_1(t)\| \leq c_1(t)\int_0^t (t-s)^{\alpha-\beta-1}\mu(K)(t-s)\lambda(w)(s)ds \tag{2.14}$$

$$[u_1]_{C^\beta([0,t];X)} \leq c_2(t)\int_0^t (t-s)^{\alpha-\beta-1}\mu(K)(t-s)\lambda(w)(s)ds \tag{2.15}$$

$$\|u_2(t)\| \leq c_3(t)\,\|Bu_o\|_Y \int_0^t s^{\alpha-\beta-1}\mu(K)(s)ds \tag{2.16}$$

$$[u_2]_{C^\beta([0,t];X)} \le c_4(t)\, \|Bu_o\|_Y \int_0^t s^{\alpha-\beta-1}\mu(K)(s)ds \tag{2.17}$$

$$\|u_3(t)\| \le c_5(t) \int_0^t (t-s)^{\alpha-\beta-1}\mu(K)(t-s)\lambda(w)(s)ds \tag{2.18}$$

$$[u_3]_{C^\beta([0,t];X)} \le c_6(t) \int_0^t (t-s)^{\alpha-\beta-1}\mu(K)(t-s)\lambda(w)(s)ds \tag{2.19}$$

$$\|v_1\|_{C^\beta([0,t];X)} \le c_7(t) \int_0^t (t-s)^{\alpha-\beta-1}\mu(K)(t-s)\lambda(w)(s)ds \tag{2.20}$$

$$\|v_2\|_{C^\beta([0,t];X)} \le c_8(t)\, \|Bu_o\|_Y \int_0^t s^{\alpha-\beta-1}\mu(K)(s)ds \tag{2.21}$$

$$\|v_3\|_{C^\beta([0,t];X)} \le c_9(t) \int_0^t (t-s)^{\alpha-\beta-1}\mu(K)(t-s)\lambda(w)(s)ds \tag{2.22}$$

Here $[u]_{C^\beta([0,t];X)}$ denotes the Hölder coefficient of order β of u in $[0,t]$. Moroever the functions c_i $(i = 1,\dots,9)$ are increasing from R_+ into itself: they depend also on α,β,M_o,M_1 and M_2 (see formulas (2.24)-(2.29) and (2.31)).

Proof. It suffices to prove estimates (2.14)-(2.19), since (2.20)-(2.22) are immediate consequences of the following estimate

$$\|v\|_{C^\beta([0,t];X)} \le (t^{1-\alpha}+t^{1-\alpha-\beta})\, \|u\|_{C([0,t];X)} + t^{1-\alpha}[u]_{C^\beta([0,t];X)} \tag{2.23}$$

where $u \in C^\beta([0,T];X)$ and $v(t) = t^{1-\alpha}u(t)$, $t \in [0,T]$.

In particular from (2.14)-(2.19) and (2.23) we infer that the functions c_{6+j} $(j = 1,2,3)$ appearing in (2.20)-(2.22) are given by the formulas

$$c_{6+j}(t) = (t^{1-\alpha}+t^{1-\alpha-\beta})c_{2j-1}(t)+t^{1-\alpha}c_{2j}(t) \quad t \in [0,T],\ j = 1,2,3 \tag{2.24}$$

In the sequel we shall use the shorthands $\lambda(t) = \lambda(w)(t)$, $\mu(t) = \mu(K)(t)$ and the fact that functions $t \to \int_0^t (t-s)^{\alpha-\beta-1}\mu(t-s)\lambda(s)ds$ and $t \to \int_0^t s^{\alpha-\beta-1}\lambda(s)ds$ are non-decreasing.

From the Definition (2.8) of u_3 we immediately get (2.18) with

$$c_5(t) = t^{\beta} \tag{2.25}$$

To prove (2.19) let us suppose $0 \leqq t_1 < t_2 \leqq t$: we have

$$\|u_3(t_2)-u_3(t_1)\| = \Big\| \int_0^{t_1} (t_2-s)^{\alpha-1}[K(t_2-s)-K(t_1-s)]w(s)ds +$$

$$+ \int_0^{t_1} [(t_2-s)^{\alpha-1}-(t_1-s)^{\alpha-1}] K(t_1-s)w(s)ds +$$

$$+ \int_{t_1}^{t_2} (t_2-s)^{\alpha-1}K(t_2-s)w(s)ds \Big\| \leqq I_1 + I_2 + I_3$$

where

$$I_1 = (t_2-t_1)^{\beta} \int_0^{t_1} (t_2-s)^{\alpha-1}\mu(t_2-s)\lambda(s)ds \leqq (t_2-t_1)^{\beta}\int_0^{t_2} s^{\alpha-1}\mu(s)\lambda(t_2-s)ds \leqq$$

$$\leqq (t_2-t_1)^{\beta}t^{\beta} \int_0^{t} s^{\alpha-\beta-1}\mu(s)\lambda(t-s)ds$$

$$I_2 = (1-\alpha)^{-1} \int_0^{t_1} \Big(\int_{t_1-s}^{t_2-s} \tau^{\alpha-2}d\tau\Big)\mu(t_1-s)\lambda(s)ds \leqq$$

$$\leqq (1-\alpha)^{-1} \int_0^{t_1} \Big[(t_1-s)^{\alpha-\beta-1} \int_{t_1-s}^{t_2-s} \tau^{\beta-1}d\tau\Big]\mu(t_1-s)\lambda(s)ds \leqq$$

$$\leqq (1-\alpha)^{-1}\beta^{-1}(t_2-t_1)^{\beta} \int_0^{t_1} (t_1-s)^{\alpha-\beta-1}\mu(t_1-s)\lambda(s)ds \leqq$$

$$\leqq (1-\alpha)^{-1}\beta^{-1}(t_2-t_1)^{\beta} \int_0^{t} s^{\alpha-\beta-1}\mu(s)\lambda(t-s)ds$$

$$I_3 = \int_{t_1}^{t_2} (t_2-s)^{\alpha-1}\mu(t_2-s)\lambda(s)ds \leqq (t_2-t_1)^{\beta} \int_{t_1}^{t_2} (t_2-s)^{\alpha-\beta-1}\mu(t_2-s)\lambda(s)ds \leqq$$

$$\leqq (t_2-t_1)^{\beta}\int_0^{t_2} s^{\alpha-\beta-1}\mu(s)\lambda(t_2-s)ds \leqq (t_2-t_1)^{\beta} \int_0^{t} s^{\alpha-\beta-1}\mu(s)\lambda(t-s)ds$$

Whence (2.19) follows by setting

$$c_6(t) = t^{\beta} + (1-\alpha)^{-1}\beta^{-1} + 1. \tag{2.26}$$

From Definition (2.7) of u_2 we obtain (2.16) with

$$c_3(t) = M_o t^{\beta} \tag{2.27}$$

Likewise for $0 \leqq t_1 < t_2 \leqq t$ we obtain

$$\|u_2(t_2)-u_2(t_1)\| = \Big\| \int_0^{t_1} e^{As}(t_2-s)^{\alpha-1}[AK(t_2-s)-AK(t_1-s)]Bu_o ds +$$

$$+ \int_0^{t_1} e^{As}[(t_2-s)^{\alpha-1}(t_1-s)^{\alpha-1}]AK(t_1-s)Bu_o ds +$$

$$+ \int_{t_1}^{t_2} e^{As}(t_2-s)^{\alpha-1}AK(t_2-s)Bu_o ds\Big\| \leqq$$

$$\leqq M_o \|Bu_o\| \gamma\{(t_2-t_1)^{\beta} \int_0^{t_1} (t_2-s)^{\alpha-1}\mu(t_2-s)ds+$$

$$+ (1-\alpha)^{-1} \int_0^{t_1} (\int_{t_1-s}^{t_2-s} \tau^{\alpha-2}d\tau)\mu(t_1-s)ds +$$

$$+ \int_{t_1}^{t_2} (t_2-s)^{\alpha-1}\mu(t_2-s)ds\} \leqq M_o \|Bu_o\| \gamma(t_2-t_1)^{\beta}[t^{\beta}+(1+\alpha)^{-1}\beta^{-1}+1]$$

$$\times \int_0^t (t-s)^{\alpha-\beta-1}\mu(t-s)ds$$

Hence we get (2.17) with

$$c_4(t) = M_o[t^{\beta} + (1-\alpha)^{-1}\beta^{-1}+1]. \tag{2.28}$$

Finally, using (2.19) and (2.18), from the definition (2.6) of u_1 we get

$$\|u_1(t)\| = \Big\| \int_0^t Ae^{A(t-s)}[u_3(s)-u_3(t)]ds + e^{At}u_3(t)-u_3(t)\Big\| \leqq$$

$$\leqq M_1 c_6(t) \int_0^t (t-s)^{\alpha-\beta-1}\mu(t-s)\lambda(s)ds \int_0^t (t-s)^{\beta-1}ds +$$

$$(M_o+1)c_5(t) \int_0^t (t-s)^{\alpha-\beta-1}\mu(t-s)\lambda(s)ds$$

i.e. (2.14) with

$$c_1(t) = M_1\beta^{-1}c_6(t)t^\beta + (M_0+1)c_5(t). \tag{2.29}$$

Moreover for $0 \leqq t_1 < t_2 \leqq t$ we have

$$\begin{aligned}
u_1(t_2)-u_1(t_1) &= A\int_0^{t_2} e^{A(t_2-s)}[u_3(s)-u_3(t_2)]ds + A\int_0^{t_2} e^{A(t_2-s)}u_3(t_2)ds + \\
&- A\int_0^{t_1} e^{A(t_1-s)}[u_3(s)-u_3(t_1)]ds - A\int_0^{t_1} e^{A(t_1-s)}u_3(t_1)ds = \\
&= \int_0^{t_2} Ae^{A(t_2-s)}[u_3(s)-u_3(t_2)]ds - \int_0^{t_1} Ae^{A(t_1-s)}[u_3(s)-u_3(t_1)]ds + \\
&+ (e^{At_2}-I)[u_3(t_2)-u_3(t_1)] + (e^{At_2}-e^{At_1})u_3(t_1).
\end{aligned} \tag{2.30}$$

Since $u_3 \in C^\beta([0,T];X)$ and $u_3(0) = 0$, from (4.18) in [3] we deduce the following inequality

$$\left\| \int_0^{t_2} Ae^{A(t_2-s)}[u_3(s)-u_3(t_2)]ds - \int_0^{t_1} Ae^{A(t_1-s)}[u_3(s)-u_3(t_1)]ds \right\| \leqq$$

$$\leqq (t_2-t_1)^\beta \left(M_2\int_0^{+\infty}(s+1)^{-1}s^{\beta-1}ds + 2M_0+M_1\beta^{-1}\right)[u_3]_{C^\beta([0,t];X)}.$$

Moreover from the inequality

$$\|(e^{At_2}-I)[u_3(t_2)-u_3(t_1)]\| \leqq (M_0+1)(t_2-t_1)^\beta[u_3]_{C^\beta([0,t];X)}$$

and from (2.18) and (2.25) we infer

$$\|(e^{At_2}-e^{At_1})u_3(t_1)\| = \left\|\int_{t_1}^{t_2} Ae^{A\sigma}u_3(t_1)d\sigma\right\| \leqq M_1\int_{t_1}^{t_2}\sigma^{-1}t_1{}^\beta d\sigma\int_0^{t_1}(t_1-s)^{\alpha-\beta-1}\mu(t_1-s)\times$$

$$\times\, \lambda(s)ds \leqq$$

$$\leqq M_1\int_{t_1}^{t_2}\sigma^{\beta-1}d\sigma\int_0^{t_1}s^{\alpha-\beta-1}\mu(s)\lambda(t_1-s)ds \leqq M_1\beta^{-1}(t_2-t_1)^\beta\int_0^t s^{\alpha-\beta-1}\mu(s)\lambda(t-s)ds$$

Finally from (2.30) we derive that

$$[u_1]_{C^\beta([0,t];X)} \leq (M_2 \int_0^{+\infty} (s+1)^{-1} s^{\beta-1} ds + 3M_0 + M_1\beta^{-1}+1)[u_3]_{C^\beta([0,t];X)} +$$

$$+ \quad M_1\beta^{-1} \int_0^t s^{\alpha-\beta-1} \mu(s)\lambda(t-s)ds.$$

Taking (2.19) into account we get (2.15) with

$$c_2(t) = (M_2 \int_0^{+\infty} (s+1)^{-1} s^{\beta-1} ds + 3M_0 + M_1\beta^{-1}+1)c_6(t) + M_1\beta^{-1} \tag{2.31}$$

□

We prove now that $\lambda(w)$ and $\mu(K)$ satisfy a system of integral inequalities. To this purpose we introduce the functions ρ_1,ρ_2: $[0,T] \to [0,+\infty[$ so defined

$$\rho_1(t) = \|w_0\|_{C([0,t];X)} \tag{2.32}$$

$$\rho_2(t) = \|\Psi K_0\|_{C^\beta([0,t];K)} \tag{2.33}$$

the linear operator Ψ being defined in assumption (H9).

<u>THEOREM 2.2</u>. The pair of fucntions $(\lambda(w),\mu(K))$ defined by equations (2.12), (2.13) is a solution to the following system of integral inequalities:

$$\lambda(w)(t) \leq \rho_1(t) + c_{10}(T) \int_0^t (t-s)^{\alpha-\beta-1}\mu(K)(t-s)\lambda(w)(s)ds +$$

$$+ c_{11}(T)\,\|Bu_0\|_Y \int_0^t s^{\alpha-\beta-1}\mu(K)(s)ds \tag{2.34}$$

$$\mu(K)(t) \leq \rho_2(t) + c_{12}(T) \int_0^t (t-s)^{\alpha-\beta-1}\mu(K)(t-s)\lambda(w)(s)ds +$$

$$+ c_{13}(T)\,\|Bu_0\|_Y \int_0^t s^{\alpha-\beta-1}\mu(K)(s)ds \tag{2.35}$$

The functions c_i $(i = 10,\ldots,13)$ are increasing from R_+ to itself and depend also on α,β,M_0,M_1,M_2, $\|BA^{-1}\|_{L(X)}$ and $\|\Psi\|_{L(K)}$ $\|\Phi\|_{L(X;K)}$ (see (2.36)-(2.39)).

<u>Proof</u>. From (2.2), (2.14) and (2.16) we easily get the inequalities

$$\lambda(w)(t) \leq \rho_1(t) + \|BA^{-1}\|_{L(X)} \{\|u_1\|_{C([0,t];X)} + \|u_2\|_{C([0,t];X)}\} \leq$$

$$\leq \rho_1(t) + \|BA^{-1}\|_{L(X)} c_1(t) \int_0^t (t-s)^{\alpha-\beta-1}\mu(K)(t-s)\lambda(w)(s)ds +$$

$$+ \|BA^{-1}\|_{L(X)} c_3(t) \|Bu_o\|_Y \int_0^t s^{\alpha-\beta-1}\mu(K)(s)ds.$$

Whence, choosing

$$c_{10}(T) = \|BA^{-1}\|_{L(X)} c_1(T) \tag{2.36}$$

$$c_{11}(T) = \|BA^{-1}\|_{L(X)} c_3(T), \tag{2.37}$$

we immediately derive inequality (2.34).

Similarly from (2.3), (2.20)-(2.22) we get

$$\mu(K)(t) \leq \rho_2(t) + \|\Psi\|_{L(K)} \|\Phi\|_{L(X;K)} \{\|v_1\|_{C^\beta([0,t];X)} + \|v_2\|_{C^\beta([0,t];X)} +$$

$$+ \|v_3\|_{C^\beta([0,t];X)}\} \leq \rho_2(t) + \|\Psi\|_{L(K)} \|\Phi\|_{L(X;K)}\{[c_7(t)+c_9(t)] \times$$

$$\times \int_0^t (t-s)^{\alpha-\beta-1}\mu(K)(t-s)\lambda(w)(s)ds\} + \|\Psi\|_{L(K)} \|\Phi\|_{L(X;K)} c_8(t) \|Bu_o\|_Y \int_0^t s^{\alpha-\beta-1}\mu(K)(s)ds$$

Such inequalities imply (2.35), if we choose $c_{12}(T)$ and $c_{13}(T)$ as follows:

$$c_{12}(T) = \|\Psi\|_{L(K)} \|\Phi\|_{L(X;K)}[c_7(T) + c_9(T)] \tag{2.38}$$

$$c_{13}(T) = \|\Psi\|_{L(K)} \|\Phi\|_{L(X;K)} c_8(T). \tag{2.39}$$

□

In the sequel we shall need the following results about integral inequalities.

PROPOSITION 2.3. Let $f,g \in L^\infty(]0,T[)$ be a pair of a.e. non-negative functions such that

$$f(t) \leq c \int_0^t (t-s)^{\beta-1} f(s)ds+g(t) \text{ for a.e. } t \in]0,T[\tag{2.40}$$

for some positive constants c and β. Then we have

$$f(t) \leq g(t) + \int_0^t k(t-s)g(s)ds \text{ for a.e. } t \in]0,T[\tag{2.41}$$

$$k(t) = \sum_{n=1}^{\infty} c^{n-1}\Gamma(\beta)^n\Gamma(n\beta)^{-1}t^{n\beta-1}, \tag{2.42}$$

Γ denoting the gamma function.

Proof. See e.g. Lemma 1.1. of [1].

PROPOSITION 2.4. Let $f \in L^{\infty}_{loc}([0,T[)$ be a non-negative solution to the integral inequality

$$f(t) \leq c \int_0^t s^{\gamma-1}f(s)ds+g(t) \text{ for a.e. } t \in]0,T[\tag{2.43}$$

where c and γ are positive constants and $g \in L^{\infty}(]0,T[)$. Then

$$f(t) \leq c \int_0^t s^{\gamma-1}g(s)\exp[\frac{c}{\gamma}(t^{\gamma}-s^{\gamma})]ds + g(t) \text{ for a.e. } t \in]0,T[\tag{2.44}$$

In particular $f \in L^{\infty}(]0,T[)$ and satisfies the estimate

$$\|f\|_{L^{\infty}(]0,T[)} \leq \exp(\frac{c}{\gamma}t^{\gamma}) \|g\|_{L^{\infty}(]0,T[)} \quad \text{for a.e. } t \in]0,T] \tag{2.45}$$

Proof. Setting

$$\phi(t) = \int_0^t s^{\gamma-1}f(s)ds \qquad t \in [0,T[,$$

from (2.43) we deduce the following inequalities

$$f(t) \leq c\phi(t) + g(t) \quad \text{for a.e. } t \in]0,T[\tag{2.46}$$

and

$$\phi'(t) \leq ct^{\gamma-1}\phi(t) + t^{\gamma-1}g(t) \quad \text{for a.e. } t \in]0,T[.$$

From this latter we get

$$\frac{d}{dt}[\phi(t)\exp(-\frac{c}{\gamma}t^\gamma)] \leqq t^{\gamma-1}g(t)\exp(-\frac{c}{\gamma}t^\gamma) \text{ for a.e. } t \in]0,T[.$$

Integrating over [0,t] we get

$$\phi(t) \leqq \exp(\frac{c}{\gamma}t^\gamma)\int_0^t s^{\gamma-1}g(s)\exp(-\frac{c}{\gamma}s^\gamma)ds \text{ for any } t \in [0,T[\tag{2.47}$$

From (2.46) and (2.47) we immediately derive (2.44). In particular it implies

$$\begin{aligned} 0 \leqq f(t) &\leqq [c\exp(\frac{c}{\gamma}t^\gamma)\int_0^t s^{\gamma-1}\exp(-\frac{c}{\gamma}s^\gamma)ds+1]\,\|g\|_{L^\infty(]0,t[)} = \\ &= \exp(\frac{c}{\gamma}t^\gamma)\,\|g\|_{L^\infty(]0,t[)} \quad \text{for a.e. } t \in]0,T[\end{aligned} \tag{2.48}$$

Hence $f \in L^\infty(]0,T[)$ and satisfies (2.45). □

Now we will show that the solution (w,K) of (2.2), (2.3) is bounded in some interval $[0,\tau]$ ($\tau \in]0,T]$) if w or K is bounded there.

THEOREM 2.5. If a solution (w,K) of (2.2), (2.3) exists in the interval $[0,\tau[$, then

$$\sup_{0\leqq t<\tau} \lambda(w)(t) < +\infty \iff \sup_{0\leqq t<\tau} \mu(K)(t) < +\infty \tag{2.49}$$

Moroever if $\mu(K)(t) \leqq M'$ for any $t \in [0,T]$

$$\lambda(w)(t) \leqq C(M',T,\|Bu_0\|_{Y,\alpha,\beta}) \quad \text{for a.e. } t \in]0,T[\tag{2.50}$$

Proof. We use again the shorthands $\lambda(w)(t) = \lambda(t)$ and $\mu(K)(t) = \mu(t)$. Suppose that

$$\sup_{0\leqq t<\tau} \lambda(t) = M < +\infty:$$

then from (2.35) we get

$$\mu(t) \leqq \rho_2(T) + [Mc_{12}(T) + \|Bu_o\|_\gamma c_{13}(T)] \int_0^t s^{\alpha-\beta-1}\mu(s)ds$$

for a.e. $t \in]0,T[$.

From Proposition 2.4, we deduce $\mu \in L^\infty(]0,\tau[)$. Conversely if

$$\sup_{0 \leqq t < \tau} \mu(t) = M' < +\infty,$$

then from (2.34) we obtain

$$\lambda(t) \leqq \rho_1(T) + c_{10}(T)M' \int_0^t (t-s)^{\alpha-\beta-1}\lambda(s)ds +$$

$$c_{11}(T)\,\|Bu_o\|_\gamma M'(\alpha-\beta)^{-1}T^{\alpha-\beta} \text{ for a.e. } t \in]0,T[.$$

Now by virtue of Proposition 2.3 we easily deduce that $\lambda \in L^\infty(]0,\tau[)$ and satisfies (2.50). □

3. PROOF OF THEOREM 1.1.

In addition to the notations of Theorem 1.1. let us also set

$$w_i = Au_i' \quad (i = 1,2) \tag{3.1}$$

and let us define w_{oi} by formula (2.4) with v_o and f' replaced respectively by $v_{oi} = Au_{oi} + f_i(o)$ and f_i'. Similarly K_{oi} is defined by formula (2.5) with G'', v_o and f' replaced respectively by G_i'', v_{oi} and f_i'.
As (w_i,K_i) $(i=1,2)$ are solutions to two systems analogous to (2.2), (2.3), by difference we get

$$\begin{aligned} w_2-w_1 = {} & w_{o2}-w_{o1} + BA^{-1}[u_1(K_2-K_1,w_2)+u_1(K_1,w_2-w_1)+u_2(K_2-K_1,u_{o2}) + \\ & + u_2(K_1,u_{o2}-u_{01})] \end{aligned} \tag{3.2}$$

$$\begin{aligned} K_2-K_1 = {} & \Psi(K_{o2}-K_{o1})-\Psi\Phi[v_1(K_2-K_1,w_2)+v_1(K_1,w_2-w_1)+v_2(K_2-K_1,u_{o2}) + \\ & + v_2(K_1,u_{o2}-u_{o1}) + v_3(K_2-K_1,w_2) + v_3(K_1,w_2-w_1)]. \end{aligned} \tag{3.3}$$

Using such equations we derive (via Lemma 2.1) that the pair of functions $(\lambda(w_2-w_1), \mu(K_2-K_1))$ (see Definitions (2.12), (2.13)) is a solution to the following system of linear integral inequalities

$$\lambda(w_2-w_1)(t) \leqq \|w_{o2}-w_{o1}\|_{C([0,t];X)} + \|BA^{-1}\|_{L(X)}\{c_1(T)\bar{M} \times\left[\int_0^t s^{\gamma-1}\mu(K_2-K_1)(s)ds + \int_0^t (t-s)^{\gamma-1}\lambda(w_2-w_1)(s)ds\right] + c_3(T)\tilde{M}\int_0^t s^{\gamma-1}\mu(K_2-K_1)(s)ds + c_3(T)\bar{M}\frac{T^\gamma}{\gamma}\|B(u_{o2}-u_{o1})\|_Y \tag{3.4}$$

$$\mu(K_2-K_1)(t) \leqq \|\Psi(K_{o2}-K_{o1})\|_{C^\beta([0,t];K)} + \|\Psi\|_{L(K)}\|\Phi\|_{L(X;K)} \times \times\Big\{c_7(T)\bar{M}\left[\int_0^t s^{\gamma-1}\mu(K_2-K_1)(s)ds + \int_0^t (t-s)^{\gamma-1}\lambda(w_2-w_1)(s)ds\right] + + c_8(T)\tilde{M}\int_0^t s^{\gamma-1}\mu(K_2-K_1)(s)ds + c_8(T)\bar{M}\frac{T^\gamma}{\gamma}\|B(u_{o2}-u_{o1})\|_Y + + c_9(T)M\left[\int_0^t s^{\gamma-1}\mu(K_2-K_1)(s)ds + \int_0^t (t-s)^{\gamma-1}\lambda(w_2-w_1)(s)ds\right]\Big\} \tag{3.5}$$

where

$$\gamma = \alpha - \beta \tag{3.6}$$

We recall that the positive constants $\tilde{M}$ and $\bar{M}$ are so related to Bu_{oj}, $\lambda(w_j)$, $\mu(K_j)$ $(j = 1,2)$ (see Hypotheses (1.6), (1.7) and estimate (2.51)):

$$\max\{\|Bu_{o1}\|_Y, \|Bu_{o2}\|_Y\} \leqq \tilde{M} \tag{3.7}$$

$$\max\{\lambda(w_1)(T), \lambda(w_2)(T), \mu(K_1)(T), \mu(K_2)(T)\} \leqq \bar{M} \tag{3.8}$$

Introduce then the functions $\rho, \nu: [0,T] \to [0, +\infty[$ and the positive constants c_{14}, c_{15}, c_{16} so defined

$$\rho(t) = \lambda(w_2-w_1)(t) + \mu(K_2-K_1)(t) \tag{3.9}$$

$$\nu(t) = \|w_{o2}-w_{o1}\|_{C([0,t];X)} + \|K_{o2}-K_{o1}\|_{C^\beta([0,t];K)} + \|B(u_{o2}-u_{o1})\|_Y \tag{3.10}$$

$$c_{14} = \|BA^{-1}\|_{L(X)}c_1(T)\bar{M} + \|\Psi\|_{L(K)}\ \|\Phi\|_{L(X;K)}\bar{M}[c_7(T)+c_9(T)] \tag{3.11}$$

$$c_{15} = \|BA^{-1}\|_{L(X)}[c_1(T)\bar{M} + c_3(T)\tilde{M}] +$$

$$+ \|\Psi\|_{L(K)}\ \|\Phi\|_{L(X;K)}[c_7(T)\bar{M} + c_8(T)\tilde{M} + c_9(T)\bar{M}] \tag{3.12}$$

$$c_{16} = 1 + \|\Psi\|_{L(K)} + [\ \|BA^{-1}\|_{L(X)}c_3(T) +$$

$$+ \|\Psi\|_{L(K)}\ \|\Phi\|_{L(X;K)}c_8(T)]\bar{M}^{-1}T^\gamma \tag{3.13}$$

Adding (3.4) and (3.5) we get

$$\rho(t) \leqq c_{14}\int_0^t (t-s)^{\gamma-1}\rho(s)ds+c_{15}\int_0^t s^{\gamma-1}\rho(s)ds+c_{16}\nu(t) \quad t \in]0,T] \tag{3.14}$$

By virtue of Proposition 2.3, setting

$$K(t) = \sum_{n=1}^{+\infty} c_{14}^{n-1}\Gamma(\gamma)^n\Gamma(n\gamma)^{-1}t^{n\gamma-1}, \tag{3.15}$$

we get

$$\rho(t) \leqq \int_0^t K(t-s)[c_{15}\int_0^s \sigma^{\gamma-1}\rho(\sigma)d\sigma+c_{16}\nu(s)]ds + c_{15}\int_0^t s^{\gamma-1}\rho(s)ds +$$

$$c_{16}\nu(t) = c_{16}[\nu(t) + \int_0^t K(t-s)\nu(s)ds] + c_{15}\int_0^t K_1(t-s)s^{\gamma-1}\rho(s)ds, \tag{3.16}$$

where

$$K_1(t) = 1 + \int_0^t K(\sigma)d\sigma \tag{3.17}$$

Note that K_1 depends on c_{14} and is a bounded non-decreasing function of t. Setting

$$\pi(t) = \nu(t) + \int_0^t K(t-s)\nu(s)ds \qquad t \in]0,T] \tag{3.18}$$

$$c_{17} = c_{15}K_1(T) \tag{3.19}$$

from (3.15), (3.16), (3.17) we deduce the integral inequality

$$\rho(t) \leqq c_{16}\pi(t) + c_{17}\int_0^t s^{\gamma-1}\rho(s)ds \qquad t \in]0,T] \tag{3.20}$$

Using Proposition 2.4 we get the estimate

$$\|\rho\|_{C([0,T])} \leqq c_{18}\|\pi\|_{C([0,T])} \tag{3.21}$$

where

$$c_{18} = c_{16}\exp(\frac{1}{\gamma}\, c_{17}T^{\gamma}). \tag{3.22}$$

Whence and from (3.18) we infer

$$\|\rho\|_{C([0,T])} \leqq c_{19}\|\nu\|_{C([0,T])} \tag{3.23}$$

with

$$c_{19} = c_{18}(1 + \int_0^T K(t)dt). \tag{3.24}$$

Taking definitions (3.9) and (3.10) into account, estimate (3.23) can be so rewritten

$$\|w_2-w_1\|_{C([0,T];X)} + \|K_2-K_1\|_{C^{\beta}([0,T];K)} \leqq c_{19}\{\|w_{o2}-w_{o1}\|_{C([0,T];X)} + \\ + \|K_{o2}-K_{o1}\|_{C^{\beta}([0,T];K)} + \|B(u_{o2}-u_{o1})\|_{Y}\} \tag{3.25}$$

In order to derive (2.5) we need to estimate $\|u_2-u_1\|_{C^1([0,T];Y)}$, $\|w_{o2}-w_{o1}\|_{C([0,T];X)}$ and $\|K_{o2}-K_{o1}\|_{C^{\beta}([0,T];K)}$. To this purpose let us observe that from the definitions of w_1 and w_2 we deduce the equation

$$u_2(t)-u_1(t) = u_{o2}-u_{o1} + A^{-1}\int_0^t [w_2(s)-w_1(s)]ds \qquad t \in [0,T].$$

Whence we get the estimate

$$\|u_2-u_1\|_{C^1([0,T];Y)} \leqq \|u_{o2}-u_{o1}\|_Y + (1+T) \times \\ \times (1 + \|A^{-1}\|_{L(X)}) \|w_2-w_1\|_{C([0,T];X)}. \tag{3.26}$$

As

$$g_2(t)-g_1(t) = e^{At}(v_{o2}-v_{o1}) + \int_0^t e^{A(t-s)}[f_2'(s)-f_1'(s)]ds$$

we deduce from estimate (4.61) in [3] that

$$\|A(g_2-g_1)\|_{C^\beta([0,T];X)} \leq c_{20}(T)(\|f_2-f_1\|_{C^{1+\beta}([0,T];X)} + \\ + \|A(v_{o2}-v_{o1})+f_2'(0)-f_1'(0)\|_{D_A(\beta,+\infty)}), \tag{3.27}$$

where

$$c_{20}(T) = M_o(3+T^\beta)+M_1\beta^{-1}(2+T^\beta)+M_2\int_0^{+\infty}(s+1)^{-1}s^{\beta-1}ds+T^\beta+1+M_o+\beta^{-1} \tag{3.28}$$

As

$$w_{o2}(t)-w_{o1}(t) = BA^{-1}A[g_2(t)-g_1(t)] \qquad t \in [0,T], \tag{3.29}$$

from this and (3.27) we infer

$$\|w_{o2}-w_{o1}\|_{C[0,T];X)} \leqq \|BA^{-1}\|_{L(X)}c_{20}(T) \{\|f_2-f_1\|_{C^{1+\beta}([0,T];X)} + \\ + \|A(v_{o2}-v_{o1}) + f_2'(0)-f_1'(0)\|_{D_A(\beta,+\infty)}\} \tag{3.30}$$

Finally from the formula

$$K_{o2}(t)-K_{o1}(t) = t^{1-\alpha}[G_2''(t)-G_1''(t)]-t^{1-\alpha}\Phi A[g_2(t)-g_1(t)]-t^{1-\alpha}\Phi[f_2'(t)-f_1'(t)]$$

we get

$$\|K_{o2}-K_{o1}\|_{C^{\beta}([0,T];K)} \leq \|t^{1-\alpha}(G_2''-G_1'')\|_{C^{\beta}([0,T];K)} +$$

$$+ (T^{1-\alpha}+T^{1-\alpha-\beta}) \|\Phi\|_{L(X;K)}\{\|A(g_2-g_1)\|_{C^{\beta}([0,T];X)} + \|f_2'-f_1'\|_{C^{\beta}([0,T];X)}\}$$

Whence, using (3.27), we obtain the estimate

$$\|K_{o2}-K_{o1}\|_{C^{\beta}([0,T];K)} \leq c_{21}(T) \{\|t^{1-\alpha}(G_2''-G_1'')\|_{C^{\beta}([0,T],K)} + \\ + \|f_2'-f_1'\|_{C^{\beta}([0,T];X)} + \|A(v_{o2}-v_{o1}) + f_2'(0)-f_1'(0)\|_{D_A(\beta,+\infty)}\} \tag{3.31}$$

with

$$c_{21}(T) = 1 + (T^{1-\alpha}+T^{1-\alpha-\beta}) \|\Phi\|_{L(X;K)} [c_{20}(T)+1] \tag{3.32}$$

Now from (3.25), (3.30) and (3.31) we infer

$$\|w_2 - w_1\|_{C([0,T];X)} + \|K_2-K_1\|_{C^{\beta}([0,T];K)} \leq c_{19}[\|BA^{-1}\|_{L(X)}c_{20}(T)+c_{21}(T)+1] \times \\ \times \{\|t^{1-\alpha}(G_2''-G_1'')\|_{C^{\beta}([0,T];K)} + \|f_2-f_1\|_{C^{1+\beta}(0,T];X)} + \|A(v_{o2}-v_{o1}) + \\ f_2'(0)-f_1'(0)\|_{D_A(\beta,+\infty)} + \|B(u_{o2}-u_{o1})\|_Y\}. \tag{3.33}$$

Finally from (3.26) and (3.33) we derive the desired estimate

$$\|u_2-u_1\|_{C^1([0,T];Y)} + \|K_2-K_1\|_{C^{\beta}([0,T];K)} \leq \|u_{o2}-u_{o1}\|_Y + \\ + c_{19}[(1+T)(1+\|A^{-1}\|_{L(X)}) + 1] \times \\ \times [\|BA^{-1}\|_{L(X)}c_{20}(T)+c_{21}(T)+1] \{\|t^{1-\alpha}(G_2''-G_1'')\|_{C^{\beta}([0,T];K)} + \\ + \|f_2-f_1\|_{C^{1+\beta}([0,T];X)} + \|A(v_{o2}-v_{o1})+f_2'(0)-f_1'(0)\|_{D_A(\beta,+\infty)} + \\ \|B(u_{02}-u_{o1})\|_Y\}. \quad \square$$

4. MAXIMAL AND GLOBAL SOLUTIONS

An immediate corollary of Theorem 1.1. states that two solutions (u_1,K_1) and (u_2,K_2) to problem (1.2)-(1.4) (with T replaced respectively by T_1 and T_2) must coincide in $[0,\min(T_1,T_2)]$. Taking advantage of this property we can show that each (local) solution can be uniquely extended to a maximal (i.e. no further continuable) solution. To this purpose it will suffice to prove that it can be locally extended.

THEOREM 4.1. Let $(w,K) \in C([0,\sigma];X) \times C^{\beta}([0,\sigma];K)$ $(0 < \sigma < T)$ be a solution to system (2.2), (2.3). Then there exists $\varepsilon \in (0,\min(T-\sigma,\sigma))$ and a pair of functions $(\bar{w},\bar{K}) \in C([0,\sigma+\varepsilon];X) \times C^{\beta}([0,\sigma+\varepsilon];K)$ such that $(\bar{w},\bar{K})$ is a solution to (2.2), (2.3) and coincides with (w,K) in $[0,\sigma]$.

Proof. Suppose first that $(\bar{w},\bar{K})$ is a pair of functions enjoying the properties stated in the theorem. Then from (2.8) we derive the equations

$$u_3(\bar{K},\bar{w})(t) = \int_0^{t-\sigma} (t-s)^{\alpha-1}\bar{K}(t-s)w(s)ds + \int_{t-\sigma}^{\sigma} (t-s)^{\alpha-1}K(t-s)w(s)ds + \\ + \int_{\sigma}^{t} (t-s)^{\alpha-1}K(t-s)\bar{w}(s)ds \qquad t \in [\sigma,\sigma+\varepsilon]. \tag{4.1}$$

Similarly from Definition (2.7) we get the equation

$$u_2(\bar{K},u_o)(t) = \int_0^{\sigma} e^{A(t-s)}s^{\alpha-1}AK(s)Bu_o ds + \int_{\sigma}^{t} e^{A(t-s)}s^{\alpha-1}A\bar{K}(s)Bu_o ds \qquad t \in [\sigma,\sigma+\varepsilon]. \tag{4.2}$$

Finally from (2.6) and (4.1) we infer the equation

$$u_1(\bar{K},\bar{w})(t) = A\int_0^{\sigma} e^{A(t-\rho)}u_3(K,w)(\rho)d\rho + A\int_{\sigma}^{t} e^{A(t-\rho)}u_3(\bar{K},\bar{w})(\rho)d\rho = \qquad t \in [\sigma,\sigma+\varepsilon].$$

$$= A\int_0^{\sigma} e^{A(t-\rho)}d\rho \int_0^{\rho} (\rho-s)^{\alpha-1}K(\rho-s)w(s)ds + \\ + A\int_{\sigma}^{t} e^{A(t-\rho)}d\rho \int_{\rho-\sigma}^{\sigma} (\rho-s)^{\alpha-1}K(\rho-s)w(s)ds + \\ + A\int_{\sigma}^{t} e^{A(t-\rho)}d\rho \int_0^{\rho-\sigma} (\rho-s)^{\alpha-1}\bar{K}(\rho-s)w(s)ds + \\ + A\int_{\sigma}^{t} e^{A(t-\rho)}d\rho \int_{\sigma}^{\rho} (\rho-s)^{\alpha-1}K(\rho-s)\bar{w}(s)ds \qquad t \in [\sigma,\sigma+\varepsilon].$$

Recall now that $(\bar{w},\bar{K})$ is a solution to system (2.2), (2.3) extending to $[0,\sigma+\varepsilon]$ the solution (w,K) existing in $[0,\sigma]$. From equations (4.1)-(4.3) we deduce that $(\bar{w},\bar{K})$ is a solution in $[\sigma,\sigma+\varepsilon]$ to the following linear system of Volterra integral equations

$$\bar{w}(t) = \bar{w}_0(t) + B\,\{\int_\sigma^t d\rho \int_\sigma^\rho s^{\alpha-1}e^{A(t-\rho)}\bar{K}(s)w(\rho-s)ds + \tag{4.4}$$

$$+ \int_\sigma^t e^{A(t-\rho)}\,d\rho \int_\sigma^\rho (\rho-s)^{\alpha-1}K(\rho-s)\bar{w}(s)ds + \int_\sigma^t s^{\alpha-1}e^{A(t-s)}\bar{K}(s)Bu_0 ds\}$$

$$\bar{K}(t) = \Psi(\bar{K}_0(t))-t^{1-\alpha}\Psi\Phi\{A\int_\sigma^t d\rho \int_\sigma^\rho s^{\alpha-1}e^{A(t-\rho)}\bar{K}(s)w(\rho-s)ds +$$

$$+ A\int_\sigma^t e^{A(t-\rho)}d\rho \int_\sigma^\rho (\rho-s)^{\alpha-1}K(\rho-s)\bar{w}(s)ds + A\int_\sigma^t s^{\alpha-1}e^{A(t-s)}\bar{K}(s)Bu_0 ds +$$

$$+ \int_\sigma^t s^{\alpha-1}\bar{K}(s)w(t-s)ds + \int_\sigma^t (t-s)^{\alpha-1}K(t-s)\bar{w}(s)ds \qquad t \in [\sigma,\sigma+\varepsilon]. \tag{4.5}$$

The functions $\bar{w}_0$ and $\bar{K}_0$ are so defined in $[\sigma,\sigma+\varepsilon]$

$$\bar{w}_0(t) = w_0(t) + B\,\{\int_0^\sigma e^{A(t-\rho)}d\rho \int_0^\sigma (\rho-s)^{\alpha-1}K(\rho-s)w(s)ds + \tag{4.6}$$

$$+ \int_\sigma^t e^{A(t-\rho)}d\rho\int_{\rho-\sigma}^\sigma (\rho-s)^{\alpha-1}K(\rho-s)w(s)ds + \int_0^\sigma s^{\alpha-1}e^{A(t-s)}K(s)Bu_0 ds\}$$

$$\bar{K}_0(t) = K_0(t)-t^{1-\alpha}\Psi\Phi\,\{A\int_0^\sigma e^{A(t-\rho)}d\rho \int_0^\rho (\rho-s)^{\alpha-1}K(\rho-s)w(s)ds +$$

$$+ A\int_\sigma^t e^{A(t-\rho)}d\rho\int_{\rho-\sigma}^\sigma (\rho-s)^{\alpha-1}K(\rho-s)w(s)ds+A\int_0^t s^{\alpha-1}e^{A(t-s)}K(s)Bu_0 ds +$$

$$+ \int_{t-\sigma}^\sigma (t-s)^{\alpha-1}K(t-s)w(s)ds \qquad t \in [\sigma,\sigma+\varepsilon]. \tag{4.7}$$

Suppose on the contrary that $(\tilde{w},\tilde{K}) \in C([\sigma,\sigma+\varepsilon];X) \times C^\beta([\sigma,\sigma+\varepsilon];K)$ is a solution to (4.4), (4.5) for some $\varepsilon \in (0,\min(T-\sigma,\sigma))$. Then the pair of functions $(\bar{w},\bar{K})$: $[0,\sigma+\varepsilon] \to X \times K$ so defined

$$\bar{w}(t) = \begin{cases} w(t) & t \in [0,\sigma] \\ \tilde{w}(t) & t \in]\sigma,\sigma+\varepsilon] \end{cases}; \quad \bar{K}(t) = \begin{cases} K(t) & t \in [0,\sigma] \\ \tilde{K}(t) & t \in]\sigma,\sigma+\varepsilon] \end{cases} \tag{4.8}$$

belongs to $C([0,\sigma+\varepsilon];X) \times C^{\beta}([0,\sigma+\varepsilon];K)$ and is a solution to (2.2), (2.3) with $T = \sigma + \varepsilon$.

Such a property is a consequence of the following equalities

$$\tilde{w}_0(\sigma) = w_0(\sigma) + B \int_0^{\sigma} e^{A(\sigma-\rho)}d\rho \int_0^{\rho} (\rho-s)^{\alpha-1}K(\rho-s)w(s)ds + \\ + \int_0^{\sigma} s^{\alpha-1}e^{A(\sigma-s)}K(s)Bu_0 ds = w(\sigma) \tag{4.9}$$

$$\tilde{K}_0(\sigma) = K_0(\sigma)-\sigma^{1-\alpha}\Psi\Phi \{A \int_0^{\sigma} e^{A(\sigma-\rho)}d\rho \int_0^{\rho} (\rho-s)^{\alpha-1}K(\rho-s)w(s)ds + \\ + \int_0^{\sigma} s^{\alpha-1}e^{A(\sigma-s)}K(s)Bu_0 ds + \int_0^{\sigma} (\sigma-s)^{\alpha-1}K(\sigma-s)w(s)ds = K(\sigma) \tag{4.10}$$

We have thus proved that a solution (w,K) to system (2.2), (2.3) on the interval $[0,\sigma]$ can be extended to the interval $[0,\sigma+\varepsilon]$, if, and only if, we can solve system (4.4), (4.5) in $[\sigma,\sigma+\varepsilon]$. On the other hand, using Banach's contraction theorem, we can show that (4.4), (4.5) admits a (unique) solution in $[\sigma,\sigma+\varepsilon]$. To this purpose we introduce the new unknown functions $z \in C([0,\varepsilon];X)$ and $H \in C^{\beta}([0,\varepsilon];K)$ defined by the equations

$$z(t) = \bar{w}(t+\sigma); \; H(t) = \bar{K}(t+\sigma) \qquad t \in [0,\varepsilon] \tag{4.11}$$

Replacing t by $t+\sigma$ in equations (4.4), (4.5) and performing standard changes of variables in the integrals, it is easy to realize that (z,H) is a solution to the following system

$$z(t) = z_0(t) + BA^{-1}\{A \int_0^{t} e^{A(t-s)}[u_4(H,w)(s)+u_3(K,z)(s)]ds+u_2(H,u_0)(t)\} \\ t \in [0,\varepsilon] \tag{4.12}$$

$$H(t) = \Psi(H_0(t))-t^{1-\alpha}\Psi\Phi \{A \int_0^{t} e^{A(t-s)}[u_4(H,w)(s)+u_3(K,z)(s)]ds + \\ + u_2(H,u_0)(t) + u_4(H,w)(t) + u_3(K,z)(t)\} \qquad t \in [0,\varepsilon] \tag{4.13}$$

The functions $u_2(H,u_0), u_3(K,z)$ and $u_4(H,w), z_0, H_0$ are defined in $[0,\varepsilon]$

respectively by (2.7), (2.8) and the following equations:

$$u_4(H,w)(t) = \int_0^t (t-s+\sigma)^{\alpha-1} H(t-s)w(s)ds \tag{4.14}$$

$$z_o(t) = \bar{w}_o(t+\sigma) \tag{4.15}$$

$$H_o(t) = \bar{K}_o(t+\sigma) \tag{4.16}$$

Observe now that $u_4(\ ,w)$ satisfies the following estimate

$$\|u_4(H,w)\|_{C^\beta([0,\varepsilon];K)} \leq \frac{2}{\alpha}(\varepsilon^\alpha + \varepsilon^{\alpha-\beta}) \|H\|_{C^\beta([0,\varepsilon];K)} \|w\|_{C([0,\varepsilon];X)}.$$

Proceeding as in the proof of Theorem 4.1 in [2], we can prove that system (4.12), (4.13) admits a (unique) solution $(z,H) \in C([0,\varepsilon];X) \times C^\beta([0,\varepsilon];K)$ for ε small enough.

This fulfills the proof of the theorem. □

Let us prove now that, if a maximal solution is bounded, then it is a global one, i.e. it is defined on the whole of $[0,T]$.

THEOREM 4.2. Let $(w,K) \in C_{loc}([0,\tau[;X) \times C^\beta_{loc}([0,\tau[;K)$ be the maximal solution to the integral system (2.22), (2.23). If

$$\sup_{\sigma\in]0,\tau[} \|K\|_{C^\beta([0,\sigma];K)} = M' < +\infty \tag{4.17}$$

then $\tau = T$. The same conclusion holds when (4.17) is replaced by

$$\sup_{\sigma\in]0,\tau[} \|w\|_{C([0,\sigma];X)} = M < +\infty. \tag{4.18}$$

Proof. From Theorem 2.5, we deduce that there exists a positive constant $C(M')$ depending on M' such that

$$\sup_{\sigma\in]0,\tau[} \|w\|_{C([0,\sigma];X)} \leq C(M').$$

Whence and from Proposition 3.1 in [2] we infer that the function

$$u_3(K,w)(t) = \int_0^t (t-s)^{\alpha-1}K(t-s)w(s)ds \qquad t \in [0,\tau[$$

can be extended by a function of $C^\beta([0,\tau];X)$.

A similar conclusion holds for the functions

$$t \to B \int_0^t e^{A(t-s)} u_3(K,w)(s)ds = (BA^{-1})u_1(K,w)(t)$$

$$t \to (BA^{-1})u_2(K,w)(t)$$

(for the latter see the proof of Theorem 4.1. in [2]).

What we have just proved implies that the solution (w,K) in $[0,\tau[$ to the integral system (2.2), (2.3) can be extended to a solution $(\bar{w},\bar{K})$ in $C([0,\tau];X) \times C^\beta([0,\tau];K)$. If $\tau < T$, this contradicts the maximality of (w,K): hence we get $\tau = T$. □

We conclude this paper by observing that stability estimate (1.5) can be replaced by another one more suitable to deal with data f,u_0,G obtained by physical measurements. This will be the subject of a future paper.

References

[1] A. Lorenzi: Determination of a time-dependent coefficient in a quasilinear parabolic equation. Ricerche di Mat. (Napoli) 32 (1983), 263-284.

[2] A. Lorenzi - E. Sinestrari: An inverse problem in the theory of materials with memory. (Preprint).

[3] E. Sinestrari: On the abstract Cauchy problem of parabolic type in spaces of continuous functions. J. Math. An. Appl. 107 (1985), 16-66.

Alfredo Lorenzi
Dipartimento di Matematica
Università di Milano
Via Saldini 50
20133 Milan
Italy.

Eugenio Sinestrari
Dipartimento di Matematica
Università di Roma I
Piazzale Aldo Moro 2
00185 Rome
Italy.

J. MILOTA

Asymptotic behaviour of parabolic equations with infinite delay

1. INTRODUCTION

In this contribution we will present the background of a geometric theory for an abstract equation with infinite delay

$$\dot{u}(t) = Au(t) + Lu_t + f. \tag{E}$$

Here A stands for a generator of an analtyic semigroup e^{At} in a Banach space X, L is a linear continuous operator from an appropriate phase space Y into X. We denote by u_t the translation of u given by $u_t(s) = u(t+s)$ for $s \in \mathbf{R}^- = (-\infty, 0)$, $t \in \mathbf{R}^+ = [0,\infty)$. As we are interested in the asymptotic behaviour of solutions of (E) we endow the space Y with the quality of fading memory. In order to avoid undue details we choose a concrete representation for Y whose elements will be strongly measurable functions on $\mathbf{R}^-$ with values in X^α (the domain of the fractional power $(-A)^\alpha$ with the graph norm). We fix some $\gamma > 0$ and define

$$\|\varphi\|_Y = \int_{-\infty}^{0} e^{\gamma s} \|\varphi(s)\|_\alpha ds.$$

For further purposes we put $\varphi(s) = 0$ for $s \in \mathbf{R}^+$.

Our approach is based on the variation of parameters formula for (E). A solution of (E) which satisfies the initial conditions

$$u(0) = x, \quad u_0 = \varphi \tag{IC}$$

(i.e. $u(s) = \varphi(s)$ for $s \in \mathbf{R}^-$), can be written (under some assumptions - see Section 2) in the form

$$u(t) = R(t)x + \int_0^t R(t-s)[L\varphi_s + f(s)]ds,$$

where R is the so-called resolvent operator. It is possible to show that R has properties very similar to those of the semigroup e^{At}. We can therefore investigate a solution operator $T(t) : (x,\varphi) \to (u(t),u_t)$ for a homogeneous equation (E) with help of the formula (VP). For example, T(t) is a C_0-semigroup on the space $Z = X^\alpha \times Y$ provided $\alpha < 1$. In spite of undue restrictions, the semigroup approach yields in a quite straightforward way results on the behaviour of solutions of a nonlinear equation (E) near a stationary point. Considering the nonlinear equation as a perturbation of the linear one, we will prove the existence of a stable, an unstable and a central manifold (Section 4). To do this, we need sharp asymptotic estimates for linear manifolds which will be derived in Section 3. The semigroup approach requires the very restrictive assumption $\alpha < 1$ (see e.g. [9]) and therefore this cannot be used in many important applications. It makes a direct use of the formula (VP) more desirable. We will give some remarks in the last section.

The author wishes to thank Prof. O. Diekmann and Prof. J. Prüss for helpful discussions during the Trento meeting, Prof. A. Lunardi for sending the preprints [4], [5], and Dr. H. Petzeltová. This contribution is mainly based on the joint work [12].

2. VARIATION OF PARAMETERS FORMULA

We will use the following two notions of a solution to the initial value problem (E), (IC): A function $u : (-\infty,T) \to X^\alpha$ is called a mild solution on the interval $[0,T)$ if u satisfies the initial condition (IC), belongs to the space $C([0,T),X^\alpha)$ and

$$u(t) = e^{At}x + \int_0^t e^{A(t-s)}(Lu_s + f)ds \tag{1}$$

holds for $t \in [0,T)$. A mild solution is said to be a strong solution if it is differentiable and satisfies (E) on $(0,T)$.

Many papers (see e.g. [3], [8], [10], [13] and references given there) are devoted to the existence and properties of a resolvent operator for a special case of (E), in which L is an integral operator. The standard approach is based on the Laplace transform. Suppose f = 0 in (E), u is a mild solution on $\mathbb{R}^+$ with the Laplace transform $\hat{u}(\lambda)$. We have formally

$$[\lambda - A - B(\lambda)]\,\hat{u}(\lambda) = x + \widehat{L\varphi_{\bullet}}(\lambda),$$

where $B(\lambda)x = L(s \to e^{\lambda s}x)$. Under the assumption

$$\left.\begin{array}{l}\text{Either } \alpha < 1 \text{ or} \\ \|B(\lambda)\|_{L(X^{\alpha},X)} \leq \text{const.}\ |\lambda|^{-\varepsilon} \text{ for } \operatorname{Re}\lambda > d \text{ with} \\ \text{some fix } \varepsilon > 0.\end{array}\right\} \qquad \text{(H)}$$

the inverse Laplace transform to $D(\lambda) = [\lambda - A - B(\lambda)]^{-1}$ there exists. Its properties are stated in the following theorem which is a slight modification of Theorems 1,2 in [12].

THEOREM 1. Under the above assumptions:

(i) There exists a resolvent R such that the function $R(t)x$ is a strong solution of

$$\dot{u}(t) = Au(t) + Lu_t,\ u(0) = x,\ u_0 = 0,$$

provided $x \in X^{\alpha}$.

(ii) If $\alpha < 1$, $f \in L^p(\mathbf{R}^+,X)$ where $p > (1-\alpha)^{-1}$ or $p \geq 1$ for $\alpha = 0$, $x \in X^{\alpha}$, $\varphi \in Y$, then the function u given by the formula (VP) is a mild solution of (E), (IC) on $\mathbf{R}^+$. Moreover, u is a strong solution provided $f(t)$, φ_t are locally Hölder continuous on $(0,+\infty)$.

(iii) If $\alpha = 1$, $f \in L^{\infty}(\mathbf{R}^+,X)$, $x \in X^1$, $\varphi \in Y$, and $f(t)$, φ_t are locally Hölder continuous on $(0,+\infty)$, then u is a mild solution.

COROLLARY. Let $\alpha < 1$ and $Z = X^{\alpha} \times Y$. Let u be a (unique) mild solution of the initial value problem

$$\dot{u}(t) = Au(t) + Lu_t,\ u(0) = x,\ u_0 = \varphi. \qquad (2)$$

Then the operators $T(t) : (x,\varphi) \to (u(t),u_t)$, $t \in \mathbf{R}^+$, form a C_0-semigroup on the space Z.

3. SOLUTION SEMIGROUP

The semigroup $T(t)$ allows us to rewrite (VP) in a more convenient form for further purposes

$$z(t) = (u(t),u_t) = T(t)(x,\varphi) + \int_0^t T(t-s)F(s)ds, \tag{3}$$

where $F = (f,0)$. Our aim is to apply this formula to prove the existence of local invariant manifolds for equations with a small nonlinear term. As the first step, we will prove the existence of a stable and an unstable manifold for the linear equation. Although the resolvent R has the same asymptotic and regularity properties as e^{At} has, the semigroup $T(t)$ loses almost all qualities of e^{At}, e.g. $T(t)$ is neither analytic, nor eventually differentiable, nor eventually compact. But we can get information on the spectrum of $T(t)$ by means of the spectrum of its generator B.

PROPOSITION 1. Let B be the generator of the semigroup $T(t)$. Then $\mathcal{D}(B) = \{(x,\varphi) \in Z;\ x \in X^1,\ \dot{\varphi} \in Y,\ \varphi(0-) = x,\ Ax + L\varphi \in X^{\alpha}\}$ is the domain of B and $B(x,\varphi) = (Ax + L\varphi, \dot{\varphi})$ for $(x,\varphi) \in \mathcal{D}(B)$.

Proof. This is an easy calculation by employing the Laplace transform of $T(t)$ and the variation of parameters formula (VP).

PROPOSITION 2. Let A have a compact resolvent. Then the continuous spectrum of B is empty, the residual spectrum $R_\sigma(B) = \{\lambda \in \mathbb{C};\ \mathrm{Re}\,\lambda \leqq -\gamma\}$. Further, λ belongs to the point spectrum $P_\sigma(B)$ if and only if the characteristic equation

$$\lambda x - Ax - B(\lambda)x = 0 \tag{4}$$

has a nontrivial solution.

Proof. We find $(x,\varphi) \in \mathcal{D}(B)$ such that

$$x - Ax - L\varphi = y \tag{5}$$

$$\lambda\varphi - \dot{\varphi} = \psi \tag{6}$$

hold. The equation (6) has a unique solution

$$\varphi(s) = e^{\lambda s} [x + \int_s^0 e^{-\lambda\sigma}\psi(\sigma)d\sigma]. \tag{7}$$

If Re $\lambda \leqq -\gamma$, then φ is not in Y, when $x \neq - \int_{-\infty}^0 e^{-\lambda\sigma}\psi(\sigma)d\sigma$.

This follows that such λ belongs to $R_\sigma(B)$. Let Re $\lambda > -\gamma$. Then the function (7) belongs to Y and the equation (5) has the form

$$[\lambda - A - B(\lambda)]\, x = y + \widehat{L\psi_{\cdot}}(\lambda) = \xi.$$

The assumption on a resolvent of A implies that the embedding of X^1 into X^α is compact (see e.g. [7]). We choose some $\lambda_1 \in \rho(A)$ and rewrite (5) into the form

$$(I-K)x = (\lambda_1 - A)^{-1}\xi .$$

Here $K = (\lambda_1 - A)^{-1}[\lambda_1 - \lambda + B(\lambda)]$ is a compact operator from X into X^α. Therefore, either $1 \notin \sigma(K)$ (and then $\lambda \in \rho(B)$) or $1 \in P_\sigma(K)$. In this case the characteristic equation (4) has a nontrivial solution, i.e. $\lambda \in P_\sigma(B)$.

Now, we are coming to the spectrum of T(t). The spectral mapping theorem does not hold for T(t). Hence we consider T(t) as a perturbation of $T_0(t)$, which is the solution semigroup in Z for the equation (2) with L = 0. We will use the notion of an essential spectrum in the sense of F. Browder, i.e. $\lambda \in \sigma(B)$ - ess (B) if and only if λ is an isolated eigenvalue of a finite multiplicity. Let r_e denotes the radius of an essential spectrum.

<u>PROPOSITION 3.</u> Let A have a compact resolvent. Then for any $\varepsilon > 0$ there is a constant c such that

$$r_e(T(t)) = r_e(T_0(t)) \leqq c\, e^{(-\gamma + \varepsilon)t}. \tag{8}$$

<u>Proof.</u> As A has a compact resolvent, e^{At} is compact. By the construction, R is also compact. The explicit expression yields the compactness of $T(t) - T_0(t)$. The left equality follows. Choose $d \in (-\gamma, -\gamma + \varepsilon)$ such that $\sigma_0^+ = \{\lambda \in \sigma(B_0);\ \text{Re}\ \lambda > d\}$, $\sigma_0^- = \{\lambda \in \sigma(B_0);\ \text{Re}\ \lambda > d\}$ are spectral sets

for the generator B_0 of $T_0(t)$. The spectral projections, which correspond to these sets, can be calculated. The explicit expressions for P_0^+, P_0^- imply that $P_0^+(Z)$ is finite dimensional and $\|P_0^- T_0(t)\| \leq ce^{dt}$, $t \in \mathbb{R}^+$. The inequality in (8) follows.

COROLLARY. Let A have a compact resolvent, let $i\mathbb{R} \cap \sigma(B) = \emptyset$. Then $\sigma^+ = \{\lambda \in \sigma(B);\ \mathrm{Re}\ \lambda > 0\}$, $\sigma^- = \{\lambda \in \sigma(B);\ \mathrm{Re}\ \lambda < 0\}$ are spectral sets and σ^+ is finite. If P^+, P^- denote the corresponding projections, then $Z^+ = P^+(Z)$ is finite dimensional. Moreover, there are positive constants c, δ such that

$$\|P^- T(t)\| \leq ce^{-\delta t}, \quad t \in \mathbb{R}^+, \tag{9}$$

$$\|P^+ T(t)\| \leq ce^{\delta t}, \quad t \in \mathbb{R}^-. \tag{10}$$

Proof. Fix some $\varepsilon > 0$. If $\lambda \in \sigma(B)$, $\mathrm{Re}\ \lambda > -\gamma + \varepsilon$, then there is $t_0 > 0$ such that $e^{\mathrm{Re}\lambda t_0} > ce^{(-\gamma+\varepsilon)t_0}$, where c is a constant from (8). Hence $e^{\lambda t_0} \notin \mathrm{ess}\ T(t_0)$. This means that λ is an isolated point of $\sigma(B)$ and it has a finite multiplicity. Especially, σ^+ is finite and Z^+ has a finite dimension. Further, $B^+ = B/Z^+$ generates a group on Z^+. The spectral mapping theorem implies (10). Let $B^- = B/Z^-$, where $Z^- = P^-(Z)$. Then B^- is a generator of a C_0-semigroup $T^-(t)$ on Z^- and $\sigma(T^-(t)) \subset \mathrm{ess}\ T(t) \cup \{e^{\lambda t};\ \lambda \in \sigma^-\}$. By (8), the estimate (9) follows.

REMARK. The estimates (9), (10) show that the subspace Z^- is a stable manifold for the equation (2) and Z^+ is an unstable manifold.

4. INVARIANT MANIFOLDS FOR NONLINEAR EQUATIONS

The following method of construction of a stable manifold for a nonlinear equation is rather standard and goes back to E. Coddington, N. Levinson [2]. It was also used by D. Henry [7] for parabolic equations and O. Diekmann and S. van Gils [6] for integral equations. Therefore we will present the main steps only.

Suppose that a nonlinear term $F : Z \to Z$ satisfies

$F \in C^1(Z)$, $F(0) = 0$, $DF(0) = 0$, DF is uniformly continuous and bounded (BUC) on bounded subsets of Z. (H1)

LEMMA 1. Let $G \in BUC(\mathbb{R}^+,Z)$. Then $z \in BUC(\mathbb{R}^+,Z)$ satisfies the equation

$$z(t) = T(t-s)z(s) + \int_s^t T(t-\sigma)G(\sigma)d\sigma$$

for all $0 \leqq s \leqq t$, if and only if there is $\xi \in Z^-$ such that

$$z(t) = T(t)\xi + \Phi(G)(t),$$

where

$$\Phi(G)(T) = \int_0^t T(t-\sigma)P^-G(\sigma)d\sigma - \int_t^{+\infty} T(t-\sigma)P^+G(\sigma)d\sigma.$$

Let the map $\mathcal{F}:BUC(\mathbb{R}^+,Z) \times Z^- \to BUC(\mathbb{R}^+,Z)$ be defined by

$$\mathcal{F}(z,\xi) = z(.) - \Phi(F(z)) - T(.)\xi.$$

LEMMA 2. A function $z \in BUC(\mathbb{R}^+,Z)$ is a mild solution on $\mathbb{R}^+$ of the equation

$$\dot{z}(t) = Bz(t) + F(z(t)) \tag{11}$$

if and only if $\mathcal{F}(z,P^-z(0)) = 0$.

The proofs of both Lemmas are straightforward. Using the implicit function theorem for the equation $\mathcal{F}(z,\xi) = 0$, we obtain the following theorem. The part concerning an unstable manifold requires only small changes in the procedure described above.

THEOREM 2. Let A have a compact resolvent, let (H1) be satisfied and let $i\mathbb{R} \cap \sigma(B) = 0$. Then there exist two C^1-manifold S and U with properties:

(i) S, resp. U, is diffeomorphic to a neighbourhood of 0 in the subspace Z^-, resp. Z^+.

(ii) Z^- (Z^+) is tangent to S(U) at the origin.

(iii) For any $\xi \in S$ $(\xi \in U)$ there is a mild solution z of the equation (11) which satisfies the initial condition $z(0) = \xi$. This solution is defined on $\mathbb{R}^+$ $(\mathbb{R}^-)$ and $\lim_{t\to+\infty} z(t) = 0$ $(\lim_{t\to-\infty} z(t) = 0)$.

(iv) There is $\varepsilon > 0$ such that the trajectory of a solution z belongs to S(U) provided $\|\xi\| < \varepsilon$, $\xi \in S$ ($\xi \in U$).

A construction of a central manifold is more sophisticated. There are many proofs of the central manifold theorem, also for infinite dimensional cases, but a semigroup is supposed to be eventually diferentiable. The following procedure is a modification of those in [1], [15].

We assume $\sigma^0 = \sigma(B) \cap i\mathbb{R} \neq \emptyset$ and let P^0 be the corresponding spectral projection. For the sake of simplicity we suppose $\sigma^+ = \emptyset$, so that $Z = Z^0 \oplus Z^-$, $Z^0 = P^0(Z)$, $Z^- = P^-(Z)$. We will need a stronger assumption on F, namely

$$F \in C^2(Z),\ F(0) = 0,\ DF(0) = 0,\ D^2F \text{ is bounded on bounded sets.} \tag{H2}$$

As usually, we cut off F with help of some $\omega \in C^\infty(\mathbb{R}^+)$ which has a compact support and $\omega(t) = 1$ for $t \in [0,1]$. We put $G(u^0 + u^-,\varepsilon) = \varepsilon^{-1}\omega(\|u^0\|)F(\varepsilon(u^0 + u^-))$ and investigate the equation

$$\dot{u} = Bu + G(u,\varepsilon). \tag{12}$$

We are looking for a central manifold in the form $u^- = \varphi(u^0)$, where φ is an appropriate smooth function. Decompose (12) by applying the spectral projections P^0, P^-. Let $u^0(t,\varphi,\xi)$ be a global solution (on $\mathbb{R}$) of the equation

$$\dot{u}^0(t) = Bu^0(t) + G^0(u^0(t) + \varphi(u^0(t))),\ u^0(0) = \xi, \tag{13}$$

where $G^0 = P^0G$. We wish to find a function φ such that $v(t) = \varphi(u^0(t,\varphi,\xi))$ is a bounded mild solution on $\mathbb{R}$ of the equation

$$\dot{v}(t) = Bv(t) + G^-(u^0(t) + \varphi(u^0(t))),\ v(0) = \varphi(\xi),$$

i.e.

$$v(t) = T(t-s)v(s) + \int_s^t T(t-\sigma)G^-(u^0(\sigma) + \varphi(u^0(\sigma)))d\sigma \tag{14}$$

is true for $-\infty < s \leq t < +\infty$. As in Lemma 1, $v \in Z^-$ is a bounded solution of (14) on $\mathbb{R}^-$ if and only if

$$v(t) = \int_{-\infty}^{t} T(t-\sigma)G^-(u^o(\sigma) + \varphi(u^o(\sigma)))d\sigma. \tag{15}$$

Especially,

$$\varphi(\xi) = \int_{-\infty}^{0} T(-\sigma)G^-(u^o(\sigma,\varphi,\xi) + \varphi(u^o(\sigma,\varphi,\xi)))d\sigma \tag{16}$$

is an equation for determining of φ. Note that any solution of (14) is bounded on $\mathbb{R}^+$ because of (H2), (9) and the Gronwall inequality. The equation (13) is an ordinary differential equation (Z^o is finite dimensional) and therefore (H2) implies that $u^o(t,\varphi,\xi)$ is differentiable with respect to ξ and $\frac{\partial u^o}{\partial \xi}$ is Lipschitzian. The function $u^o(t,\varphi,\xi)$ is also Lipschitzian with respect to φ with a constant which can be arbitrarily small for $\varepsilon \to 0$. The equation (16) can be solved by the contraction principle in the space $M_{q,K} = \{\varphi \in C^1(Z^o,Z^-); \varphi(0) = 0, D\varphi(0) = 0, \|\varphi\|, \|D\varphi\|$ are bounded on Z^o by the constant K and φ, $D\varphi$ are Lipschitzian on Z^o with the Lipschitz constant q$\}$. As (13) is an autonomous equation, $u^-(t) = \varphi(u^o(t))$ satisfies (15) on $\mathbb{R}^-$. Because of cutting, the equation (11) is equivalent to (12) only in a neigbhourhood of the origin. Therefore the following theorem is merely local.

THEOREM 3. Let A have a compact resolvent and let (H2) be satisfied. If $\{\lambda \in \sigma(B); \mathrm{Re}\,\lambda \geq 0\} = i\mathbb{R} \cap \sigma(B) \neq \emptyset$ then there is a finite dimensional C^1-maniofld C such that

(i) C is diffeomorphic to a neighbourhood of 0 in Z^o.

(ii) C is tangent to Z^o at the origin.

(iii) C is locally invariant for the equation (11).

(iv) There exists $\delta > 0$ such that any solution z of (11), with $\|z(t)\| \leq \delta$ for all $t \in \mathbb{R}$, must have $z(t) \in C$ for $t \in \mathbb{R}$.

5. CONCLUSION

It is well known that the central manifold theorem is one of the possible starting points to bifurcations of the Hopf type. The existence of a branch of periodic solutions to a parabolic integrodifferential equation has been recently proved by G. da Prato, A. Lunardi [4]. Their method is based on Fourier series and therefore it needs strong smoothness assumptions on a nonlinear term. Theorem 3 was used for finding periodic solutions of the special equation - the so called Volterra population model - in [11].

The core of our approach to Theorem 2 consists of an investigation of the linear equation for which the origin is a hyperbolic point. In accordance with ordinary differential equations one can expect also the existence of bounded solutions for bounded forcing terms. There are two important contributions to this question, namely [5] and [14]. The authors have used directly the variation of parameters formula to obtain results which are valid also for $\alpha = 1$. We believe that their method can be modified for nonlinear equations to yield theorems similar to Theorems 2,3. Our semigroup roundabout excludes not only the case $\alpha = 1$ but restricts also a nonlinear term f requiring that f maps Z into X^{α}.

References

[1] J. Carr, Applications of Centre Manifold Theory, Applied Math. Sci. No. 35, Springer 1981.

[2] E.A. Coddington and N. Levinson, Theory of Ordinary Differential Equations. McGraw-Hill, New York 1955.

[3] G. Da Prato and M. Iannelli, Existence and regularity for a class of integrodifferential equations of parabolic type. J. Math. Anal. Appl. 112 (1985), 36-55.

[4] G. Da Prato and A. Lunardi, Hopf bifurcation for nonlinear integro-differential equations in Banach spaces with infinite delay. To appear in Indiana Math. J.

[5] G. Da Prato and A. Lunardi, Solvability on the real line of a class of linear Volterra integrodifferential equations of parabolic type. Università di Pisa, 1987.

[6] O. Diekmann and S.A. van Gils, Invariant manifolds for Volterra integral equations of convolution type. J. Differential Equations 54 (1984), 139-180.

[7] D. Henry, Geometric Theory of Semilinear Parabolic Equations. Lecture Notes in Mathematics No. 840, Springer 1981.

[8] R.C. Grimmer and F. Kappel, Series expansions for resolvents of Volterra integrodifferential equations in Banach spaces. SIAM J. Math. Anal. 15 (1984), 595-604.

[9] K. Kunish and W.Schappacher, Necessary conditions for differential equations with delay to generate C_0-semigroup. J. Differential Equations 50 (1983), 49-79.

[10] A. Lunardi, Laplace transform methods in integrodifferential equations. J. Integral Equations 10 (1985), 185-211.

[11] J. Milota, Stability in populations models with infinite delay. Conference on Mathematical Biology, Oberwolfach 1987.

[12] H. Petzeltová and J. Milota, Resolvent operator for abstract functional differential equations with infinite delay. Numerical Funct. Anal. Optim. 9 (1987), 779-807.

[13] J. Prüss, On linear Volterra equations of parabolic type in Banach spaces. Trans. Amer. Math. Soc. 300 (1987), 1-31.

[14] J. Prüss, Bounded solutions of Volterra equations. To appear in SIAM J. Math. Anal. 1987.

[15] J. Sijbrand, Studies in nonlinear stability and bifurcation theory. Thesis, University Utrecht 1981.

Jaraslav Milota
Matematicko-fyzikální fakulta,
Universita Karlova
Sokolovská 83
18600 Praha 8
Czechoslovakia.

E. MITIDIERI & M. TOSQUES

Nonlinear integrodifferential equations in Hilbert space: the variational case

0. INTRODUCTION

In this paper we study a class of evolution equations of the form

$$u' + \partial^{-}f(u) \ni F(u) \tag{0.1}$$

where $f:H \to \mathbb{R} \cup \{+\infty\}$ is a given lower semicontinuous function, H is a real Hilbert space, $\partial^{-}f$ is the subdifferential of f (see the next Section) and F is a given continuous map

$$F: C([0,T];H) \to L^2([0,T];H).$$

Equations of the form (0.1) have attracted great attention in the past few years, mainly because they represent abstract models of some physical problems (see [4], [6], [7], [8], [13],...,[16]).

In this paper we do not discuss the possible applications of (0.1) to concrete physical situations, we refer the interested reader to [16] for some information in this direction.

The aim of this paper is to prove a local existence result for (0.1) in the case in which $f:H \to \mathbb{R} \cup \{+\infty\}$ is <u>not</u> a convex function (i.e. $\partial^{-}f$ is <u>not</u> a monotone operator).

The main ingredient in this paper is the use of a special class of function called "functions with monotone subdifferential of order two" (abbreviated MS(2)) which is a noteworthy generalization of the class of convex functions (see [3] for relevant properties) and a compactness argument.

Some existence, uniqueness and regularity results for Volterra integral equations associated with non linear operators, more general than $\partial^{-}f$, are given in [12].

The paper is organized in the following way: the first section contains some preliminaries and the main results, the second one shows some applications

1. PRELIMINARIES AND MAIN RESULTS

Let H be a real Hilbert space with norm $\|\cdot\|$ and inner product $(\cdot,\cdot)$. Throughout the paper Ω will denote an open subset of H. If $r > 0$ and $u \in H$ we will set $B(u,r) = \{v \in H : \|u-v\| < r\}$. Let $g: \Omega \to \mathbf{R} \cup \{+\infty\}$, then the set $D(g) = \{v \in \Omega: g(v) \in \mathbf{R}\}$ will be called the domain of g. If $u \in \Omega$ we can define (see also [3]):

$$\partial^- g(u) = \begin{cases} \emptyset & , \text{ if } u \in \Omega - D(g) \\ \{\alpha \in H: \liminf\limits_{v \to u} \dfrac{g(v)-g(u)-(\alpha,v-u)}{\|v-u\|} \geq 0\} & , \text{ otherwise.} \end{cases}$$

We will set $D(\partial^- g) = \{u \in \Omega: \partial^- g(u) \neq \emptyset\}$. It is clear that $D(\partial^- g) \subset D(g)$. It is not difficult to see that $\partial^- g(u)$ is a closed convex set for every $u \in D(\partial^- g)$.

Now we introduce a class of functions which will play an important role in the rest of the paper (see [3]).

(1.1) DEFINITION. Let $f:\Omega \to \mathbf{R} \cup \{+\infty\}$ be a lower semicontinuous function. Then we say that f has a monotone subdifferential of order two (in short $f \in MS(2)$) if there exists a continuous function $\chi: D(f)^2 \times \mathbf{R}^2 \to \mathbf{R}^+$ such that

$$(\alpha-\beta,u-v) \geq -\chi(u,vf(u),f(v))(1+ \|\alpha\|^2 + \|\beta\|^2)\cdot \|u-v\|^2 \tag{1.2}$$

for every $u,v \in D(\partial^- f)$ and $\alpha \in \partial^- f(u)$, $\beta \in \partial^- g(v)$.

It follows readily from the above definition that:

a) every lower semicontinuous convex function belongs to MS(2)

b) perturbations of lower semicontinuous convex functions by functions whose gradient is Lipschitz continuous, belong to MS(2).

Other remarkable concrete examples of functions of class MS(2) can be found in [3], [9], [10] and [11].

(1.3) DEFINITION. Let $f: \Omega \to \mathbf{R} \cup \{+\infty\}$ be a function. Let $I \in \mathbf{R}$ be an interval with $\mathring{I} \neq \emptyset$ and $h \in L^1(I;H)$. We say (see [17]) that a function

$u: I \to \Omega$ is a <u>strong solution</u> of

$$u' + \partial^- f(u) \ni h \tag{1.4}$$

on I, if the following conditions are satisfied:

a) u is continuous on I and absolutely continuous on the compact subsets of $\mathring{I}$.

b) $u(t) \in D(\partial^- f)$ a.e. on I and $u'(t)-h(t) \in -\partial^- f(u(t))$ holds a.e. on I.

c) $f \circ u$ is bounded on the compact subsets of $\mathring{I}$.

The following theorems (1.5) and (1.9) are important tools in the proof of our main result: Theorem (1.13). They can be easily deduced by the results proved in [17].

(1.5) <u>THEOREM</u>. Let $f: \Omega \to \mathbb{R} \cup \{+\infty\}$ be of class MS(2). Then for every $u_0 \in D(f)$ and every $k \geq 0$ there exists $\bar{T} > 0$ such that for every $T > 0$ and $h \in L^2(0,T;H)$ with $\|h\|_2 \leq k$, there exists a unique strong solution of the problem

$$\begin{aligned} &u' + \partial^- f(u) \ni h \text{ on } [0,\min(T,\bar{T})], \\ &u(0) = u_0 \end{aligned} \tag{1.6}$$

Moreover $u \in H^{1,2}(0,\min(T,\bar{T});H)$ and $f \circ u \in AC(0,\min(T,\bar{T});\mathbb{R})$.

Furthermore if $[0,T_0[$ is the maximal interval such that, for every $t \in [0,T_0[$, u is a strong solution of (1.4) on $[0,t]$ and $u \in H^{1,2}(0,t;H)$, then

$$\limsup_{t \to T_0^-} \max(-f(u(t)),d(u(t),\partial\Omega)^{-1}) = +\infty, \text{ if } T_0 < T \tag{1.7}$$

and we have, for every $t \in [0,T_0[$:

$$\int_0^t \|u'(s)\|^2 ds \leq 2(f(u_0)-f(u(t))) + \int_0^t \|h(s)\|^2 ds \tag{1.8}$$

(1.9) THEOREM. Let $f:\Omega \to \mathbb{R} \cup \{+\infty\}$ be of class $MS(2)$.

Let $a,b \in \mathbb{R}^+$ with $a < b$ and $h_i \in L^1(a,b;H)$, $(i = 1,2)$.

Let $u_i (i = 1,2)$ denote a strong solution of (1.4) on $[a,b]$ corresponding to h_i, i.e. u_i solves

$$u_i' + \partial^- f(u_i) \ni h_i \quad \text{on } [a,b].$$

Then for every $t \in [a,b]$ we have

$$\begin{aligned} \|u_1(t)-u_2(t)\| \leq & \left(\|u_1(a)-u_2(a)\| + \int_a^t \|h_1(s)-h_2(s)\| \, ds \right) . \\ & \cdot \exp \left(\int_a^t \chi(u_1(s),u_2(s),f(u_1(s)),f(u_2(s))) . \right. \\ & \left. \cdot (1 + \|u_1'(s)\|^2 + \|u_2'(s)\|^2) ds \right) \end{aligned} \tag{1.10}$$

(with the convention that $0.\infty = \infty$).

We are now ready to state the following definition.

(1.11) DEFINITION. Let $\Lambda \subset H$ be an open set.

Let $f: \Omega \to \mathbb{R} \cup \{+\infty\}$ be of class $MS(2)$ and let $I \subset \mathbb{R}$ such that $\mathring{I} \neq \emptyset$. If $F: C(I;\Lambda) \to L^1_{loc}(\mathring{I};\Lambda)$ is a given map, then we say that a function $u: I \to \Omega \cap \Lambda$ is a strong solution of

$$u' + \partial^- f(u) \ni F(u) \text{ on } I, \tag{1.12}$$

if u is a strong solution of (1.4) corresponding to $h = F(u)$.

Throughout the paper we shall use the following conventions.

(1.13) CONVENTIONS

a) Let $T \in \mathbb{R}^+$ and $F: C([0,T];\Lambda) \to L^1_{loc}(0,T;H)$ be a given map. Let $T' \in]0,T]$ and $v \in C([0,T']; \ )$ then $F(v)(t) = F(\bar{v})(t)$ a.e. on $[0,T]$ where $\bar{v}(t) = v(t)$ if $0 \leq t \leq T'$, $\bar{v}(t) = v(T')$ if $T' \leq t \leq T$.

b) Let $0 < T' \leq T$ and let $u:[0,T'[\to D(f) \cap \Lambda$ be a continuous function.

Then $[0,T'[$ is the maximal interval of existence of u as strong solution of $u' + \partial^- f(u) \ni F(u)$ if for every $\bar{T} \in [0,T'[$, u is a strong solution of (1.12) on $[0,\bar{T}]$ and for every $\bar{T} \in [T',T]$, u cannot be extended to a continuous function $\bar{u}$ on $[0,\bar{T}]$ such that $\bar{u}$ is a strong solution of (1.12) on $[0,\bar{T}]$.

The main result of this paper is the following:

(1.14) THEOREM. Let Ω, Λ be open sets of H, and

$$F: C([0,T];\Lambda) \to L^2(0,T;H) (0 < T \leq +\infty)$$

be a continuous map. Let $f: \Omega \to \mathbf{R} \cup \{+\infty\}$ be of class MS(2), such that the following condition holds:

$$\begin{aligned}&\text{for every } \hat{u}_0 \in D(f), \text{ there exists } r > 0 \text{ such that}\\&\overline{B(\hat{u}_0,r)} \subset \Omega \text{ and the set}\\&L(c) = \{v: f(v) \leq c,\ v \in \overline{B(\hat{u}_0,r)}\}\\&\text{is compact for every } c \in \mathbf{R}.\end{aligned} \tag{1.15}$$

Then, for every $u_0 \in D(f) \cap \Lambda$, there exists $T_0 \in]0,T]$ and a strong solution $u: [0,T_0] \to D(f) \cap \Lambda$ of (1.12) on $[0,T_0]$, such that:

$$u(0) = u_0,\ u \in H^{1,2}(0,T_0;H),\ f \circ u \in AC(0,T_0;\mathbf{R}) \tag{1.16}$$

and

$$\int_0^t \|u'(s)\|^2 ds \leq 2(f(u_0)-f(u(t))) + \int_0^t \|F(u)(s)\|^2 ds, \text{ for every } t \in [0,T_0] \tag{1.17}$$

Proof. The proof is based on the application of the Schauder fixed point theorem.

Let $u_0 \in D(f) \cap \Lambda$. Since f is lower semicontinuous and F is continuous, we can suppose that r, given by (1.15), verifies also the following properties:

$$f(v) \geq f(u_0)-1 \text{ for every } v \in \overline{B(u_0,r)} \tag{1.18}$$

and

$$\|F(v)\|_2 \quad M \quad \text{for every } v \in C([0,T];B(u_0,r)). \tag{1.19}$$

Applying Theorem (1.5) with the choice $k = M$, we can find $T > 0$ ($\bar{T} \leq T$) such that: for every $v \in C([0,T];\overline{B(u_0,r)})$ there exists a unique strong solution $u_v:[0,\bar{T}] \to D(f)$ of (1.6) corresponding to $h_v = F(v)$, i.e. u_v solves

$$u_v' + \partial^- f(u_v) \ni F(v) \text{ on } [0,\bar{T}] \tag{1.20}$$

and $u_v(0) = u_0$, $u_v \in H^{1,2}(0,\bar{T};H)$, $f \circ u_v$ is absolutely continuous and

$$\int_0^t \|u_v'(s)\|^2 ds \leq 2(f(u_0)-f(u_v(t))) + \int_0^t \|F(u_v)(s)\|^2 ds, \text{ for every } t \in [0,\bar{T}]. \tag{1.21}$$

Let $T_0 \in R^+$ such that

$$T_0 \leq \min \{\frac{r^2}{2+M^2}\}, \bar{T} , \tag{1.22}$$

By setting $K = \{v \in C([0,T_0];\overline{B(u_0,r)}): v(0) = u_0\}$ and by a) of (1.13), one easily sees that the map $\mathcal{F}:K \to H^{1,2}(0,\bar{T};H)$ defined by $\mathcal{F}(v) = u_v$ (u_v being the strong solution to (1.20)) is well defined. Let $J:H^{1,2}(0,\bar{T};H) \to C([0,T_0];H)$ be the usual continuous injection. We claim that $J(\mathcal{F}(K)) \subseteq K$ and $J \circ \mathcal{F}$ is continuous.

First of all we have that for every $t \in [0,T_0]$

$$\mathcal{F}(v)(t)-u_0\| \leq \int_0^t \|\mathcal{F}(v)'(s)\| ds \leq \sqrt{t}\sqrt{2(f(u_0)-f(\mathcal{F}(v(t))))+\int_0^t \|F(v)(s)\|^2 ds}$$

which implies, by using (1.18), (1.19) and (1.22) that, for every $t \in [0,T_0]$

$$\mathcal{F}(v)(t) \in \overline{B(u_0,r)} \tag{1.23}$$

therefore $J(\mathcal{F}(K)) \subseteq K$.

To prove the continuity of $J \circ F$, we observe that, by (1.10), the following condition holds:

For every $v_1, v_2 \in K$ and $t \in [0,T_0]$ we have

$$\|F(v_1)(t)-F(v_2)(t)\| \leq \sqrt{t}\, \|F(v_1)-F(v_2)\|_2 \cdot$$
$$\cdot \exp\Big(\int_0^t \chi(F(v_1(s),F(v_2)(s),f(F(v_1)(s)),f(F(v_2)(s))) \cdot \qquad (1.24)$$
$$\cdot (1 + \|F(v_1)'(s)\|^2 + \|F(v_2)'(s)\|^2)ds\Big).$$

Furthermore by (1.18) and (1.23) we have,

$$f(F(v_i(t))) \geq f(u_0)-1 \text{ for } i = 1,2, \text{ for every } t \in [0,T_0] \qquad (1.25)$$

and, by using (1.19) and (1.21)

$$f(F(v_1(t))) \leq f(u_0) + \frac{1}{2}\int_0^t \|F(v_i)(s)\|^2 ds \quad f(u_0) + \frac{1}{2}M^2, \qquad (1.26)$$
$$\text{for } i = 1,2 \text{ and every } t \in [0,T_0].$$

Since χ is continuous we can suppose, unless of decreasing r that

$$L = \sup \{\chi(x_1,x_2,s_1,s_2): \|x_i-u_0\| \leq r,\ f(u_0)-1 \leq s_i \leq f(u_0) + \frac{1}{2}M^2, \qquad (1.27)$$
$$i = 1,2\} < +\infty.$$

Finally by (1.24)-(1.27) one has,

$$\|F(v_1)-F(v_2)\|_\infty \leq \sqrt{T_0}\, \|F(v_1)-F(v_2)\|_2 \cdot$$
$$\cdot \exp\Big(L(T_0 + \sum_{i=1}^{2} (2(f(u_0)+f(F(v_i)(T_0)) + \int_0^{T_0} \|F(v_i)\|^2 ds))\Big) \leq$$
$$\leq \|F(v_1)-F(v_2)\|_2 \sqrt{T_0} \exp(L(T_0 + 4 + M^2)),$$

which, together with the continuity of F, gives that $J \circ F$ is continuous. To apply Schauder theorem we need to show that the family:

$$\{J \circ F(v) : v \in K\}$$

is equicontinuous. Indeed, if $t_1, t_2 \in [0,T_0]$ and $t_1 \leq t_2$ we have that

$$\|F(v)(t_2)-F(v)(t_1)\| \leq \int_{t_1}^{t_2} \|F(v)'(s)\| ds \leq$$

$$\leq \sqrt{t_2-t_1}\ (2(f(F(v)(t_1))-f(F(v)(t_2))) + \int_{t_1}^{t_2} \|F(v)(s)\|^2 ds)^{1/2}$$

$$\leq \sqrt{t_2-t_1}\ \sqrt{2\ (\frac{M^2}{2}\ L\ 1) + M^2}$$

for every $v \in K$ which implies the required equicontinuity.

Now, by an estimate similar to (1.26) and the compactness hypotheses (1.15) it follows that there exists $c > 0$ such that $F(v)([0,T_0]) \subset \{u \in \overline{B(u_0,r)}: f(u) \leq c\}$, for every $v \in K$ (the latter set being compact). Therefore by the Ascoli-Arzela theorem the operator

$$\Theta = J \circ F : K \to K$$

is completely continuous. An application of the Schauder fixed point principle gives the existence of $u \in K$ such that $\Theta(u) = u$. That is, u is a strong solution of (1.12) which satisfies (1.16) - (1.17). □

The following result characterizes the maximal interval of existence of a strong solution of the equation (1.4).

(1.29) <u>THEOREM</u>. Suppose that the hypotheses of Theorem (1.14) hold and that, in addition, F satisfies the following property:

$$\begin{aligned} &\text{for every } v,w \in C([0,T];A) \text{ such that} \\ &v(t) = w(t) \text{ for every } t \in [0,\bar{T}]\ (\bar{T} \leq T), \text{ we have} \\ &F(v)(t) = F(w)(t) \text{ a.e. on } [0,\bar{T}] \end{aligned} \tag{1.30}$$

Let $u : [0,T'[\to D(f) \cap A$ $(T' < T)$ be a continuous function. If $[0,T'[$ is the maximal interval of existence of u, as strong solution of $u'+\partial^- f(u) \ni F(u)$ and $u \in H^{1,2}(0,t;H)$ for every $t \in [0,T'[$, then:

$$\limsup_{t \to T'^-} \max\{-f(u(t)), \|F(u)\|_{L^2(0,t;H)}, d(u(t),\partial\Omega)^{-1}, d(u(t),\partial\Lambda)^{-1}) =$$

$$= +\infty. \tag{1.31}$$

<u>Proof</u>. If (1.31) does not hold then there exists a sequence $(t_n) \subset [0,T'[$ such that: $\lim_n t_n = T'$,

$$m = \inf_n\{f(u(t_n))\} > -\infty, \ M = \sup_n \{\|F(u)\|_{L^2(0,t_n;H)}\} < +\infty, \tag{1.32}$$

and

$$\inf_n \{d(u(t_n),\partial\Omega), \ d(u(t_n)\partial\Lambda)\} > 0 \tag{1.33}$$

By (1.32) and (1.17) it follows that, for every n:

$$\int_0^{t_n} \|u'(s)\|^2 ds \leq 2(f(u_o)-f(u(t_n)) + \|F(u)\|^2_{L^2(0,t_n;H)} \leq 2(f(u_o)-m)+M^2 \tag{1.34}$$

Therefore there exists $\lim_{t \to T'^-} u(t) = \bar{u}$, and by (1.33) $\bar{u} \in \Omega \cap \Lambda$. Using the lower semicontinuity of f and (1.34) we get in particular that $\bar{u} \in D(f)$. Let

$$\tilde{F}:\{v \in C([0,T-T'];\Lambda):v(0) = \bar{u}\} \to L^2(0,T-T';H)$$

be defined by $\tilde{F}(v) = F(\tilde{v})$ where $\tilde{v}(t) = u(t)$, if $t \in [0,T']$ and $\tilde{v}(t) = v(t-T')$, it $t \in [T',T]$. Using Theorem (1.14), we get that there exists a function $u:[0,\varepsilon] \to D(f)$ which is a strong solution of the problem

$$\begin{cases} w' + \partial^- f(w) \ni \tilde{F}(w) \text{ on } [0,\varepsilon], \\ w(0) = \bar{u}. \end{cases}$$

Let $\bar{u}:[0,T'+\varepsilon] \to D(f)$ be the continuous function defined by $\bar{u}(t) = u(t)$, if $t \in [0,T']$ and $\bar{u}(t) = \tilde{u}(T'+\varepsilon-t)$ if $t \in [T'+\varepsilon]$. Since, by (1.30), $F(\bar{u})(t) = F(u)(t)$ a.e. on $[0,T']$, we get that $\bar{u}$ is a strong solution of (1.12) on $[0,T'+\varepsilon]$, which extends u on $[0,T'+\varepsilon]$, against the definition of T'. □

2. SOME APPLICATIONS

As an example of the application of the abstract results (1.14) and (1.29), we consider the following situation. Let $T \in \mathbb{R}^+$ and let $\Lambda \subset H$ be an open set. Let $G:[0,T] \times \Lambda \to H$ be a Caratheodory mapping such that there exists $\alpha:[0, +\infty[\to [0, +\infty[$ continuous and nondecreasing such that for every $(t,x) \in [0,T] \times \Lambda$ the following growth condition holds

$$\|G(t,x)\| \leq k_1(t)\alpha(\|x\|) + k_2 \tag{2.1}$$

with $k_1 \in L^2(0,T;R)$ and $k_2 \in \mathbb{R}$.

Let $a \in L^2(0,T;R)$ and consider the following maps

$$F_i: C([0,T];\Lambda) \to L^2(0,T;H),\ i = 1,2$$

defined by:

$$F_1(v)(t) = G(t,v(t)),\quad F_2(v)(t) = \int_0^t a(t-s)G(s,v(s))ds$$

for every $v \in C([0,T];\Lambda)$ and $t \in [0,T]$.

An application of Theorem (1.14) and (1.29) gives:

(2.2) THEOREM. Let $\Omega \subset K$ be an open set.

Let $f: \Omega \to \mathbb{R} \cup \{+\infty\}$ be of class MS(2) satisfying (1.15). Then for every $u_0 \in D(f)$, there exists $T_{0,i} (\leq T)$ and a strong solution $u_i:[0,T_{0,i}] \to D(f) \cap \Lambda$ of

$$u_i' + \partial^- f(u_i) \ni F_i(u_i) \text{ on } [0,T_{0,i}] \quad (i = 1,2) \tag{2.3}$$

such that (1.16) and (1.17) hold.

If $[0,T_i']$ $(i = 1,2)$ is the maximal interval of existence of u_i $(i = 1,2)$, as strong solution of

$$w_i' + \partial^- f(w_i) \ni F_i(w_i) \qquad (i = 1,2), \tag{2.4}$$

such that $u_i \in H^{1,2}(0,t;H)$ $(i = 1,2)$, for every $t \in [0,T_i'[$, and if $T_i' < T$, then

$$\limsup_{t\to T_i'^-} \max\{ \|u_i\|_{L^\infty(0,t;H)}, -f(u_i(t)), d(u_i(t),\partial\Omega)^{-1}, d(u_i(t),\partial\Lambda)^{-1}\} = +\infty. \tag{2.5}$$

Proof. Since F_i $(i = 1,2)$ is a continuous map, then there exists $T_{0,i}$ which satisfies the properties of Theorem (1.14). Furthermore if $\limsup_{t\to T_i'^-} \|u_i\|_{L^\infty(0,t;H)} < +\infty$, $(i = 1,2)$ then by (2.1) we get $\limsup_{t\to T_i'^-} \|F(u_i)\|_{L^2(0,t;H)} < +\infty$, $(i = 1,2)$. Since (1.30) is certainly true then (2.5) follows from (1.31). □

By requiring stronger hypotheses on f, we can obtain the following Corollary

(2.6) COROLLARY. Suppose that the hypotheses listed in the proceding theorem are satisfied. If there exists $c \geq 0$ such that

$$f(v) \geq -c(1 + \|v\|^2), \text{ for every } v \in D(f) \tag{2.7}$$

and

$$\alpha(t) \leq t \text{ for every } t \in [0,+\infty[, \tag{2.8}$$

then the thesis of Theorem (2.2) holds and (2.5) can be substituted with

$$\liminf_{t\to T_i'^-} \min \{d(u_i(t),\partial\Omega), d(u_i(t),\partial\Lambda)\} = 0. \tag{2.9}$$

(2.10) REMARK. If $f:\Omega \to \mathbf{R} \cup \{+\infty\}$ is a lower semicontinuous function satisfying the following condition:

$$\begin{gathered}\text{there exists a continuous function} \\ X:D(f) \times \mathbf{R} \to \mathbf{R}^+ \\ \text{such that for every } v \in D(f),\ u \in D(\partial^- f) \text{ and } \alpha \in \partial^- f(u) \text{ one has} \\ f(v) \geq f(u) + (\alpha,v-u)-X(x,f(u))(1+ \|\alpha\|^2) \|v-u\|^2\end{gathered} \tag{2.11}$$

then f $\in$ MS(2) and (2.7) holds.

For interesting properties of functions satisfying (2.11) we refer the interested reader to [3], [9] and the references therein.

PROOF OF COROLLARY (2.6). Suppose that (2.7) and (2.8) hold. By uniqueness and (1.8) we get that for every $t_o, t \in [0, T_i'[$ $(i = 1,2)$ $t_o \le t$ we have

$$\int_{t_0}^{t} \|u_i'(s)\|^2 ds \le 2(f(u_i(t_o)-f(u_i(t)))+2\int_{t_0}^{t} \|F(u_i)(s)\|^2 ds \le$$

$$\le 2(f(u_i(t_o))+c+c\,\|u_i(t)\|^2)+2(\sup_{s\in[t_o,t]}\|u(s)\|^2)\int_0^T k_1^2(s)ds+2k_2^2T.$$

Then,

$$\sup_{s\in[t_o,t]} \|u(s)-u(t_o)\|^2 \quad (t-t_o)(2f(u(t_o))+2c+4+4(c+\int_0^T k_1^2(s)ds).$$

$$\cdot \sup_{s\in[t_o,t]} \|u(s)-u(t_o)\|^2+4\,(c+\int_0^T k_1^2(s)ds)\,\|u_i(t_o)\|^2+2k_2^2T).$$

Therefore if we choose $t_o \in [0,T_i'[$ such that:

$$8c(T_i-t_o)\int_0^T k_1^2(s)ds < 1$$

we get that there exists $\bar{c} > 0$ such that $\sup_{t_o \le s \le T_i'} \|u_i(s)-u_i(t_o)\|^2 \le \bar{c}$,

which implies that $\sup\{\|u_i(t)\| : 0 \le t < T_i'\} < +\infty$. Therefore by (2.7) and (2.9), it follows that (2.5) holds. □

References

[1] M. Degiovanni, A. Marino and M. Tosques, General properties of (p,q)-convex functions and (p,q)-monotone operators, Ricerche di Matematica, 32 (1983), 285-319.

[2] M. Degiovanni, A. Marino and M. Tosques, Evolution equations associated with (p,q)-convex functions and (p,q)-monotone operators, Ricerche di Matematica, 33 (1984), 81-112.

[3] M. Degiovanni, A. Marino and M. Tosques, Evolution equations with lack of convexity, Nonlinear Anal. Th. Math. Appl., (12), 9 (1985), 1401-1443.

[4] V. Barbu, Nonlinear Semigroups and Differential Equations in Banach Spaces (Noordhoff, Leyden, 1976).

[5] H. Brezis, Operateurs maximaux monotones et semigroupes de contractions dans les espaces de Hilbert (North-Holland, Amsterdam, 1973).

[6] G. Gripenberg, On some integral and integro-differential equations in a Hilbert space, Annali di Matematica Pura e Applicata (IV) 118 (1978), 181-198.

[7] M.G. Crandall, S.O. Londen, J.A. Nohel, An abstract Volterra integro-differential equations, MRC. Technical Summary Report 1684, The University of Wisconsin.

[8] M.G. Crandall, J.A. Nohel, An abstract functional differential equation and a related nonlinear Volterra equation, Israel J. of Math. 29 (1978), 313-328.

[9] A. Marino, Evolution equations and multiplicity of critical points with respect to an obstacle, (Bologna 1985), Res. Notes in Math., Pitman, in press.

[10] A. Marino, D. Scolozzi, Autovalori dell'operatore di Laplace ed equazioni di evoluzione in presenza di un ostacolo, in "Problemi differenziali e teoria dei punti critici", Marzo (1984)-Universita di Bari.

[11] G. Chobanov, A. Marino, D. Scolozzi, Una classe di problemi vincolati. Autovalore dell'operatore di Laplace rispetto ad un ostacolo e relativa equazione di evoluzione, preprint n. 166, Dipartimento di Matematica, Pisa 1986.

[12] E. Mitidieri, M. Tosques, Volterra integral equations associated with a class of nonlinear operators in Hilbert spaces, Ann. Fac. Sci. Toulose Math. (5), VIII, (1986).

[13] E. Mitidieri, I.I. Vrabie, Existence for nonlinear functional differential equations, Hiroshima Math. Journal 17, 3 (1987).

[14] W. Fitzgibbon, Integrodifferential equations in Banach spaces, in "Integral and functional differential equations", Lecture Notes in Pure and Applied Math. M. Dekker Inc. Vol. 67 (1981), 173-184 (Editors T.L. Herdmann, S.M. Roukin III and H.W. Steck).

[15] I.I. Vrabie, Compactness methods for nonlinear evolutions, Pitman Monographs and Surveys in Pure and Applied Mathematics, Vol. 32, Longman Scientific and Technical (1987).

[16] J.A. Nohel, Nonlinear Volterra equations for heat flow in materials with memory, in "Integral and functional differential equations" Marcel Dekker, Inc. New York and Basel (1981), 3-82.

[17] M. Tosques, Quasi-autonomous evolution equations associated with a class of non linear operators, to appear.

Enzo Mitidieri
Dipartimento di Scienze Matematiche
Università degli Studi di Trieste
34100-Trieste
Italy.

Mario Tosques
Dipartimento di Matematica
Via Filippo Buonarroti, 2
56100 Pisa
Italy.

J.A. NOHEL, R.C. ROGERS & A.E. TZAVARAS

Hyperbolic conservation laws in viscoelasticity

1. INTRODUCTION

The equations of unsteady motion for nonlinear elastic materials are quasilinear systems of hyperbolic equations in which characteristic speeds are not constant. Thus weak initial waves are amplified and smooth solutions generally develop singularities in finite time. Particularly interesting situations arise when this destabilizing mechanism coexists and competes with dissipation; a simple example is provided by the quasilinear wave equation with linear first-order damping. A more subtle dissipation occurs in certain viscoelastic materials such as polymers, suspensions and emulsions which have memory; i.e. the stress at each material point depends not only on the present value of the deformation gradient (and/or velocity gradient), but on the entire temporal history of motion. These materials exhibit behaviour intermediate between that of an elastic solid and a viscous fluid. Generally, the memory fades with time: disturbances which occurred in the distant past have less influence on the present stress than those which occurred in the recent past. Our purpose is to discuss qualitative properties of the equations which model unsteady motions of such materials. In order to present some of the central ideas while avoiding serious technical difficulties, we shall restrict the discussion to a particular model problem for the motion of a one-dimensional viscoelastic material with fading memory. Many of the qualitative features in this relatively simple situation are present in more general integral and differential models in one and in several space dimensions; these are discussed in a recent monograph [37].

Motivated by the study of the growth and decay of acceleration waves by Coleman and Gurtin ([5], [6]) and by experiments ([34], [43]), it is natural to require that reasonable mathematical models for unsteady motion of viscoelastic materials should possess qualitative properties which include:

(a) Under physically reasonable assumptions, the memory should induce a weak dissipative mechanism. As a consequence, for sufficiently smooth and small data and history the equations of motion should have globally defined smooth solutions which decay as $t \to \infty$.

(b) By contrast, if the smooth data and history are chosen sufficiently large, the classical solution solutions should break down in finite time and exhibit a shock structure.

(c) For rough, bounded data, the equations of motion should possess globally defined weak solutions.

This paper is organized as follows. In Section 2, we state the model problem and related, well-understood problems, and we briefly review known results regarding properties (a) and (b) for classical solutions. We then state a new result [32] on the existence of a weak solution (in the class of bounded measurable functions) which verifies property (c) in an important special case. In Section 3, we sketch the proof of the new result using the methods of vanishing viscosity and compensated compactness; further details can be found in [32].

2. MODEL PROBLEM AND SUMMARY OF RESULTS

We study the model Cauchy problem for the system of integrodifferential equations

$$\begin{cases} w_t = v_x, \quad v_t = \sigma_x, \quad x \in \underline{R},\ t > 0 \\ w(x,0) = w_0(x), \quad v(x,0) = v_0(x), \quad x \in \underline{R}, \end{cases} \tag{VE}$$

where the function $\sigma(x,t)$ is determined by the history of $w(x,\cdot)$ through the constitutive assumption

$$\sigma(x,t) = \varphi(w(x,t)) + \int_0^t k(t-\tau)\psi(w(x,\tau))d\tau, \tag{CA}$$

which generalizes Boltzmann's law for linear viscoelaticity [3]. The given functions $\varphi(w)$, $\psi(w)$ and $k(t)$ are assumed to be smooth and, in addition,

$$\varphi'(w) > 0, \ w \in \underline{R},$$

so that the structure of (VE) is hyperbolic. The system (VE) is a model for one-dimensional motion of a viscoelastic material with fading memory. The functions $v(x,t)$, $w(x,t)$ and $\sigma(x,t)$ stand for the velocity, deformation gradient and stress, respectively, while the constitutive assumption (CA) states that the stress is a particular functional of the history of the deformation gradient; in (CA) the history of $w(x,\cdot)$ is assumed to be zero for $t < 0$ and the body force is taken to be zero. Under appropriate assumptions on the kernel k, the memory induces a weak dissipative mechanism which competes with the hyperbolic character of (VE). In the sequel, we will limit our discussion to the Cauchy problem. We remark that when x is restricted to a finite interval, boundary conditions which are compatible with the data w_0, v_0 must be adjoined to (VE).

It is useful to consider several special cases. If $k \equiv 0$, (VE) reduces to the system of conservation laws

$$\begin{cases} w_t = v_x, \ v_t = \varphi(w)_x \\ w(x,0) = w_0(x), \ v(x,0) = v_0(x), \end{cases} \tag{E}$$

which describes the motion of one-dimensional elastic materials. The solutions of (E) generally develop singularities in finite time no matter how smooth and small the data w_0, v_0 are taken ([19], [24], [18]). At the other extreme, if one formally sets $k = -\delta'$, where δ is the Dirac mass at the origin, (VE) reduces to the system

$$w_t = v_x, \ v_t = \psi(w)_{xt} + \varphi(w)_x,$$

which is parabolic and which possesses globally defined smooth solutions even for the large data w_0, v_0, whenever ψ is smooth and $\psi'(\cdot) > 0$ ([1], [21]). The theory of (VE) lies between these two extremes. In the special case $\psi \equiv \phi$, (VE) reduces to the system

$$\begin{aligned} & w_t = v_x, \\ & v_t = \varphi(w)_x + \int_0^t k(t-\tau)\varphi(w(x,\tau))_x d\tau, \quad x \in \underline{R}, \ t > 0, \\ & w(x,0) = w_0(x), \ v(x,0) = v_0(x), \qquad x \in \underline{R}, \end{aligned} \tag{2.1}$$

which, as we shall see, exhibits a close resemblance to the theory of the system

$$\begin{aligned} w_t &= v_x \\ v_t &= \varphi(w)_x - v. \end{aligned} \tag{FE}$$

(FE) describes the motion of a one-dimensional frictionally damped elastic material and enjoys properties (a), (b) and (c) which are established in [29, 38, 13] respectively. The similarity between (2.1) and (FE) is revealed by the following idea of MacCamy [22]. Let $r(t)$ be the resolvent kernel associated with k; i.e. r is the solution of the linear Volterra equation

$$r(t) + \int_0^t k(t-\tau)r(\tau)d\tau = k(t), \ t \geqq 0. \tag{r}$$

Convolving $(2.1)_2$ with $r(t)$, a simple calculation yields

$$\begin{aligned} \int_0^t k(t-\tau)\ (w(x,\tau))_x d\tau &= \int_0^t r(t-\tau)v_t(x,\tau)d\tau \\ &= r(0)v(x,t) - r(t)v_0(x) + \int_0^t r'(t-\tau)v(x,\tau)d\tau. \end{aligned} \tag{2.2}$$

Thus, for classical solutions (2.1) is equivalent to

$$w_t = v_x,\ v_t = \varphi(w)_x + F[v], \qquad x \in \underline{R},\ t > 0, \tag{2.3}$$

$$w(x,0) = w_0(x),\ v(x,0) = v_0(x), x \in \underline{R},$$

where

$$F[v](x,t) := r(0)v(x,t) - r(t)v_0(x) + \int_0^t r'(t-\tau)v(x,\tau)d\tau. \tag{2.4}$$

Under physically reasonable assumptions (e.g., $k = a'$, where $a:[0,\infty) \to \underline{R}$ is smooth, positive, decreasing and convex, $r(0) = a'(0) < 0$) the term $r(0)v(x,t)$ has a damping effect. Its effect is dominant close to equilibrium ($w \equiv 0$, $v \equiv 0$, $\phi(0) = 0$), thereby inducing property (a) ([22], [11], [39]). However, for suitably large data, the hyperbolic part of (2.1) dominates irrespective of the sign of the memory, and property (b) holds if ϕ is convex or concave ([8], [31], [35]).

To discuss the results for classical solutions of (VE) when $\psi \neq \phi$, we assume that the smooth constitutive functions ϕ, ψ defined on $\underline{R}$ satisfy

$$\phi(0) = \psi(0) = 0, \ \phi'(\cdot) > 0, \ \psi'(0) > 0.$$

To simplify the exposition, we also assume that the kernel $k = a'$, where $a : [0,\infty) \to \underline{R}$ is a smooth, positive, nonincreasing, and convex function on $[0,\infty)$; such kernels are relevant for applications in viscoelasticity, (more generally, one can assume that a is strongly positive on $[0,\infty)$, [33], [11], [12], [16], [39]). The following results regarding the behaviour of classical solutions of (VE) have been established under additional technical assumptions which are omitted:

(a) If $\phi'(0) - a(0)\psi'(0) > 0$ (i.e., the equilibrium stress modulus is positive) and if the initial data w_o, v_o are sufficiently smooth and small, it is shown in [12], [16] using delicate energy estimates and properties of Volterra kernels that (VE) has a globally defined smooth solution which decays as $t \to +\infty$. Similar global results for more general models are discussed in [15], [37; Ch. IV].

(b) By constrast, for sufficiently large, smooth data w_o, v_o and smooth kernels k irrespective of sign properties, the solutions of (VE) develop singularities in finite time; the first derivatives of w and v become infinite while w, v remain bounded. This suggests the formation of shock front. Such results are established for ϕ convex or concave by the method of characteristics in [8] and [31]. Similar results for other integral models with smooth memory kernels and for several differential models are given in [37], Chapter II. Numerical evidence of finite-time breakdown of smooth solutions for large data and for the onset of a shock structure for (VE) are provided in [26].

A parallel theory regarding properties (a) and (b) has been developed in [30], [25], [7], [23] for the simpler model Cauchy problem for the single conservation law with memory

$$w_t + \sigma_x = 0, \ w(x,0) = w_o(x), \qquad \text{(CLM)}$$

where σ is given by (CA). When $k \equiv 0$, (CLM) reduces to the Burgers equation

$$w_t + \phi(w)_x = 0, \tag{B}$$

while the analogue of (FE) is

$$w_t + \phi(w)_x + w = 0. \tag{FB}$$

Our discussion of (VE) is limited to kernels which are smooth on $[0,\infty)$. However, we remark that a result confirming property (a) for (VE) with a singular kernel ($k = a'$, a positive, decreasing, convex on $[0,\infty)$, and a' having an integrable singularity at zero) has been established by Hrusa and Renardy [17] (also see these proceedings, paper by Renardy), if x is confined to a finite interval; it is not known whether property (a) holds for the corresponding Cauchy problem. Moreover, it is not known whether property (b) holds for (VE) with such singular kernels.

The results (b) discussed above provide a motivation for studying weak solutions of (VE). So far, only special results concerning weak solutions of (VE) have appeared in the literature. Greenberg [14] (also see [37, Section 2.6) established the existence of travelling wave solutions (steady compression shocks) for a history value problem associated with (VE). Regarding unsteady weak solutions, we present a new existence result for the Cauchy problem (2.1) when the data $w_0, v_0 \in L^\infty(\underline{R})$, and not necessarily small.

We assume: the constitutive function φ satisfies

$$\begin{aligned}&\varphi : \underline{R} \to \underline{R} \text{ is a twice continuously differentiable}\\&\text{function such that } \varphi'(w) > 0,\ w \in \underline{R};\\&\varphi \text{ has a single inflection point at } w = w_i \text{ and is}\\&\text{convex on } (w_i,\infty) \text{ and concave on } (-\infty,w_i).\end{aligned} \tag{2.5}$$

The kernel k satisfies

$$k:[0,\infty) \to \underline{R},\ k \in C^1[0,\infty), \tag{2.6}$$

and the data $w_0(x)$, $v_0(x)$ satisfy

$$w_0(x),\ v_0(x) \in L^\infty(\underline{R}) \cap L^2(\underline{R}). \tag{2.7}$$

The following result establishes property (c) for the system (2.1).

THEOREM 2.1. Let the hypotheses (2.5)-(2.7) be satisfied. Given $T > 0$, there exists a weak solution $\{w(x,t),v(x,t)\}$ of (2.1) on $\underline{R} \times [0,T]$, such that

$$(w,v) \in L^\infty([0,T];L^2(\underline{R})) \cap L^\infty(\underline{R} \times [0,T).$$

The proof is sketched in Section 3 using the method of compensated compactness of Murat [27] and Tartar [40, 41, 42]; complete details of the proof will appear in [32]. This approach has been employed with success by Tartar [41] to obtain L^∞ solutions for the general, scalar Burgers equation (B), by DiPerna [13] and Rascle [36] to construct L^∞ solutions to the hyperbolic system (E), and by Dafermos [10] to obtain L^∞ solutions for (CLM); unfortunately, the elegant approach of [10] does not seem to apply to (VE) or (2.1). We also remark that Boldrini [2] used the method of compensated compactness to show that as the memory weakens (i.e. in (2.1) $k = k(\delta,t) = O(\delta)$), uniformly bounded solutions $\{w^\delta,v^\delta\}$ of (2.1) converge to a weak solution of (E) as $\delta \downarrow 0$.

Serious difficulties remain to be overcome in order to prove a result similar to Theorem 2.1 for (VE) when $\psi \not\equiv \phi$. The central issue is whether the memory term $\int_0^t k(t-\tau)\psi(w)(x,\tau))_x d\tau$ can be shown to be of lower order; such is shown to be the case if $\psi \equiv \phi$, even for weak solutions.

3. SKETCH OF THE PROOF OF THEOREM 2.1.

The weak solutions of (2.1), in the space of bounded measurable functions, will be constructed as limits of solutions of the regularized (parabolic) system

$$\begin{aligned} w_t &= v_x + \varepsilon w_{xx} \\ v_t &= \varphi(w)_x + \int_0^t k(t-\tau)\varphi(w(x,\tau))_x d\tau \\ &\quad + \varepsilon(v_{xx} + \int_0^t k(t-\tau)v_{xx}(x,\tau)d\tau),\ (x,t) \in \underline{R} \times (0,T] \qquad (3.1) \\ w(x,0) &= w_0(x),\ v(x,0) = v_0(x), \qquad x \in \underline{R}, \end{aligned}$$

as the real parameter $\varepsilon \downarrow 0$. This unconventional regularization preserves one of the main features of (2.1); namely, that the integration with respect to t offsets differentiation with respect to x and the memory terms are of lower order. Its relevance is revealed by the following calculation. Convolving $(3.1)_2$ with r(t), and using equations (r) and (2.4), we find

$$\begin{aligned}\int_0^t k(t-\tau)(\varphi(w(x,\tau))_x + \varepsilon v_{xx}(x,\tau))d\tau &= \int_0^t r(t-\tau)v_t(x,\tau)d\tau \\ &= r(0)v(x,t)-r(t)v_0(x) + \int_0^t r'(t-\tau)v(x,\tau)d\tau = F[v](x,t).\end{aligned} \tag{3.2}$$

Thus, the initial value problem (3.1) can be written in the form

$$\begin{aligned}&w_t = v_x + \varepsilon w_{xx},\ v_t = \varphi(w)_x + F[v] + \varepsilon v_{xx},\ (x,t) \in \underline{R} \times (0,T] \\ &w(x,0) = w_0(x),\ v(x,0) = v_0(x),\quad x \in \underline{R}.\end{aligned} \tag{3.3}$$

It is evident that systems (3.1) and (3.3) are equivalent for smooth data and classical solutions. For the data in $L^\infty(\underline{R}) \cap L^2(\underline{R})$, the above ideas extend to weak solutions; i.e. it is shown in [32] that (3.1) and (3.3) (also (2.1) and (2.3)) are equivalent for L^∞ solutions, using C^1-test functions which have compact support on $\underline{R} \times [0,T]$ for any $T > 0$ and which vanish at $t = T$. Moreover, (2.2) and (3.2) hold in the weak sense.

Our next objective is to obtain a-priori estimates, independent of ε, for solutions $\{w^\varepsilon,v^\varepsilon\}$ of (3.1) corresponding to initial data $w_0,v_0 \in L^\infty(\underline{R}) \cap L^2(\underline{R})$; we will employ the equivalent system (3.3). The following result is needed to complete the proof of Theorem 2.1.

<u>THEOREM 3.1</u>. Under the hypotheses (2.5)-(2.7), for each $\varepsilon > 0$, $T > 0$, the initial value problem (3.3) (respectively (3.1)) has a unique solution $\{w^\varepsilon(x,t),v^\varepsilon(x,t)\}$ defined on $R \times [0,T]$ such that $w^\varepsilon,\ v^\varepsilon \in C([0,T]; L^2(\underline{R})) \cap L^\infty(\underline{R} \times [0,T])$, $w_x^\varepsilon,v_x^\varepsilon,w_{xx}^\varepsilon,w_t^\varepsilon,v_t^\varepsilon \in C((0,T];L^2(\underline{R}))$ and, also, $w_x^\varepsilon,v_x^\varepsilon \in L^2(\underline{R}\times[0,T])$. In addition, as $\varepsilon \downarrow 0$, the families $\{w^\varepsilon(x,t),v^\varepsilon(x,t)\}_{\varepsilon>0}$ and $\{\varepsilon^{1/2}w_x^\varepsilon(x,t),\varepsilon^{1/2}v_x^\varepsilon(x,t)\}_{\varepsilon>0}$ lie in bounded sets of $L^\infty([0,T];L^2(\underline{R}) \cap L^\infty(\underline{R}))$ and $L^2(\underline{R} \times [0,T])$, respectively.

The existence, uniqueness and regularity properties of the solution $(w^\varepsilon,v^\varepsilon)$

of (3.3) are established in [32] by a fairly standard fixed-point argument on an appropriately chosen Banach space using energy methods and the L^2-theory of the heat equation. We will sketch the proof of the a-priori estimates; we emphasize that due to the nonlocal nature of the memory-term in (3.3), the L^∞-estimates cannot be obtained by finding an invariant region [4] (compare with the proof of the existence of weak solutions for (E) in [13]). To avoid a confusing notation, we now drop the superscript ε.

The a-priori estimates are deduced by using a class of exponentially growing convex entropies for (E) constructed by Dafermos [9]. Let $\{w(x,t),v(x,t)\}$ be the solution of (3.3) on $\underline{R} \times [0,T]$ satisfying the regularity properties of Theorem 3.1; this assumption justifies the calculations which follow. In the sequel, C will stand for a generic constant depending on the $L^\infty(\underline{R})$ and $L^2(\underline{R})$-norms of the initial data, on the $C^1[0,T]$-norm of $r(t)$, on properties of the function $\varphi(w)$, on T, but not on ε. Whenever the constant depends on ε it will be denoted by C_ε. The analysis of (3.3) will be based on the concept of entropy-entropy flux pairs for the elastic problem (E) (cf. Lax [20]). A smooth, convex function $\eta(w,v)$ defined on $\underline{R} \times \underline{R}$ is an entropy for (E), with corresponding entropy flux $q(w,v)$, if

$$\partial_t \eta(w(x,t),v(x,t)) + \partial_x q(w(x,t),v(x,t)) = 0 \tag{3.4}$$

for any smooth solution $\{w(x,t),v(x,t)\}$ of (E). Such pairs are generated as solutions of the system of equations

$$\begin{aligned} q_w &= -\varphi'(w)\eta_v \\ q_v &= -\eta_w, \end{aligned} \tag{3.5}$$

provided $\eta(w,v)$ is convex. Eliminating $q(w,v)$ in (3.5), we find that $\eta(w,v)$ must be a convex solution of the linear wave equation

$$\eta_{ww} = \varphi'(w)\eta_{vv}; \tag{3.6}$$

$q(w,v)$ is then determined by (3.5). A classical example of an entropy-entropy flux pair is

$$\eta(w,v) = \frac{1}{2} v^2 + \int_0^w \varphi(\xi)d\xi, \quad q(w,v) = - v\varphi(w). \tag{3.7}$$

For each entropy-entropy flux pair $\{\eta(w,v),q(w,v)\}$ for (E), we denote by $\{\bar{\eta}(w,v),\bar{q}(w,v)\}$ the parts of the pair $\{\eta,q\}$ which vanish to quadratic order in (w,v). Then $\{\bar{\eta},\bar{q}\}$ also form an entropy-entropy flux pair for (E), and a simple computation yields the identity

$$\begin{aligned}\partial_t\bar{\eta}(w,v) + \partial_x\bar{q}(w,v) &= \bar{\eta}_v(w,v)F[v] + \varepsilon\partial_x^2\bar{\eta}(w,v)\\ &\quad - \varepsilon[\eta_{ww}(w,v)w_x^2 + 2\eta_{wv}(w,v)w_xv_x + \eta_{vv}(w,v)v_x^2].\end{aligned} \tag{3.8}$$

Integrating (3.8) over $\underline{R} \times (0,t)$, $0 < t \leqq T$, we obtain

$$\begin{aligned}&\int_{-\infty}^{\infty} \bar{\eta}(w(x,t),v(x,t))dx\\ &\quad + \varepsilon\int_0^t \int_{-\infty}^{\infty} (\eta_{ww}(w,v)w_x^2 + 2\eta_{wv}(w,v)w_xv_x + \eta_{vv}(w,v)v_x^2)dxd\tau\\ &= \int_{-\infty}^{\infty} \bar{\eta}(w_0(x),v_0(x))dx + \int_0^t \int_{-\infty}^{\infty} \bar{\eta}_v(w,v)F[v]dxd\tau, \quad 0 \leqq t \leqq T.\end{aligned} \tag{3.9}$$

For the first estimate, we employ the pair (3.7) and use (3.9) to deduce

$$\begin{aligned}&\int_{-\infty}^{\infty} [\tfrac{1}{2} v^2(x,t) + \Phi(w(x,t))]dx + \varepsilon \int_0^t \int_{-\infty}^{\infty}(\varphi'(w)w_x^2 + v_x^2)dxd\tau\\ &= \int_{-\infty}^{\infty} [\tfrac{1}{2} v_0^2(x,t) + \Phi(w_0(x,t))]dx\\ &\quad + \int_0^t \int_{-\infty}^{\infty} vF[v]dxd\tau, \ 0 \leqq t \leqq T,\end{aligned} \tag{3.10}$$

where

$$\Phi(w) := \int_0^w (\varphi(\xi) - \varphi(0))d\xi. \tag{3.11}$$

Assumption (2.5) implies that

$$\varphi'(w) \geqq \varphi'(w_i) > 0, \text{ and } \Phi(w) \geqq \frac{\varphi'(w_i)}{2} w^2 \geqq 0, \text{ for every } w \in \underline{R}. \tag{3.12}$$

Using (2.4) and the Cauchy-Schwarz inequality, a simple calculation yields

$$\int_0^t \int_{-\infty}^{\infty} vF[v]dxd\tau \leqq C(1 + \int_0^t \int_{-\infty}^{\infty} v^2(x,\tau)dxd\tau),\ 0 \leqq t \leqq T. \tag{3.13}$$

Combining (3.10) with (3.12) and (3.13), gives

$$\int_{-\infty}^{\infty} v^2(x,t)dx \leqq C + C\int_0^t \int_{-\infty}^{\infty} v^2 dxd\tau, \tag{3.14}$$

whence by Gronwall's inequality, (3.10) and (3.12),

$$\begin{aligned}&\int_{-\infty}^{\infty} [v^2(x,t) + w^2(x,t)]dx \\ &\qquad + \varepsilon \int_0^t \int_{-\infty}^{\infty} (w_x^2 + v_x^2)dxd\tau \leqq C,\ 0 \leqq t \leqq T,\end{aligned} \tag{3.15}$$

where C is independent of ε. Thus $\{w(x,t),v(x,t)\}$ lies in a bounded set of $L^\infty([0,T];L^2(\underline{R}))$ and $\{\varepsilon^{1/2}w_x(x,t),\varepsilon^{1/2}v_x(x,t)\}$ lies in a bounded set of $L^2(\underline{R} \times [0,T])$, independent of ε.

The proof of the L^∞-estimates independent of ε is more subtle. For this purpose, we extend the development of Dafermos [9]. The following facts are proved in [9, Section 2]: For $\varphi(w)$ as in (2.5), the wave equation (3.6) admits a class of solutions $\{\eta^{(k)}(w,v)\}_{k>0}$ on $\underline{R} \times \underline{R}$ which are strictly convex and grow exponentially at infinity. These solutions have the form

$$\eta^{(k)}(w,v) = Y^{(k)}(w)\cosh kv,\ 0 < k < \infty, \tag{3.16}$$

where $Y^{(k)}(w)$ is the solution of the initial value problem

$$\begin{aligned}&Y^{(k)''}(w) = k^2\varphi'(w)Y^{(k)}(w), \\ &Y^{(k)}(w_i) = 1,\ Y^{(k)'}(w_i) = 0,\quad 0 < k < \infty.\end{aligned} \tag{3.17}$$

The functions $Y^{(k)}(w)$ satisfy the estimates

$$Y^{(k)}(w) \geqq \cosh[k\sqrt{\varphi'(w_i)}(w - w_i)],\ -\infty < w < \infty,\ 0 < k < \infty, \tag{3.18}$$

$$|Y^{(k)'}(w)| \leqq k\sqrt{\varphi'(w)}Y^{(k)}(w),\quad -\infty < w < \infty,\ 0 < k < \infty \tag{3.19}$$

and

$$Y^{(k)}(w) \leqq \exp[k \int_{w_i}^{w} \sqrt{\varphi'(\xi)}d\xi], \ -\infty < w < \infty, \ 0 < k < \infty. \tag{3.20}$$

We will estimate the solution $\{w(x,t),v(x,t)\}$ of (3.3) by monitoring the evolution of

$$\int_{-\infty}^{\infty} \bar{\eta}^{(k)}(w(x,t),v(x,t))dx.$$

In view of the convexity of $\bar{\eta}^{(k)}(w,v)$, (3.9) yields

$$\begin{aligned}\int_{-\infty}^{\infty} \bar{\eta}^{(k)}(w(x,t),v(x,t))dx &\leqq \int_{-\infty}^{\infty} \bar{\eta}^{(k)}(w_0(x),v_0(x))dx \\ &+ \int_0^t \int_{-\infty}^{\infty} \bar{\eta}_v^{(k)}(w(x,\tau),v(x,\tau))F[v](x,\tau)dxd\tau, \quad 0 \leqq t \leqq T,\end{aligned} \tag{3.21}$$

where, by (3.16),

$$\bar{\eta}^{(k)}(w,v) = \eta^{(k)}(w,v) - Y^{(k)}(0) - Y^{(k)\prime}(0)w \geqq 0, \tag{3.22}$$

and

$$\bar{\eta}_v^{(k)}(w,v) = k(\tanh kv)\eta^{(k)}(w,v). \tag{3.23}$$

Using (3.23), (3.22), (3.19), the Cauchy-Schwarz inequality and (2.4), a lengthy calculation shows that the last integral in (3.21) can be estimated as follows:

$$\begin{aligned}&\left|\int_0^t \int_{-\infty}^{\infty} \bar{\eta}_v^{(k)}(w,v)F[v]dxd\tau\right| \\ &\quad = \left|\int_0^t \int_{-\infty}^{\infty} k(\tanh kv)(\bar{\eta}^{(k)}(w,v) + Y^{(k)}(0) \right. \\ &\qquad \left. + Y^{(k)\prime}(0)w)F[v]dxd\tau\right| \\ &\quad \leqq Ck^2Y^{(k)}(0) \int_0^t \int_{-\infty}^{\infty} (|w| + |v|)|F[v]|dxd\tau \\ &\qquad + k \int_0^t \int_{-\infty}^{\infty} \bar{\eta}^{(k)}(w,v)|F[v]|dxd\tau\end{aligned} \tag{3.24}$$

$$\leq Ck^2\gamma^{(k)}(0)\left[\int_0^t\int_{-\infty}^{\infty}(w^2+v^2)dxd\tau+\int_{-\infty}^{\infty}v_0^2(x)dx\right]$$

$$+\ Ck\int_0^t M(\tau)\left[\int_{-\infty}^{\infty}\bar{\eta}^{(k)}(w(x,\tau),v(x,\tau))dx\right]d\tau,$$

where

$$M(t) := \sup_{x\in \underline{R}}\ \{1+|v(x,t)|+\int_0^t|v(x,\tau)|d\tau\},\ 0\leq t\leq T. \tag{3.25}$$

Combining (3.21), (3.24), (3.15), and then using the Gronwall equality yields the estimate

$$\int_{-\infty}^{\infty}\bar{\eta}^{(k)}(w(x,t),v(x,t))dx$$

$$\leq \{\int_{-\infty}^{\infty}\bar{\eta}^{(k)}(w_0(x),v_0(x))dx \tag{3.26}$$

$$+\ Ck^2\gamma^{(k)}(0)\}\exp\{Ck\int_0^t M(\tau)d\tau\},\quad 0\leq t\leq T.$$

Raising both sides of (3.26) to the 1/k power, letting $k\to\infty$ and then taking the logarithm of both sides of the resulting inequality (see [32] for justification of this procedure), we conclude

$$\sup_{x\in R}\ \{1+|w(x,t)|+|v(x,t)|\}\leq C+C\int_0^t M(\tau)d\tau,\ 0\leq t\leq T. \tag{3.27}$$

Substituting (3.25) into (3.27) yields

$$S(t)\leq C+C(1+T)\int_0^t S(\tau)d\tau, \tag{3.28}$$

where $S(t) := \sup_{x\in \underline{R}}\ \{1+|w(x,t)|+|v(x,t)|\}$. Finally, integrating (3.28) we conclude

$$\sup_{x\in \underline{R}}\ \{1+|w(x,t)|+|v(x,t)|\}\leq\ M,\ 0\leq t\leq T, \tag{3.29}$$

where the constant M depends on T but not on ε. Therefore, the solution $\{w(x,t),v(x,t)\}$ of (3.3) is bounded on $\underline{R}\times[0,T]$, uniformly in $\varepsilon>0$. This completes the sketch of the proof of the a-priori estimates in Theorem 3.1. □

Our final task in proving Theorem 2.1 is to use Theorem 3.1 to justify passage to the limit as $\varepsilon \downarrow 0^+$ in the Cauchy problem (3.1). For $\varepsilon > 0$, let $\{w^\varepsilon(x,t), v^\varepsilon(x,t)\}$ be the solution of (3.1) on $Q_T := \underline{R} \times [0,T]$, with regularity properties as in Theorem 3.1. By virtue of (3.29) the family of functions $\{w^\varepsilon(x,t), v^\varepsilon(x,t)\}_{\varepsilon>0}$ is uniformly bounded in $L^\infty(Q_T)$. There exist functions $w(x,t)$, $v(x,t)$ and $\bar{\varphi}(x,t)$ in $L^\infty(Q_T)$ such that, along a subsequence, $w^\varepsilon(x,t) \overset{*}{\rightharpoonup} w(x,t)$, $v^\varepsilon(x,t) \overset{*}{\rightharpoonup} v(x,t)$, and $\varphi(w^\varepsilon(x,t)) \longrightarrow \bar{\varphi}(x,t)$ in L^∞-weak star as $\varepsilon \downarrow 0$. The objective is to show that $\{w(x,t), v(x,t)\}$ is a solution of (2.1) in the sense of distributions for $(x,t) \in Q_T$.

The major obstacle to overcome is that, in general, nonlinear functions are not continuous under weak star convergence, and it does not follow that $\bar{\varphi}(x,t) = \varphi(w(x,t))$. Under weak-star convergence, such composite weak limits have been characterized by Tartar [40, 41, 42] as expected values of a family of probability measures $\nu_{(x,t)}$, called the Young measures. The relevant issue here is whether the Young measure reduces to a Dirac mass. Using the theory of compensated compactness and a class of entropy-entropy flux pairs introduced by Lax [20], DiPerna [13] proved the following key result which will be employed in the sequel.

<u>PROPOSITION 3.2.</u> (DiPerna). Let $\{w^\varepsilon, v^\varepsilon\}: Q_T \to \underline{R}$ be a collection of functions such that

$$\|w^\varepsilon\|_{L^\infty(Q_T)} + \|v^\varepsilon\|_{L^\infty(Q_T)} \leq C,$$

where C is a constant independent of ε. Suppose also that for any smooth entropy-entropy flux pair $\{\eta(w,v), q(w,v)\}$ of (E) with φ satisfying (2.5),

$$\partial_t \eta(w^\varepsilon(x,t), v^\varepsilon(x,t)) + \partial_x q(w^\varepsilon(x,t), v^\varepsilon(x,t))$$

lies in a compact set of $H^{-1}_{loc}(Q_T)$. Then there exists a subsequence $\{w^{\varepsilon'}, v^{\varepsilon'}\}$ and functions $w, v \in L^\infty(Q_T)$ such that

$$w^{\varepsilon'}(x,t) \to w(x,t), \; v^{\varepsilon'}(x,t) \to v(x,t), \quad \text{a.e. for } (x,t) \in Q_T$$

as $\varepsilon' \downarrow 0$.

We apply Proposition 3.2 to the family $\{w^\varepsilon(x,t),v^\varepsilon(x,t)\}_{\varepsilon>0}$ of solutions of (3.1) (which also satisfy (3.3)). A straightforward calculation using (3.3) and (3.5) yields

$$\begin{aligned}\partial_t\eta(w^\varepsilon,v^\varepsilon) &+ \partial_x q(w^\varepsilon,v^\varepsilon)\\ &= \varepsilon^{1/2}\partial_x(\varepsilon^{1/2}\eta_w(w^\varepsilon,v^\varepsilon)w_x^\varepsilon + \varepsilon^{1/2}\,\eta_v(w^\varepsilon,v^\varepsilon)v_x^\varepsilon) \qquad (3.30)\\ &- \varepsilon[\eta_{ww}(w^\varepsilon,v^\varepsilon)(w_x^\varepsilon)^2 + 2\eta_{wv}(w^\varepsilon,v^\varepsilon)w_x^\varepsilon v_x^\varepsilon + \eta_{vv}(w^\varepsilon,v^\varepsilon)(v_x^\varepsilon)^2]\\ &+ \eta_v(w^\varepsilon,v^\varepsilon)F[v^\varepsilon] := I_1 - I_2 + I_3.\end{aligned}$$

The a-priori estimates (3.15) and (3.29) imply that the family I_1 converges to 0 and is thereby compact in $H^{-1}(Q_T)$, while, the family I_2 is bounded in $L^1(Q_T)$. In addition,

$$\partial_t\eta(w^\varepsilon,v^\varepsilon) + \partial_x q(w^\varepsilon,v^\varepsilon) - I_3 \qquad (3.31)$$

lies in a bounded set of $W^{-1,\infty}(Q_T)$. Using a lemma of Murat [28] we deduce that (3.31) lies in a compact set of $H^{-1}_{loc}(Q_T)$. Finally, since I_3 lies in a bounded set of $L^2(Q_T)$ and thus in a compact set of $H^{-1}_{loc}(Q_T)$, the left side of (3.30) also lies in a compact set of $H^{-1}_{loc}(Q_T)$. Thus, Proposition 3.2 implies that

$$w^{\varepsilon'}(x,t) \to w(x,t),\ v^{\varepsilon'}(x,t) \to v(x,t) \text{ a.e. in } Q_T,$$

along a subsequence $\varepsilon' \downarrow 0$, and permits passage to the limit $\varepsilon \downarrow 0$ in (3.1) in the sense of distributions. The pair of functions $\{w(x,t),v(x,t)\}$ belongs to $L^\infty([0,T];L^2(\underline{R})) \cap L^\infty(Q_T)$ and is a weak solution of (2.1) in Q_T. This completes the sketch of the proof of Theorem 2.1. □

Acknowledgements

This paper was prepared at the Center for the Mathematical Sciences, University of Wisconsin-Madison. The research of the first author was partially supported by the United States Army under Grant No. DAAL03-87-K-0036 and by National Science Foundation Grants DMS-8620303 and DMS-8712058; the second author was partially supported by the United States Airforce under Grant AFOSR-87-0191 and by the National Science Foundation under Grant No. DMS-210950; the third author was partially supported by the National Science Foundation under Grant No. DMS-8501397, and by the United States Airforce under Grant AFOSR-87-0191.

References

[1] G. Andrews, On the existence of solutions to the equation $u_{tt} = u_{xxt} + \sigma(u_x)_x$, J. Diff. Eq. 35 (1980), 200-231.

[2] J.L. Boldrini, Is elasticity the proper asymptotic theory for materials with memory? Brown Univ. Ph.D. Thesis '85, LCDS Report #85-9.

[3] L. Boltzmann, Zur Theorie der elastischen Nachwirkung, Ann. Physik 7 (1876), Ergänzungsband, 624-625.

[4] K.N. Chueh, C.C. Conley and J.A. Smoller, Positively invariant regions for systems of nonlinear diffusion equations, Indiana Univ. Math. J. 26 (1977), 372-411.

[5] B.D. Coleman and M.E. Gurtin, Waves in materials with memory II. On the growth and decay of one dimensional acceleration waves, Arch. Rational Mech. Anal. 19 (1965), 239-265.

[6] B.D. Coleman, M.E. Gurtin and I.R. Herrera, Waves in materials with memory I, Arch. Rational Mech. Anal. 19 (1965), 1-19.

[7] C.M. Dafermos, Dissipation in materials with memory, Viscoelasticity and Rheology, Proceedings of Mathematics Research Center Symposium, October 1984, A.S. Lodge, J.A. Nohel and M. Renardy, coeditors, Academic Press, Inc., New York (1985), 221-234.

[8] C.M. Dafermos, Development of singularities in the motion of materials with fading memory, Arch. Rational Mech. Anal. 91 (1986), 193-205.

[9] C.M. Dafermos, Estimates for conservation laws with little viscosity, SIAM J. Math. Anal. 18 (1987), 409-421.

[10] C.M. Dafermos, Solutions in L^∞ for a conservation law with memory, Analyse Mathématique et Applications, Gauthier Villars (to appear).

[11] C.M. Dafermos and J.A. Nohel, Energy methods for nonlinear hyperbolic Volterra integrodifferential equations, Comm. PDE 4 (1979), 219-278.

[12] C.M. Dafermos and J.A. Nohel, A nonlinear hyperbolic Volterra equation in viscoelasticity, Amer. J. Math. Supplement (1981), 87-116.

[13] R.J. DiPerna, Convergence of approximate solutions to conservation laws, Arch. Rational Mech. Anal. 82 (1983), 27-70.

[14] J.M. Greenberg, The existence of steady shock waves in nonlinear materials with memroy, Arch. Rat. Mech. Anal. 24 (1967), 1-21.

[15] W.J. Hrusa, A nonlinear functional differential equation in Banach space with applications to materials with fading memory, Arch. Rational Mech. Anal. 84 (1983), 99-137.

[16] W.J. Hrusa and J.A. Nohel, The Cauchy problem in one-dimensional nonlinear viscoelasticity, J. Diff. Eq. 59 (1985), 388-412.

[17] W.J. Hrusa and M. Renardy, On a class of quasilinear partial integrodifferential equations with singular kernels, J. Diff. Eq. 64 (1986), 195-220.

[18] S. Klainerman and A. Majda, Formation of singularities including the nonlinear vibrating string, Comm. Pure Appl. Math. 33 (1980), 241-263.

[19] P.D. Lax, Development of singularities of solutions of nonlinear hyperbolic partial differential equations, J. Math. Phys. 5 (1964), 611-613.

[20] P.D. Lax, Shock waves and entropy, Contributions to Functional Analysis, ed. E.A. Zarantonello, Academic Press, New York, 1971, 603-634.

[21] R.C. MacCamy, Existence, uniqueness and stability of solutions of the equation $u_{tt} = \sigma(u_x)_x + \lambda(u_x)u_{xt}$, Indiana Univ. Math. J. 20 (1970), 231-238.

[22] R.C. MacCamy, A model for one-dimensional nonlinear viscoelasticity, Quart. Appl. Math. 35 (1977), 22-33.

[23] R.C. MacCamy, A model Riemann problem for Volterra equations, Arch. Rational Mech. Anal. 82 (1983), 71-86.

[24] R.C. MacCamy and V.J. Mizel, Existence and nonexistence of solutions of quasilinear wave equations, Arch. Rational Mech. Anal. 25 (1967), 299-320.

[25] R. Malek-Madani and J.A. Nohel, Formation of singularities for a convservation law with memory, SIAM J. Math. Anal. 16 (1985), 530-540.

[26] P. Markowich and M. Renardy, Lax-Wendroff methods for hyperbolic history value problems, SIAM J. Numer. Anal. 21 (1984), 24-51.

[27] F. Murat, Compacité par compensation, Ann. Scuola Norm. Sup. Pisa, Sci. Fis. Mat. 5 (1978), 489-507.

[28] F. Murat, L'injection du cone positif de H^{-1} dans $W^{-1,q}$ est compacte pour tout $q < 2$, J. Math. Pures et Appl. 60 (1981), 309-322.

[29] T. Nishida, Nonlinear hyperbolic equations and related topics in fluid mechanics, Publ. Math. Orsay (1978), 46-60.

[30] J.A. Nohel, A nonlinear conservation law with memory, Volterra and Functional Differential Equations, Kenneth B. Hannsgen, Terry L. Herdman, Harlan W. Stech and Robert L. Wheeler, eds., Marcel Dekker, Inc., New York (1982), 91-123.

[31] J.A. Nohel and M. Renardy, Development of singularities in nonlinear viscoelasticity, Amorphous Polymers and Non-Newtonian Fluids, C. Dafermos, J.L. Ericksen and D. Kinderlehrer (Eds.), IMA Volumes in Mathematics and its Applications, Springer Verlag, Vol. 6 (1987),139-152.

[32] J.A. Nohel, R.C. Rogers and A. Tzavaras, Weak solutions for a non-linear system in viscoelasticity, Comm. PDE 13 (1988), 97-127.

[33] J.A. Nohel and D.F. Shea, Frequency domain methods for Volterra equations, Adv. Math. 22 (1976), 278-304.

[34] J.W. Nunziato, E.K. Walsh, K.W. Schuler and L.M. Barker, Wave propagation in nonlinear viscoelastic solids, Handbuch der Physik Vol VIa/4, Ed. C. Truesdell, Springer-Verlag Berlin (1974), 1-108.

[35] M.A. Rammaha, Development of singularities of smooth solutions of nonlinear hyperbolic Volterra equations, Comm. PDE 12 (1987), 243-262.

[36] M. Rascle, Un résultat de" compacité par compensation à coefficients variables". Application à l'élasticité nonlinéaire, Compt. Rend. Acad. Sci. Paris, Série I, 302 (1986), 311-314.

[37] M. Renardy, W.J. Hrusa and J.A. Nohel, Mathematical Problems in Viscoelasticity, Longman Scientific & Technical, Pitman Monographs and Surveys in Pure and Applied Mathematics 35 (1987), 273 pp.

[38] M. Slemrod, Instability of steady shearing flows in a nonlinear viscoelastic fluid, Arch. Rational Mech. Anal. 68 (1978), 211-225.

[39] O. Staffans, On a nonlinear hyperbolic Volterra equation, SIAM J. Math. Anal 11 (1980), 793-812.

[40] L. Tartar, Une nouvelle méthode de resolution d'équations aux dérivées partielles nonlinéaires, Lecture Notes in Math., vol. 665, Springer-Verlag (1977), 228-241.

[41] L. Tartar, Compensated compactness and applications to partial differential equations, in Nonlinear Analysis and Mechanics: Heriot-Watt Symposium, vol. IV, Research Notes in Mathematics, 39, R.J. Knops, Ed., Pitman Publ. Inc., 1979.

[42] L. Tartar, The compensated compactness method applied to systems of conservation laws, Systems of Nonlinear Partial Differential Equations, J.M. Ball, ed., Reidel Publishing Co., Holland (1983), 263-285.

[43] E.K. Walsh, Wave propagation in viscoelastic solids, in: A.S. Lodge, M. Renardy and J.A. Noehl, Viscoelasticity and Rheology, Academic Press 1985.

J.A. Nohel, R.C. Rogers* and A.E. Tzavaras
Department of Mathematics and Center for the Mathematical Sciences
University of Wisconsin-Madison
Madison
Wisconsin 53705
USA.

* Rogers has moved to Department of Mathematics, Virginia Polytechnic Institute and State University, Blackburg, VA 24601-4097, USA.

J. PRÜSS

Regularity and integrability of resolvents of linear Volterra equations

1. INTRODUCTION

Let X be a Banach space, A a closed linear operator in X with dense domain D(A), and $a \in BV_{loc}(R_+)$ a scalar kernel locally of bounded variation on $R_+ = [0,\infty)$. In this paper we consider the problem

$$(P)\quad \begin{aligned} u'(t) &= \int_0^t da(\tau)Au(t-\tau) + f(t), \quad t \geq 0 \\ u(0) &= u_0, \end{aligned}$$

where $f \in C(R_+,X)$. The solutions of (P) are given by the variation of parameters formula

$$(S)\quad u(t) = S(t)u_0 + \int_0^t S(t-\tau)\, f(\tau)d\tau, \ t \geq 0,$$

whenever the resolvent S(t) exists in the sense of

DEFINITION 1.1. A strongly continuous family $\{S(t)\}_{t\geq 0} \subset \mathcal{B}(X)$ of bounded linear operators in X is called a resolvent for (P) if S(0) = I (identity operator), S(t) commutes with A, and the resolvent equation

$$(R)\quad S'(t)x = \int_0^t da(\tau)AS(t-\tau)x, \ t \geq 0, \ x \in D(A)$$

is satisfied.

Representation formula (S) exhibits the central role of the resolvent S(t) in the linear theory.

Problem (P) arises in several different fields such as thermodynamics, electrodynamics, continuum mechanics and population biology. Let us briefly indicate two applications to the theory of linear viscoelasticity; cp. Christensen [4] or Pipkin [18] or Renardy, Hrusa and Nohel [26] for more details and background material.

EXAMPLE 1.1. Linear one-dimensional viscoelasticity.

Consider a linear viscoelastic material undergoing simple shearing. If x denotes the distance from the shearing plane, $v(t,x)$ the displacement and $b(t,x)$ the body forces involved, conservation of momentum yields the equation

$$v_{tt}(t,x) = b(t,x) + \sigma_x(t,x). \tag{1.1}$$

The stress $\sigma(t,x)$ in turn is related to the strain $v_x(t,x)$ in the material by a constitutive equation which reflects the memory properties of the material.

$$\sigma(t,x) = \int_0^\infty da(\tau)v_{xt}(t-\tau,x). \tag{1.2}$$

Inserting (1.2) into (1.1) we obtain the following equation for the velocity field $u(t,x) = v_t(t,x)$

$$u_t(t,x) = \int_0^t da(\tau)u_{xx}(t-\tau,x) + f(t,x), \tag{1.3}$$

where the inhomogeneity $f(t,x)$ incorporates the body forces $b(t,x)$ as well as the strain history past $t = 0$

$$f(t,x) = b(t,x) + \int_t^\infty da(\tau)u_{xx}(t-\tau,x). \tag{1.4}$$

Thus we have obtained (P) with $A = (d/dx)^2$ in a suitable function space X.

EXAMPLE 1.2. Isotropic and incompressible materials.

Consider now the three-dimensional case; again conservation of momentum yields

$$u_t(t,x) = b(t,x) + \mathrm{div}_x\sigma(t,x) \tag{1.5}$$

but this time the stress tensor is related to the strain tensor

$$\varepsilon(t,x) = [\mathrm{grad}_x v(t,x) + (\mathrm{grad}_x v(t,x))^T]/2 \tag{1.6}$$

according to the constitutive law

$$\sigma(t,x) = -p(t,x)I + 2\int_0^\infty da(\tau)\varepsilon_t(t-\tau,x). \tag{1.7}$$

Here the pressure p(t,x) is determined only implicitly by the incompressibility condition

$$\mathrm{div}_x u(t,x) = 0. \tag{1.8}$$

Inserting (1.6) and (1.7) into (1.5) we arrive at

$$\begin{aligned} u_t(t,x) &= \int_0^t da(\tau)\Delta_x u(t-\tau,x) - \mathrm{grad}_x p(t,x) + f(t,x) \\ \mathrm{div}_x u(t,x) &= 0. \end{aligned} \tag{1.9}$$

As usual, projecting onto the space of divergence-free L^2-functions, the pressure can be eliminated and we obtain a problem of the form (P), where A now becomes the so-called Stokes operator which is selfadjoint and dissipative.

The kernel da(t) in the constitutive equations (1.2) and (1.7) is called <u>stress relaxation modulus</u> and in general is of the form

(a) $a(t) = a_0 + a_\infty t + \int_0^t a_1(s)ds,$

with Newtonian viscosity $a_0 \geqq 0$, elasticity modulus $a_\infty \geqq 0$, and $a_1(t) \geqq 0$ is nonincreasing, $\lim_{t\to\infty} a_1(t) = 0$ and is of positive type (Second Law of Thermodynamics). It is generally believed that $a_1(t)$ is even completely monotonic on $(0,\infty)$; cp. Bland [1].

There are many papers dealing with (P) and more general problems, e.g. Carr and Hannsgen [2], [3], Clement and Nohel [5], [6], Dafermos [7], [8], Desch, Grimmer and Schappacher [10], Grimmer and Prüss [14], Miller and Wheeler [16], Da Prato and Iannelli [20], Noren [17], to mention a few which are related to this work. Recently, in Prüss [24] we presented a new approach to (P) which works under the following hypothesis.

(A) A is the generator of a strongly continuous cosine family T(t) in X;

$a(t)$ is of the form (a) with a_0, $a_\infty \geqq 0$, $a_1(t) > 0$ nonincreasing, $\lim_{t\to\infty} a_1(t)=0$ and $\log a_1(t)$ convex.

Obviously, these assumptions are motivated by Examples 1.1 and 1.2 above. This paper relies strongly on this approach, therefore we present a brief explanation of the method and a summary of the results obtained so far in Section 2. The remainder of this paper is then devoted to regularity and integrability properties of the resolvent $S(t)$ for (P).

2. THE RAYLEIGH PROBLEM AND RESOLVENTS

It is well known that existence of resolvents for (P) can be characterized in terms of the formal Laplace transform $H(\lambda) = \hat{S}(\lambda)$ of the resolvent $S(t)$. A simple computation yields

$$H(\lambda) = \frac{1}{\lambda} (I-a(\lambda)A)^{-1}. \tag{2.1}$$

The following result is due to Da Prato and Iannelli [20].

THEOREM 2.1. Let A be closed linear and densely defined, $a \in BV_{loc}(\mathbb{R}_+)$ with $\int_0^\infty e^{-\omega t}|da(t)| < \infty$ for some $\omega \in \mathbb{R}$.

Then (P) admits a resolvent $S(t)$ such that

$$|S(t)| \leqq Me^{\omega t} \quad \text{for } t \geqq 0 \tag{2.2}$$

holds for some $M \geqq 1$ if and only if

(A1) $\hat{a}(\lambda) \neq 0$ and $1/\hat{a}(\lambda) \in \rho(A)$ for all $\lambda > \omega$;

(A2) $H(\lambda) = \lambda^{-1}(I-\hat{a}(\lambda)A)^{-1}$ satisfies the estimates

$$|H^{(n)}(\lambda)| \leqq M\cdot n!\cdot(\lambda-\omega)^{-(n+1)} \quad \text{for all } \lambda > \omega,\ n \in \mathbb{N}_0. \tag{2.3}$$

In general it is a difficult task to verify the estimates in (A2); however, in our case when assumption (A) holds this is quite easy, as we shall see now.

For this purpose let (A) be satisfied. It is well known that the Laplace transform of the cosine family T(t) is given by

$$\hat{T}(z) = \int_0^\infty T(\tau)e^{-z\tau}d\tau = z(z^2-A)^{-1}, \quad \text{Re } z > \omega_0; \tag{2.4}$$

with $z = [\hat{a}(\lambda)]^{-1/2}$ this identity yields

$$H(\lambda) = \int_0^\infty T(\tau)h(\lambda,\tau)d\tau \tag{2.5}$$

where

$$h(\lambda,\tau) = \lambda^{-1}[\hat{a}(\lambda)]^{-1/2} \exp(-\tau[\hat{a}(\lambda)]^{-1/2}). \tag{2.6}$$

Since $|T(\tau)| \leqq Me^{\omega_0 t}$ holds for all $\tau \geqq 0$ and some constants $M \geqq 1$ and $\omega_0 \geqq 0$, representation formula (2.5) easily yields conditions (A1) and (A2), provided we know that $h(\lambda,\tau)$ is completely monotonic in λ for each fixed $\tau > 0$; cp. Prüss [24] for the details of the proof of this assertion.

Thus we are led to study the function $h(\lambda,\tau)$. It turns out that $h(\lambda,\tau)$ is related to the solution of the following model problem which is known as the Rayleigh problem in viscoelasticity.

$$\begin{aligned} w_t(t,x) &= \int_0^t da(\tau)w_{xx}(t-\tau,x), \quad t,x \geqq 0 \\ w(t,0) &= 1 \qquad\qquad , \quad t > 0 \\ w(0,x) &= 0 \qquad\qquad , \quad x > 0 \end{aligned} \tag{2.7}$$

This problem represents a model for the following physical situation: A viscoelastic fluid is resting on a plate and is filling the half space $x \geqq 0$. Everything is at rest up to time $t = 0$, but then the plate is suddenly moved to the right with constant speed 1. The solution $w(t,x)$ then describes the evolution of the velocity field in the fluid with time, in particular $w(t,x)$ yields the propagation of a jump discontinuity at the boundary into the medium.

A simple computation leads to

$$\hat{w}(\lambda,x) = \lambda^{-1}\exp(-x[a(\lambda)]^{-1/2}), \tag{2.8}$$

in other words, the relation between $h(\lambda,\tau)$ and $\hat{w}(\lambda,\tau)$ is given by

$$h(\lambda,\tau) = -\frac{\partial}{\partial\tau}\hat{w}(\lambda,\tau). \tag{2.9}$$

Therefore, by Bernstein's Theorem, $h(\cdot,\tau)$ is completely monotonic if and only if $w(t,\tau)$ is nonincreasing in τ. This is one of the properties of the solution $w(t,x)$ of the Rayleigh problem which are collected in

THEOREM 2.2. Suppose (A) is satisfied and let $w(t,x)$ denote the solution of the Rayleigh problem (2.7). Then

(i) $w(t,x)$ is nondecreasing in t and $\lim_{t\to\infty} w(t,x) = 1$;
$w(t,x)$ is nonincreasing in x and $\lim_{x\to\infty} w(t,x) = 0$

(ii) The speed of propagation is $\mu = [a_\infty + a_1(0)]^{1/2}$, $\mu = \infty$ in case $a_0 > 0$, i.e.
$w(t,x) = 0$ if and only if $t \leqq x/\mu$.

(iii) If $\mu = \infty$ then $w(t,x)$ is continuous on $\mathbf{R}_+ \times \mathbf{R}_+ \setminus \{(0,0)\}$.

Suppose also $\mu < \infty$ and $\dot{a}_1(t)$ absolutely continuous on $(0,\infty)$. Then

(iv) If $\kappa = -\dot{a}_1(0) = \infty$ then $w(t,x)$ is continuous on $\mathbf{R}_+ \times \mathbf{R}_+ \setminus \{(0,0)\}$.

(v) If $\mu < \infty$ and $\kappa < \infty$ then

$$v(t,x) = w(t,x) - \eta(\mu t-x)\exp(-\kappa t/2\mu^2) \tag{2.10}$$

is continuous on $\mathbf{R}_+ \times \mathbf{R}_+$.

Here $\eta(t) = 1$ for $t > 0$, $\eta(t) = 0$ for $t \leqq 0$ denotes Heaviside's function. The proof of Theorem 2.2 is contained in Prüss [24], Section 3. Theorem 2.2(i) not only yields existence of the resolvent $S(t)$ for (P) via Theorem 2.1 as outlined above, it also yields the representation formula

$$S(t) = -\int_0^\infty T(\tau)d_\tau w(t,\tau), \quad t \geqq 0. \tag{2.11}$$

Furthermore, decomposition (2.10) leads to a refined version of (2.11) which gives more information about the resolvent. The following result has been obtained in Prüss [24], too.

THEOREM 2.3. Suppose (A) is satisfied and let $w(t,x)$ denote the solution of the Rayleigh problem (2.7). Then

(i) Problem (P) admits a resolvent $S(t)$ satisfying

$$|S(t)| \leqq Me^{\omega t}, \quad t \geqq 0, \tag{2.2}$$

for some $M \geqq 1$ and $\omega \in \mathbb{R}$.

(ii) $S(t)$ has the representation

$$S(t) = -\int_0^\infty T(\tau)d_\tau w(t,\tau), \quad t \geqq 0. \tag{2.11}$$

Assume in addition $\dot{a}_1(t)$ absolutely continuous on $(0,\infty)$ in case $\mu < \infty$, where μ and κ are defined as in Theorem 2. Then

$$\begin{aligned} S(t) &= T(\mu t)\exp(-\kappa t/2\mu^2) - \int_0^{\mu t} T(\tau)v_\tau(t,\tau)d\tau \\ &= S_1(t) + S_0(t), \quad t \geqq 0. \end{aligned} \tag{2.12}$$

(iii) If $\mu + \kappa = \infty$ then $S(t) = S_0(t)$ is continuous in $\mathbb{B}(X)$ for $t > 0$.

(iv) If $\mu + \kappa < \infty$ then $S_0(t)$ is continuous in $\mathbb{B}(X)$ for $t \geqq 0$.

Formula (2.12) shows that the resolvent decomposes into a damped elastic part $S_1(t)$ which is propagated with speed μ and a viscous part $S_0(t)$ which propagates with speed at most μ and which is more regular than the former. Let us mention a well known special case of (2.12), namely the case $a(t) \equiv 1$. Then we have $\mu = \infty$ and

$$v(t,x) = w(t,x) = \operatorname{erfc}(x/2\sqrt{t})$$

and so (2.12) yields

$$e^{At} = \frac{1}{\sqrt{\pi t}} \int_0^\infty T(\tau)\exp(-\tau^2/4t)d\tau, \quad t \geqq 0,$$

the well known relation between the C_0-semigroup and the cosine family generated by A; cp. e.g. Goldstein [12].

3. INTEGRABILITY PROPERTIES OF THE RESOLVENTS

One of the most important qualitative properties of the resolvent S(t) is its integrability over the halfline $\mathbb{R}_+$. If S(t) has this property then the unique bounded solution w(t) of the infinite delay equation on the line

$$w'(t) = \int_0^\infty da(\tau)Aw(t-\tau) + f(t), \ t \in \mathbb{R}, \tag{3.1}$$

with f bounded is given by

$$w(t) = \int_0^\infty S(\tau)f(t-\tau)d\tau \qquad t \in \mathbb{R}. \tag{3.2}$$

It can be shown that if f is t_0 - periodic, or is almost periodic, or asymptotically almost periodic, or has a limit for $t \to \infty$, then w exibits the same property; cp. Prüss [23] for proofs of these statements.

Since X is infinite dimensional, there are several different integrability concepts for resolvents; the most important are summarized in

DEFINITION 3.1. The strongly measurable family $\{R(t)\}_{t\geq 0} \subset \mathcal{B}(X,Z)$ of bounded linear operators is called

(i) strongly integrable if $\int_0^\infty |R(t)x|dt < \infty$ for each $x \in X$;

(ii) integrable if there is $\varphi \in L^1(\mathbb{R})$ such that

$$|R(t)| \leq \varphi(t) \quad \text{a.e. on } \mathbb{R}_+;$$

(iii) uniformly integrable if $R \in L^1(\mathbb{R}_+, \mathcal{B}(X,Z))$.

If the resolvent S(t) for (P) is strongly integrable, then the Laplace transform $H(\lambda) = \hat{S}(\lambda)$ exists and is strongly continuous for Re $\lambda \geq 0$ and even holomorphic for Re $\lambda > 0$. The resolvent equation (R) yields the identity

$$\lambda H(\lambda) = I + \lambda\hat{a}(\lambda)AH(\lambda) = I + \lambda\hat{a}(\lambda)H(\lambda)A, \quad \text{Re } \lambda > 0, \tag{3.3}$$

whenever $a \in BV_{loc}(\mathbb{R}_+)$ is of subexponential growth, i.e. $\int_0^\infty e^{-\varepsilon t}|da(t)| < \infty$

for each $\varepsilon > 0$. (3.3) yields the following necessary conditions for strong integrability of the resolvent; cp. Prüss [24] for a proof.

PROPOSITION 3.1. Let A be a closed linear densely defined operator in X and let $a \in BV_{loc}(\mathbb{R}_+)$ be of subexponential growth. Suppose (1.1) admits a strongly integrable resolvent S(t). Then

(i) $\hat{a}(\lambda)$ admits continuous extension to the set $\operatorname{Re}\lambda \geqq 0$, $\lambda \neq 0$, and $\lim_{\lambda\to 0}\lambda\hat{a}(\lambda) = a(\infty)$ exists and is nonzero, $a(\infty) = \infty$ is possible;

(ii) $\hat{a}(\lambda) \neq 0$ and $1/\hat{a}(\lambda) \in \rho(A)$ for all $\operatorname{Re}\lambda \geqq 0$, $\lambda \neq 0$;

(iii) A is invertible and $\int_0^\infty S(t)dt = H(0) = -A^{-1}/a(\infty)$;

(iv) $H(\lambda) = [\lambda a(\lambda)]^{-1}([a(\lambda)]^{-1}-A)^{-1}$ for all $\operatorname{Re}\lambda \geqq 0, \lambda \neq 0$;

Here $\rho(A) = \mathbb{C}\setminus\sigma(A)$ denotes the set of all $z \in \mathbb{C}$ with z-A invertible. Unfortunately, the converse of Proposition 3.1 is false, even in the semigroup case $a(t) = \eta(t)$. The first positive results have been obtained by Friedman and Shinbrot [11] for the parabolic case ($a_0 > 0$, A the generator of an analytic C_0-semigroup) and by Carr and Hannsgen [2] for A selfadjoint, negative definite in a Hilbert space X, $a_0 = 0$ and $a_1(t) > 0$, non-increasing, convex and $-\dot{a}_1(t)$ convex. In [24] we obtained the following converse of Proposition 3.1.

THEOREM 3.1. Assume (A), let $\dot{a}_1(t)$ be absolutely continuous on $(0,\infty)$ in case $\mu < \infty$, and define the type of T(t) as

$$\omega_0(A) = \lim_{t\to\infty} t^{-1}\log|T(t)|; \tag{3.4}$$

let A be invertible and in case $\omega_0(A) > 0$ suppose also that $1/a(\lambda) \in \rho(A)$ for all $\operatorname{Re}\lambda \geqq 0$, $\lambda \neq 0$, holds. Let μ, κ and S(t) be defined as in Section 2. Then

(i) If $\mu + \kappa = \infty$ then S(t) is uniformly integrable.

(ii) If $\mu + \kappa < \infty$ and $\omega_0(A) < \kappa/2\mu^3$ then S(t) is integrable.

In particular, if $\omega_0(A) = 0$ then S(t) is integrable if and only if $a(t) \not\equiv a_\infty t$.

Note that in case $a(t) \equiv a_\infty t$ we have $S(t) = T(\sqrt{a_\infty}\cdot t)$, which is never integrable. In (ii) the extra condition $\omega_0(A) < \kappa/2\mu^3$ is needed to prevent the set $\{1/\hat{a}(\lambda) : \mathrm{Re}\,\lambda \geqq 0\}$ from interaction with the spectrum of A at $\lambda = \infty$; this never happens for $\mu + \kappa = \infty$.

Next, let us consider the integrability of the derivatives $S'(t)$ and $S''(t)$ of $S(t)$ which are important for the regularity of $w(t)$ defined by (3.2). Carr and Hannsgen obtained integrability of $S'(t)$ in the space $\mathrm{B}(Y,X)$, where $Y = D((-A)^{1/2})$, A selfadjoint negative definite in the Hilbert space X, the kernel satisfies $a_1(t) \geqq 0$ nonincreasing, convex and $-\dot{a}_1(t)$ convex. To derive a corresponding statement under assumption (A) we have to introduce a suitable generalization of $D((-A)^{1/2})$ to cosine families in general Banach spaces; we define

$$Y = \{x \in X : T(\cdot)x \in C^1(\mathbb{R},X)\},\ |x|_y = \sup\{|T'(t)x| : t \in [0,1]\}. \tag{3.5}$$

It is well known that Y with norm $|.|_y$ is again a Banach space and we have the continuous and dense embeddings

$$X_A \overset{d}{\dashrightarrow} Y \overset{d}{\longrightarrow} X$$

where X_A denotes $D(A)$ equipped with the graph norm of A. $T'(t)$ is strongly continuous in $\mathrm{B}(Y,X)$ and in $\mathrm{B}(X_A,Y)$ and has the same exponential growth abscissa $\omega_0(A)$ as $T(t)$. Integration by parts in (2.11) yields

$$S(t) = I + \int_0^\infty T'(\tau)w(t,\tau)d\tau,\quad t \geqq 0$$

in $\mathrm{B}(Y,X)$, i.e. with (2.10) we obtain

$$S(t) = I + (T(\mu t) - I)e^{-\kappa t/2\mu^2} + \int_0^{\mu t} T'(\tau)v(t,\tau)d\tau \tag{3.6}$$

and finally by differentiation we arrive at the representation formula

$$S'(t) = \mu T'(\mu t)e^{-\kappa t/2\mu^2} - \frac{\kappa}{2\mu^2}(T(\mu t)-I)e^{-\kappa t/2\mu^2} + \int_0^{\mu t} T'(\tau)v_t(t,\tau)d\tau \tag{3.7}$$

in $\mathrm{B}(Y,X)$ for $S'(t)$, where the first two terms are absent in case $\mu + \kappa = \infty$.

(3.7) shows that $S'(t)$ has the same continuity in $\mathcal{B}(Y,X)$ as $S(t)$ in $\mathcal{B}(X)$. This yields the first part of

THEOREM 3.2. Assume (A), let $\dot{a}_1(t)$ be absolutely continuous on $(0,\infty)$ in case $\mu < \infty$, and let Y be defined by (3.5). Then $S'(t)$ is strongly continuous in $\mathcal{B}(Y,X)$ for all $t \geqq 0$; if $\mu + \kappa = \infty$ then $S'(t)$ is uniformly continuous in $\mathcal{B}(Y,X)$ for all $t > 0$. Assume in addition A invertible and $1/\hat{a}(\lambda) \in \rho(A)$ for all $\mathrm{Re}\,\lambda \geqq 0$, $\lambda \neq 0$ in case $\omega_o(A) > 0$. Then

(i) If $\mu + \kappa = \infty$ then $S'(t)$ is uniformly integrable in $\mathcal{B}(Y,X)$.

(ii) If $\mu + \kappa < \infty$ and $\omega_o(A) < \kappa/2\mu^3$ then $S'(t)$ is integrable in $\mathcal{B}(Y,X)$.

The same assertions also hold in $\mathcal{B}(X_A,Y)$.

The proof of the second part is almost identical with that of Theorem 3.1, we may therefore again refer to Prüss [24].

The story with $S''(t)$ is more difficult. Noren [17] considers the integrability of $S''(t)$ under the same conditions as Carr and Hannsgen [2], [3] for $S(t)$ and $S'(t)$ as mentioned above but he has to impose extra conditions of a technical nature. Let us first consider the case $a_o > 0$; if $S''(t)A^{-1}$ were integrable in $\mathcal{B}(X)$ then the relation

$$\hat{S}''A^{-1} = \lambda\hat{S}'A^{-1}-S'(0)A^{-1} = \lambda^2\hat{S}A^{-1}-\lambda A^{-1}-a_oI = \lambda((I-\hat{a}A)^{-1}-I)A^{-1}-a_oI =$$

$$= \lambda\hat{a}(\lambda)(I-\hat{a}(\lambda)A)^{-1}-a_oI$$

and $\hat{S}''(\lambda)A^{-1} \to 0$ in $\mathcal{B}(X)$ for $\lambda \to \infty$ would imply that $(I-\hat{a}(\lambda)A)^{-1}$ is invertible for large λ, i.e. A is bounded. Thus in case $a_o > 0$ and A is unbounded, $S''(t)A^{-1}$ is never integrable, not even locally on $\mathbb{R}_+$. Next, let $\mu < \infty$; differentation of (R) yields

$$S''(t)A^{-1} = \mu^2S(t) + \dot{a}_1 * S(t), \quad t \geqq 0, \tag{3.8}$$

hence since $\dot{a}_1 \in L^1(\mathbb{R}_+)$, $S''(t)A^{-1}$ enjoys the same continuity and integrability properties as $S(t)$ does. The difficult case is $a_o = 0$, but $\mu = \infty$; Noren [17] proved that then

$$- \int_0^1 a_1(t)\log t \, dt < \infty \tag{3.9}$$

is necessary for integrability of $S''(t)A^{-1}$ in a certain sense. In our situation this condition is also sufficient.

THEOREM 3.3. Assume (A) and let $\dot{a}_1(t)$ be absolutely continuous on $(0,\infty)$ in case $\mu < \infty$.

(i) If $\mu + \kappa = \infty$ then $S''(t)$ is uniformly continuous in $\mathcal{B}(X_A,X)$ for $t > 0$.

(ii) If $\mu < \infty$ then $S''(t)$ is strongly continuous in $\mathcal{B}(X_A,X)$ for $t \geqq 0$.

(iii) If $\mu = \infty$ then $\int_0^1 |S''(t)|_A dt < \infty$ if A is bounded, or $a_0 = 0$ and (3.9) holds.

Assume in addition A invertible and $1/\hat{a}(\lambda) \in \rho(A)$ for all Re $\lambda \geqq 0$, $\lambda \neq 0$ in case $\omega_0(A) > 0$; let $a_0 = 0$. Then

(iv) If $\mu + \kappa = \infty$ and (3.9) holds then $S''(t)A^{-1}$ is uniformly integrable.

(v) If $\mu + \kappa < \infty$ and $\omega_0(A) < \kappa/2\mu^3$ then $S''(t)A^{-1}$ is integrable.

The proof of this result is postponed to Section 6.

4. PERTURBATIONS

As an application of the integrability results of Section 3 let us now consider the perturbed problem

$$\text{(P1)} \quad \begin{aligned} &u'(t) = da * Au(t) + dB * u(t) + f(t), \quad t \geqq 0, \\ &u(0) = u_0, \end{aligned}$$

where a(t) and A are subject to Condition (A) and the perturbing kernel satisfies $B \in BV_{loc}(\mathbb{R}_+,\mathcal{B}(X_A,X))$. If (P1) admits a resolvent $S_1(t)$, it must be a solution of

$$S_1 = S + S_1 * dB * S, \tag{4.1}$$

where S(t) denotes the resolvent of (P) and the star indicates convolution; (4.1) follows directly from the variation of parameters formula. Thus we are led to solve (4.1).

In general, (4.1) is not solvable since the family $\{B(t)\}_{t\geq 0}$ consists of unbounded operators. However, if unboundedness is compensated by regularity in time in the following manner, then this defect can be removed and (4.1) becomes a wellposed problem. For this purpose we let

$$B = B_0 + da * B_1 + da * da * B_2 \tag{4.2}$$

with $B_0 \in BV_{loc}(\mathbb{R}_+, \mathcal{B}(X))$, $B_1 \in BV_{loc}(\mathbb{R}_+, \mathcal{B}(Y,X))$ and $B_2 \in BV_{loc}(\mathbb{R}_+, \mathcal{B}(X_A, X))$; w.l.o.g. we may also assume A invertible. Suppose also $a_0 = 0$ and (3.9). Then we obtain

$$\begin{aligned} dB * S &= dB_0 * S + dB_1 * (da * S) + dB_2 * (da * da * S) = \\ &= dB_0 * S + dB_1 * S'A^{-1} + d(B_2A^{-1})*(S''A^{-1} - \dot{a}), \end{aligned} \tag{4.3}$$

hence $dB * S \in L^1_{loc}(\mathbb{R}_+, \mathcal{B}(X))$ in case $\mu + \kappa = \infty$ and $dB * S$ is strongly continuous if $\mu < \infty$, by the results of Section 3. This argument yields

<u>THEOREM 4.1</u>. Assume (A), let $a_0 = 0$, $-\int_0^1 a_1(t) \log t \, dt < \infty$, and $\dot{a}_1(t)$ absolutely continuous on $(0,\infty)$ in case $\mu < \infty$. Suppose

$$B = B_0 + da * B_1 + da * da * B_2 \tag{4.2}$$

with $B_0 \in BV_{loc}(\mathbb{R}_+, \mathcal{B}(X))$, $B_1 \in BV_{loc}(\mathbb{R}_+, \mathcal{B}(Y,X))$, $B_2 \in BV_{loc}(\mathbb{R}_+, \mathcal{B}(X_A, X))$. Then there is a unique resolvent $S_1(t)$ for (P1). $S_1(t)$ is strongly continuous for $t \geq 0$ and in case $\mu + \kappa = \infty$ even uniformly continuous on $(0,\infty)$.

The proof follows with (4.3) from the Neumann series for (4.1). Some remarks are in order.

<u>REMARKS</u>: (i) In case $a_0 > 0$ a result much stronger than Theorem 4.1 holds: <u>any</u> $B \in BV_{loc}(\mathbb{R}_+, \mathcal{B}(X_A, X))$ with $B(0) = B(0+)$ is admissable and A needs only to be the generator of an analytic C_0-semigroup; cp. Prüss [22] for this result.

(ii) For $\mu < \infty$, assumptions (4.2) are equivalent to the following:

$$B = B_0 + B_3 + B_4 \text{ with } B_o \in BV_{loc}(\mathbb{R}_+,\mathcal{B}(X)),\ B_3 \in W^{1,\infty}_{loc}(\mathbb{R}_+,\mathcal{B}(Y,X)),$$

$$\dot{B}_3 \in BV_{loc}(\mathbb{R}_+,\mathcal{B}(Y,X)) \text{ and } B_4 \in W^{2,\infty}_{loc}(\mathbb{R}_+,\mathcal{B}(X_A,X)),\ \ddot{B}_4 \in BV_{loc}(\mathbb{R}_+,\mathcal{B}(X_A,X)).$$

(iii) Observe the effect of singular kernels: if $a(t) = 2\sqrt{t/\pi}$, then $\dot{a} * \dot{a} \equiv 1$ hence $da * da * B_2 = 1 * B_2$, i.e. any $B_4 \in W^{1,\infty}_{loc}(\mathbb{R}_+,\mathcal{B}(X_A,X))$ with $\dot{B}_4 \in BV_{loc}\ (\mathbb{R}_+,\mathcal{B}(X_A,X))$ is admissible.

Equation (4.1) in connection with (4.2) can be used to obtain necessary and sufficient conditions for integrability of the resolvent $S_1(t)$ of the perturbed problem. In fact, if $S \in L^1(\mathbb{R}_+,\mathcal{B}(X))$ and the B_j are of bounded variation on $\mathbb{R}_+$ in the corresponding spaces of bounded linear operators, then $dB * S \in L^1(\mathbb{R}_+,\mathcal{B}(X))$ by (4.3) and the Paley-Wiener Lemma in the Banach algebra $\mathcal{B}(X)$ applied to (4.1) yields $S_1 \in L^1(\mathbb{R}_+,\mathcal{B}(X))$ if and only if

$$\begin{aligned}&\lambda - \hat{da}(\lambda)A - \hat{dB}(\lambda) \text{ is invertible for all } \mathrm{Re}\lambda \geqq 0,\ \lambda \neq 0,\\ &\text{and } I - (B_o(\infty)/a(\infty) + B_1(\infty) + da * B_2(\infty)) \text{ invertible.}\end{aligned} \tag{4.4}$$

By Theorems 3.1, 3.2, 3.3 this applies to the case $a_o = 0$, $\mu + \kappa = \infty$ and (3.9), if $\mu + \kappa < \infty$ then in addition B_o, B_1, B_2 must be absolutely continuous, otherwise the Paley-Wiener Lemma does not apply since we need $dB * S \in L^1(\mathbb{R}_+,\mathcal{B}(X))$ rather than its integrability.

5. REGULARITY OF THE RESOLVENT

Let us return to the unperturbed problem (P) and hypothesis (A). We want to determine the extent of regularity $S(t)$ enjoys. In case $\mu + \kappa < \infty$ we know already that $S(t)$ is as regular as the cosine family generated by A, i.e. $S(t)$ is merely strongly continuous nothing more; Equation (2.13) shows this very clearly.

However, in case $\mu + \kappa = \infty$ more can be expected. To see this we proceed as follows.

(I) Assume there is a curve Γ: $-\omega(\rho) + i\rho$, $\rho \in \mathbb{R}$ in the complex plane with the following two properties:

(i) $\omega(\rho)$ is even, nondecreasing such that

$$\sup \omega'(\rho) = \omega_1 < \infty \quad \text{and} \quad \overline{\lim_{\rho\to\infty}} (\log\rho)/\omega(\rho) = t_0 < \infty$$

(ii) $H(\lambda) = \lambda^{-1}(I-\hat{a}(\lambda)A^{-1}$ admits bounded analytic extension to the set $\Omega = \{\lambda = \sigma + i\rho : \sigma \geqq -\omega(\rho)\}$ and satisfies

$$|H(\lambda)| \leqq M/\omega(\rho) \text{ on } \Gamma. \tag{5.1}$$

Then shifting the integration path to Γ in the complex inversion formula of the Laplace transform we obtain

$$S(t) = \frac{1}{2\pi i}\int_\Gamma e^{\lambda t}H(\lambda)d\lambda, \; t > t_0; \tag{5.2}$$

the integral is absolutely convergent. Even more is true, for $t > (n+1)t_0$ the resolvent $S(t)$ is n-times continuously differentiable and

$$S^{(n)}(t) = \frac{1}{2\pi i}\int_\Gamma e^{\lambda t}H(\lambda)\lambda^n d\lambda, \; t > (n+1)t_0. \tag{5.3}$$

Similarly, the identity

$$AH(\lambda) = H(\lambda)/\hat{a}(\lambda) - 1/\lambda\hat{a}(\lambda), \quad \lambda \in \Omega$$

also yields $AS(t) \in B(X)$ for $t > 3t_0$, at least, provided $\hat{a}(\lambda)$ admits analytic extension to Ω, and satisfies

$$|\hat{a}(\lambda)| \geqq c|\lambda|^{-2} \text{ for all } \lambda \in \Omega. \tag{5.4}$$

In particular, for $t_0 = 0$ we obtain $S \in C^\infty((0,\infty), B(X,X_A))$; if we even have $\overline{\lim_{\rho\to\infty}} \rho/\omega(\rho) < \infty$ then a standard estimate of (5.3) shows that $S(t)$ is even analytic for $t > 0$ in $B(X,X_A)$.

(II) To find such a function $\omega(\rho)$ assume in addition that $a_1(t)$ is completely monotonic; then $\hat{a}(\lambda)$ admits analytic extension to the set $\mathbb{C}\setminus(-\infty,0]$ and

$$1/\lambda^2\hat{a}(\lambda) = 1/[\lambda a_0 + a_\infty + \lambda\hat{a}_1(\lambda)] = \mu^{-2} + \hat{b}(\lambda), \tag{5.5}$$

where b(t) is again completely monotonic, by a well known result due to Reuter [21] and so (5.4) is satisfied.

(α) Since A generates a cosine family its Laplace transform $z(z^2-A)^{-1}$ is defined on a right halfplane $\mathrm{Re}z > \omega_0 \geqq 0$ and

$$|z(z^2-A)^{-1}| \leqq C/(\mathrm{Re}z-\omega_0), \quad \mathrm{Re}z > \omega_0 \tag{5.6}$$

holds; nothing more can be expected. Therefore analytic extension of $H(\lambda)$ depends on the mapping properties of the function $g(\lambda) = [\hat{a}(\lambda)]^{-1/2}$, we must have $\mathrm{Re}\, g(\lambda) > \omega \geqq 0$.

(β) For this purpose we introduce the following notations.

$$\psi_\sigma(\rho) = -\mathrm{Im}\, \hat{d}a(-\sigma+i\rho)/\rho, \quad \varphi_\sigma(\rho) = \mathrm{Re}\, \hat{d}a(-\sigma + i\rho) + \sigma\psi_\sigma(\rho)$$
$$\chi_0(\rho) = \varphi_0(\rho)/\psi_0(\rho), \qquad \theta_\sigma(\rho) = -\arg \hat{a}(-\sigma + i\rho). \tag{5.7}$$

Since $a_1(t)$ by assumption is completely monotonic, Bernstein's Theorem yields a nonnegative measure $d\alpha$ on $\mathbb{R}_+$ such that

$$a_\infty + a_1(t) = \int_0^\infty e^{-st}\, d\alpha(s), \quad t \geqq 0; \tag{5.8}$$

since $a_1 \in L^1_{loc}(R_+)$ holds we have in addition $\alpha(s)/s \to 0$ as $s \to \infty$. A simple computation yields

$$\hat{d}a(\lambda) = a_0 + \int_0^\infty (\lambda+s)^{-1} d\alpha(s) = \varphi_\sigma(\rho) - \sigma\psi_\sigma(\rho) - i\rho\, \psi_\sigma(\rho) \tag{5.9}$$

as well as

$$\varphi_\sigma(\rho) = a_0 + \int_0^\infty s[(s-\sigma)^2 + \rho^2]^{-1} d\alpha(s) \tag{5.10}$$

and

$$\psi_\sigma(\rho) = \int_0^\infty [s-\sigma)^2 + \rho^2]^{-1} d\alpha(s). \tag{5.11}$$

In the sequel we shall need the following properties of the function $\varphi_0, \psi_0, \chi_0$.

LEMMA 5.1. Let $a_1 \in L^1_{loc}(R_+)$ be completely monotonic on $(0,\infty)$, and let $\varphi_0, \psi_0, \chi_0$ be defined as in (5.7). Then

(i) $\lim_{\rho\to\infty} \rho^2\psi_0(\rho) = \mu^2$, $\lim_{\rho\to\infty} \rho^2\varphi_0(\rho) = \kappa$.

(ii) There is a function $\varepsilon(\rho)$, bounded as $|\rho| \to \infty$, such that

$$\rho^2\psi_0^2(\rho) \leqq \varepsilon(\rho)\varphi_0(\rho) \text{ for all } \rho \neq 0;$$

we may choose $\varepsilon(\rho)$ such that $\lim_{\rho\to\infty} \varepsilon(\rho) = 0$ in case $\mu + \kappa = \infty$.

(iii) $\lim_{\rho\to\infty} \chi_0(\rho) = \infty$ if and only if $\mu + \kappa = \infty$.

Proof. (i) is obvious from (5.10) and (5.11). To prove (ii) write

$$\begin{aligned}\rho^2\psi_0^2(\rho) &= \int_0^\infty \int_0^\infty [\rho^2/(s^2+\rho^2)][1/(t^2+\rho^2)]d\alpha(s)d\alpha(t) \\ &\leqq \rho^{-2} \int_0^N \int_0^N d\alpha(s)d\alpha(t) + 2 \int_N^\infty [1/(t^2+\rho^2)]\alpha(t)d\alpha(t) \\ &\leqq \varphi_0(\rho)\, [\alpha(N)^2/\rho^2\varphi_0(\rho) + 2 \sup_{t\geqq N} \alpha(t)/t],\end{aligned}$$

which shows existence and boundedness of $\varepsilon(\rho)$; clearly by (i), $\varepsilon(\rho) \to 0$ for $|\rho| \to \infty$ in case $\mu + \kappa = \infty$.

(iii) If $\mu + \kappa = \infty$ from (ii) we obtain

$$\chi_0(\rho) = \varphi_0(\rho)/\psi_0(\rho) \geqq \rho^2\psi_0(\rho)/\varepsilon(\rho) \to \infty \text{ , as } |\rho| \to \infty$$

since $\varepsilon(\rho) \to 0$; on the other hand, if $\mu + \kappa < \infty$ then by (i)

$$\chi_0(\rho) = \varphi_0(\rho)/\psi_0(\rho) = \rho^2\varphi_0(\rho)/\rho^2\psi_0(\rho) \to \kappa/\mu < \infty. \qquad \square$$

(γ) For $\hat{a}(\lambda)$ we obtain

$$\hat{a}(-\sigma+i\rho) = [\sigma^2+\rho^2]^{-1}\cdot[-\sigma\varphi_\sigma-(\rho^2-\sigma^2)\psi_\sigma - i\rho(\varphi_\sigma- 2\sigma\psi_\sigma)]. \tag{5.12}$$

If $0 \leqq \sigma \leqq |\rho|$ holds then the real part of $\hat{a}(-\sigma+i\rho)$ is always negative and so for Re $g(\lambda) > 0$ we must necessarily have $\varphi_\sigma \geqq 2\sigma\psi_\sigma$, analytic continuation of $H(\lambda)$ clearly breaks down if $\hat{a}(\lambda)$ becomes negative real. The elementary estimate

$$(1+\varepsilon^2)^{-1}\,[s^2+\rho^2]^{-1} \leqq [(s-\sigma)^2+\rho^2]^{-1} \leqq (1-\varepsilon)^{-1}[s^2+\rho^2]^{-1} \text{ for all}$$

$$\rho \in \mathbb{R},\ s \geqq 0,\ 0 \leqq \sigma \leqq \varepsilon|\rho|$$

yields

$$(1+\varepsilon^2)^{-1}\varphi_0(\rho) \leqq \varphi_\sigma(\rho) \leqq (1-\varepsilon)^{-1}\varphi_0(\rho)$$

$$(1+\varepsilon^2)^{-1}\psi_0(\rho) \leqq \psi_\sigma(\rho) \leqq (1-\varepsilon)^{-1}\psi_0(\rho) \tag{5.14}$$

for all $\rho \in \mathbb{R}$; $\rho \geqq 0$, $0 \leqq \sigma \leqq \varepsilon|\rho|$.

We have to discriminate two cases.

(a) $\nu = \varliminf_{\rho\to\infty} \chi_0(\rho)/\rho > 0$.

Then (5.14) yields for $\rho \in \mathbb{R}$, $0 \leqq \sigma \leqq \varepsilon|\rho|$

$$\mathrm{tg}(\pi-\theta_\sigma) = \frac{\rho(\varphi_\sigma - 2\sigma\,\psi_\sigma)}{\sigma\,\varphi_\sigma + (\rho^2-\sigma^2)\psi_\sigma} \geqq \frac{\gamma(\varepsilon)\chi_0(\rho)/\rho-2\varepsilon}{1+\chi_0(\rho)/\rho} \geqq \gamma > 0,$$

hence $g(\lambda)$ maps the sector $\Sigma = \{\lambda \in \mathbb{C} : -\varepsilon|\rho| \leqq \sigma,\ \rho \in \mathbb{R}\}$ into a closed sector of opening angle less than $\pi/2$; in this situation $\omega(\rho) = \varepsilon\rho - N\omega_0$ is a possible choice and the resolvent turns out to be analytic. Since this case has been studied in detail by Da Prato and Iannelli [20] as well as by Grimmer and Pritchard [13] we shall not discuss it here any further.

(b) $\lim_{\rho\to\infty} \chi_0(\rho)/\rho = 0$.

In this case (5.14) yields for $0 \leqq \sigma \leqq \varepsilon|\rho|$

$$\mathrm{tg}(\pi-\theta_\sigma) \leqq (1+\varepsilon^2)(1-\varepsilon)^{-2}\,\chi_0(\rho)/\rho \to 0 \text{ as } |\rho| \to \infty, \tag{5.15}$$

as well as

$$\begin{aligned}\varphi_\sigma(\rho) - 2\sigma\,\psi_\sigma(\rho) &\leqq (1-\varepsilon)^{-1}\varphi_0(\rho) - 2\sigma(1+\varepsilon^2)^{-1}\,\psi_0(\rho)\\ &\leqq (1-\varepsilon^2)^{-1}\,\psi_0(\rho)((1+\varepsilon^2)(1-\varepsilon)^{-1}\chi_0(\rho) - 2\sigma) \leqq 0\end{aligned}$$

in case $\varepsilon|\rho| \geqq \sigma \geqq (1+\varepsilon^2)(1-\varepsilon)^{-1} \chi_0(\rho)/2$ holds. Letting $\varepsilon \to 0$ this yields the asymptotic upper bound

$$\overline{\lim_{\rho\to\infty}} \omega(\rho)/\chi_0(\rho) \leqq 1/2 \tag{5.16}$$

for the growth of $\omega(\rho)$.

(δ) Conversely, assume (b) and let $0 \leqq \sigma \leqq \min \{\varepsilon|\rho|, \gamma\chi_0(\rho)/2\}$; then by (5.14)

$$\begin{aligned} tg(\pi-\theta_\sigma) &\geqq [\chi_0(\rho)/\rho][(1-\varepsilon)(1+\varepsilon^2)^{-1}-\gamma][1+\gamma\chi_0(\rho)^2/2\rho^2]^{-1} \\ &\geqq c_1(\varepsilon)\chi_0(\rho)/\rho \end{aligned}$$

for some constant $c_1(\varepsilon) > 0$, provided $(1-\varepsilon)(1+\varepsilon^2)^{-1} > \gamma$. Since $\theta_\sigma \approx \pi$ for large $|\rho|$ this yields

$$\cos \frac{\theta_\sigma}{2} = \sin (\pi/2 - \theta_\sigma/2) \geqq c_2 tg(\pi-\theta_\sigma)$$

for some constant $c_2 > 0$. (5.14) moreover yields

$$|\hat{a}(-\sigma+i\rho)| \leqq 2\psi_0(\rho)\sqrt{1+\chi_0(\rho)/\rho} = c_3 \cdot \varepsilon(\rho)\chi_0(\rho)/\rho^2$$

with $\varepsilon(\rho)$ from Lemma 5.1, and so we obtain

$$\text{Re } g(\lambda) \geqq |\hat{a}(\lambda)|^{-1/2}\cos\theta_\sigma/2 \geqq [c_3\varepsilon(\rho)\chi_0(\rho)/\rho^2]^{-1/2}c_2 tg(\pi-\theta_\sigma),$$

i.e. $\text{Re } g(\lambda) \geqq c_4(\varepsilon)[\chi_0(\rho)/\varepsilon(\rho)]^{1/2} > \omega$.

for $|\rho| \geqq N$ say, since in the case $\mu + \kappa = \infty$ we have $\chi_0(\rho) \to \infty$ and $\varepsilon(\rho) \to 0$ as $|\rho| \to\infty$, by Lemma 5.1 again. These estimates finally yield

$$|H(\lambda)| \leqq C(\varepsilon)/\chi_0(\rho), \tag{5.17}$$

for all $|\rho| \geqq N$ and $0 \leqq \sigma \leqq \min \{\varepsilon(\rho),\gamma \ \chi_0(\rho)/2\}$. Since $\varepsilon > 0$ was arbitrary we may let $\varepsilon \to 0$ and $\gamma \to 1$; as a summary we obtain the following result.

THEOREM 5.1. Assume (A) and let $a_1(t)$ be completely monotonic; let the function $\chi_0(\rho)$ be defined as in (5.7). Then

(i) If $\underline{\lim}_{\rho\to\infty} \chi_0(\rho)/\rho = \nu > 0$ then $S:(0,\infty) \to B(X,X_A)$ is analytic.

(ii) If $\lim_{\rho\to\infty} \chi_0(\rho)/\log\rho = \infty$ then $S:(0,\infty) \to B(X,X_A)$ is of class C^∞.

(iii) If $t_0 = 2\,\overline{\lim}_{\rho\to\infty} \log\rho/\chi_0(\rho) > 0$ then $S \in C^n((t_n,\infty), B(X))$ and $AS \in C^n((t_{n+2},\infty), B(X))$, where $t_n = (n+1)t_0$.

It is not difficult to prove that there are constants $c_1, c_2 > 0$ such that

$$c_1[a_\infty/2\rho^2 + \int_0^{1/\rho} t\,a_1(t)dt] \leq \psi_0(\rho) \leq c_2[a_\infty/2\rho^2 + \int_0^{1/\rho} t\,a_1(t)dt]$$

and

$$c_1[-\int_0^{1/\rho} t\,\dot{a}_1(t)dt] \leqq \varphi_0(\rho) \leqq c_2[-\int_0^{1/\rho} t\,\dot{a}_1(t)dt]$$

hold, where we assumed $a_0 = 0$; this gives the asymptotic relation

$$\chi_0(\rho) \infty - \int_0^{1/\rho} t\,\dot{a}_1(t)dt/[a_\infty/2\rho^2 + \int_0^{1/\rho} t\,a_1(t)dt] \text{ as } \rho \to \infty . \qquad (5.18)$$

Therefore we obtain the following examples for Theorem 5.1.

(i) $a_0 > 0$ or $a_0 = 0$ but $a_1(t) \sim ct^{\alpha-1}$, $\alpha \in (0,1)$

(ii) $a_0 = 0$, $\mu < \infty$ and $-\dot{a}_1(t) \sim ct^{\alpha-1}$, $\alpha \in (0,1)$

(iii) $a_0 = 0$, $\mu < \infty$ and $\dot{a}_1(t) \sim c \log t$.

Similar results have been obtained for the Equation (1.3) on the line by Hrusa and Renardy [15] and for certain systems of first order equations in one space dimension by Desch and Grimmer [9].

6. PROOF OF THEOREM 3.3

An easy computation shows that formally we ahve $\hat{S}''(\lambda) = G(\lambda)$ where

$$G(\lambda) = \lambda(1/\hat{a}(\lambda)-A)^{-1}A - a_0A, \quad \lambda > 0; \qquad (6.1)$$

to prove Theorem 3.3 we therefore try to recover $G(\lambda)$ as the Laplace transform of a function with values in $B(X_A,X)$. This is not possible directly, hence we use the formal relations

$$(tS'')^\wedge(\lambda) = -G'(\lambda), \quad (t^2S'')^\wedge(\lambda) = G''(\lambda), \ \lambda > 0;$$

(a) Let us consider the case $a_0 > 0$ first. Another computation yields

$$G'(\lambda) = (1/\hat{a}-A)^{-1}A + [\lambda\hat{a}'/\hat{a}](1/\hat{a}-A)^{-2}A, \ \lambda > 0; \tag{6.2}$$

hence by (2.1) we see that $G'(\lambda)$ is the transform of a $B(X_A,X)$ - valued continuous function, provided $\lambda\hat{a}'/\hat{a}$ is the transform of a measure. This is indeed the case as the following identity shows;

$$\lambda\hat{a}'/\hat{a} = \lambda\widehat{da}'/\widehat{da} - 1 = [-a_\infty/\lambda + \lambda\hat{a}_1']/\widehat{da} - 1, \ \lambda > 0;$$

note that by virtue of $a_0 > 0$ we have $1/\widehat{da} = \widehat{dr}$ for some measure dr. Thus there is $U \in C(R_+, B(X_A,X))$ such that $\hat{U}(\lambda) = -G'(\lambda)$ for all sufficiently large $\lambda > 0$; an easy computation then shows $U(t) = tS''(t)$ for all $t > 0$, by uniqueness of the Laplace transform, i.e. $S''(t)$ is $B(X_A,X)$-continuous for $t > 0$.

(b) Now let us assume $a_0 = 0$ as well as $\mu = \infty$ by the remarks in front of Theorem 3.3; we may also assume $a_\infty = 0$, replace $a_1(t)$ by $a_1(t) + a_\infty$ otherwise. Unfortunately, we do not know whether $\lambda\hat{a}'/\hat{a}$ is the transform of a measure in this case, and so we are forced to consider the second derivative of $G(\lambda)$ which is given by

$$\begin{aligned} G''(\lambda) &= \lambda^3\hat{a}''\,[\lambda^{-1}(I-\hat{a}A)^{-1}]^2A \\ &\quad + 2[-\hat{a}'/\hat{a}^{3/2}]\cdot[\hat{a}^{-1/2}(\hat{a}^{-1}-A)^{-1}]\lambda\hat{a}[\lambda^{-1}(I-\hat{a}A)^{-1}]A \\ &\quad - [\lambda(\hat{a}')^2/\hat{a}^{5/2}]\cdot[\hat{a}^{-1/2}(\hat{a}^{-1}-A)^{-1}]\lambda\hat{a}[\lambda^{-1}(I-\hat{a}A)^{-1}]A \\ &\quad + 2[\lambda(\hat{a}')^2/\hat{a}^{5/2}]\cdot[\hat{a}^{-1/2}(\hat{a}^{-1}-A)^{-1}]^3A \\ &= \{G_1(\lambda) + 2G_2(\lambda) - G_3(\lambda) + G_4(\lambda)\}A, \ \lambda > 0. \end{aligned} \tag{6.3}$$

Thus to prove the first assertions of Theorem 3.3 in this case, it suffices to show that there are $R_i \in C((0,\infty), \mathbf{B}(X))$ with $\hat{R}_i(\lambda) = G_i(\lambda)$ for $\lambda > 0$, and $t^{-2}R_i \in L^1([0,1], \mathbf{B}(X))$ if in addition (3.9) holds, $i = 1,\ldots,4$; note that (3.9) is equivalent to $\int_0^1 a(t)t^{-1}dt < \infty$.

(c) This is straightforward and easy for $G_1(\lambda)$. In fact, we have

$$\lambda^3\hat{a}'' = (6a_1 + 6t\dot{a}_1 + t^2a_1)^\wedge(\lambda) = \hat{a}_2(\lambda),\ \lambda > 0,$$

hence $R_1 = a_2 * S * S$ is $\mathbf{B}(X)$ - continuous on $\mathbb{R}_+$ since $S(t)$ has this property on $(0,\infty)$ by Theorem 2.3. Moreover, we have

$$|a_2(t)| \leqq a_2(t) - 12\dot{a}_1(t), \quad t \geqq 0$$

hence with $C = \sup\{|S(t)|: 0 \leqq t \leqq 1\}$ we obtain

$$\int_0^1 |R_1(t)|t^{-2}dt \leqq C^2\int_0^1 |a_2|*1 * 1 \cdot t^{-2}dt \leqq C^2\int_0^1[a_2*t-12t\dot{a}_1*t]t^{-2}dt$$

$$\leqq C^2\int_0^1 [14a(t)/t + a_1(t)]dt < \infty,$$

by (3.9) and (A).

(d) In principle, G_2, G_3, G_4 will be treated as G_1, however things are somewhat difficult here. We shall need the following properties of the kernel $a(t)$.

<u>LEMMA 6.1.</u> Assume $a_0 = a_\infty = 0$, $\mu = \sqrt{a_1(0)} = \infty$ and let $\omega \geqq 0$. Then

(i) There are nonnegative kernels k, $l \in L^1_{loc}(\mathbb{R}_+)$ such that

$$\hat{k}(\lambda) = 1/\lambda[\hat{a}(\lambda)]^{1/2}, \quad \hat{l}(\lambda) = -\hat{a}'(\lambda)/[\hat{a}(\lambda)]^{3/2},\ \lambda > 0.$$

(ii) There is a nonnegative kernel $b \in L^1_{loc}(\mathbb{R}_+)$ such that $\hat{b}(\lambda) = [\hat{a}(\lambda)]^{1/2}$, $\lambda > 0$.

(iii) There is $m \in BV_{loc}(\mathbb{R}_+)$ such that $\widehat{dm}(\lambda) = [1-\omega\hat{b}(\lambda)]^{-1}$, $\lambda > 0$.

(iv) For each $\alpha \in (0,1)$ there are nondecreasing kernels k_α, $l_\alpha \in BV_{loc}(\mathbb{R}_+)$

such that

$$\widehat{dk}_\alpha(\lambda) = [\lambda^2\hat{a}(\lambda)]^{-\alpha}, \ \widehat{dl}_\alpha(\lambda) = \frac{d}{d\lambda}[\lambda^{1-2\alpha}/\hat{a}(\lambda)^\alpha], \ \lambda > 0.$$

Proof. (i) is contained in Proposition 2 of Prüss [24] and will not be proved here. (ii) follows from (i) by virtue of

$$[\hat{a}(\lambda)]^{1/2} = \lambda\hat{a}(\lambda)/\lambda[\hat{a}(\lambda)]^{1/2} = \hat{a}_1(\lambda)\hat{k}(\lambda) = (a_1 * k)^\wedge(\lambda), \ \lambda > 0. \tag{6.4}$$

(iii) follows immediately from (ii).

To prove (iv) assume first $a_1(0) < \infty$; then

$$[\lambda^2\hat{a}(\lambda)]^{-\alpha} = [\lambda\hat{a}_1(\lambda)]^{-\alpha} = [\mu^2-(-\hat{\dot{a}}_1(\lambda))]^{-\alpha} = (\mu^2-x)^{-\alpha} \circ (-\hat{\dot{a}}_1(\lambda))$$

is the composition of the completely monotonic function $-\hat{\dot{a}}_1(\lambda)$ (note that $a_1(t)$ is non-increasing) and the absolutely monotonic function $(\mu^2-x)^{-\alpha}$, hence completely monotonic. If $a_1(0) = \infty$, we approximate by $a_{1n}(t) = a_1(t+1/n)$, to see $[\lambda^2\hat{a}(\lambda)]^{-\alpha}$ is always completely monotonic, hence by Bernstein's Theorem is represented as the Laplace transform of a nonnegative measure. Finally, to prove the last assertion, note that by the complex variable formula for $\alpha \in (0,1)$,

$$z^{-\alpha} = \pi^{-1} \sin\alpha\pi \int_0^\infty (z+r)^{-1} r^{-\alpha} dr, \quad z \in \mathbb{C} \setminus (-\infty,0],$$

we have

$$\lambda^{1-2\alpha}\hat{a}^{-\alpha} = \pi^{-1} \sin\alpha\pi \int_0^\infty \lambda^{1-2\alpha}(\hat{a}+r)^{-1} r^{-\alpha} dr, \quad \lambda > 0.$$

Changing the integration variable to $s = r\lambda^2$ we obtain

$$\lambda^{1-2\alpha}\hat{a}^{-\alpha} = \pi^{-1} \sin \alpha\pi \int_0^\infty [\hat{a}_1 + s/\lambda]^{-1} s^{-\alpha} ds, \quad \lambda > 0,$$

and differentiation yields

$$\frac{d}{d\lambda}[\lambda^{1-2\alpha}\hat{a}^{-\alpha}] = \pi^{-1} \sin \alpha\pi \int_0^\infty [-\hat{c}_s'/\hat{c}_s^2] s^{-\alpha} ds, \ \lambda > 0, \tag{6.5}$$

where $\hat{c}_s(\lambda) = \hat{a}_1(\lambda) + s/\lambda$ is the transform of the positive and log - convex function $a_1(t)+s$. Thus, as shown in Proposition 1 of Prüss [24], $-\hat{c}_s'/\hat{c}_s^2$ is completely monotonic for all $s \geqq 0$, and so that last assertion of Lemma 6.1 follows from Bernstein's Theorem again. Note that there are no problems in interchanging differentiation and integration in Equation (6.5). □

For future reference we define $b_1 \in L^1_{loc}(\mathbb{R}_+)$ by means of

$$b_1(t) = b * dm(t), \quad t > 0. \tag{6.6}$$

The second tool we need to handle G_2, G_3, G_4 is concerned with the integral resolvent R of (P) which in terms of transforms is defined by means of the relation

$$\hat{R}(\lambda) = [\hat{a}(\lambda)]^{-1/2}(1/\hat{a}(\lambda)-A)^{-1}, \quad \lambda > 0. \tag{6.7}$$

We prove

<u>PROPOSITION 6.1</u>. Suppose $\mu = \infty$. Then there is $R \in C((0,\infty), \mathcal{B}(X))$ such that (6.7) holds. Moreover, if the cosine family $T(t)$ generated by A satisfies $|T(t)| \leqq Me^{\omega t}$ on $\mathbb{R}_+$, then

$$|R(t)| \leqq Mb_1(t), \quad t > 0 \tag{6.8}$$

holds. In particular, $R * R \in C(\mathbb{R}_+,\mathcal{B}(X))$.

<u>Proof</u>. Define $R(t)$ according to

$$R(t) = \int_0^\infty T(\tau)w_t(t,\tau)d\tau, \quad t > 0;$$

since w_t is measurable and

$$|R(t)| \leqq M\int_0^\infty e^{\omega\tau}\, w_t(t,\tau)d\tau = Mb_1(t), \quad t > 0,$$

holds, $R(t) \in \mathcal{B}(X)$ for each $t > 0$ and (6.8) is satisfied. Taking Laplace transforms in the weak sense we see that (6.7) holds. On the other, for $x \in D(A)$ (6.7) yields

$$\hat{R}(\lambda)x = \hat{b}(\lambda)[x + \hat{da}(\lambda)\hat{S}(\lambda)Ax]$$

i.e.

$$R(t)x = b(t)x + b * da * S(t)Ax \tag{6.9}$$

and so $R(t)x$ is continuous on $(0,\infty)$ for each $x \in D(A)$; the Banach-Steinhaus Theorem implies that $R(t)$ is strongly continuous on $(0,\infty)$ since $|R(t)|$ is locally bounded on $(0,\infty)$. To show that $R(t)$ is even $\mathcal{B}(X)$ - continuous, we consider

$$(tR(t))^{\wedge}(\lambda) = -\hat{R}'(\lambda) = 2^{-1}\hat{1}(\lambda)(\hat{R}(\lambda)^2 - \widehat{da}\hat{S}(\lambda))$$

i.e.

$$2tR(t) = 1 * R * R(t) - 1 * da * S(t)$$

is $\mathcal{B}(X)$-continuous on $(0,\infty)$. The last assertion finally follows from this continuity of $R(t)$ and the fact that $b_1 * b_1 = a * dm * dm$ is bounded at zero. □

(e) Consider now $G_2(\lambda)$; with 1 from Lemma 6.1 and R from Proposition 6.1 we obtain

$$G_2(\lambda) = \hat{1}(\lambda)\hat{R}(\lambda)\hat{a}_1(\lambda)\hat{S}(\lambda), \quad \lambda > 0,$$

i.e. $G_2(\lambda) = \hat{R}_2(\lambda)$ where

$$R_2(t) = 1 * a_1 * R * S(t), \quad t \geqq 0.$$

It is clear that $R_2 \in C(\mathbb{R}_+,\mathcal{B}(X))$ since $S(t)$ is $\mathcal{B}(X)$ - continuous for $t > 0$ and bounded at zero. We have $1(t)$, $a_1(t) \geqq 0$, hence by (6.8) we obtain

$$\int_0^1 |R_2(t)|t^{-2}dt \leqq C \int_0^1 \ell * a * b \cdot t^{-2}dt = C \int_0^1 a(t)t^{-1}dt < \infty$$

by (3.9).

(f) Next we consider $G_3(\lambda)$; Lemma 6.1 (iv) with $\alpha = 1/4$ gives

$$\lambda(\hat{a}')^2/\hat{a}^{5/2} = [-\lambda^{1/2/4}\hat{a}'/\hat{a}^{1+1/4}]^2 = 16[\widehat{dl}_\alpha - 2^{-1}\widehat{dk}_\alpha]^2 = 16\,\widehat{dr}_\alpha^2$$

and so

$$G_3(\lambda) = 16\,\widehat{dr}_\alpha^2 \cdot \hat{a}_1(\lambda)\cdot\hat{R}(\lambda)\hat{S}(\lambda), \quad \lambda > 0,$$

i.e. $G_3(\lambda) = \hat{R}_3(\lambda)$ with

$$R_3(t) = 16a_1 * dr_\alpha * dr_\alpha * R*S(t) \text{ for } t \geqq 0$$

which obviously belongs to $C(\mathbb{R}_+, \mathbb{B}(X))$. Since dl_α, $dk_\alpha \geqq 0$, by Proposition 6.1 we get

$$\int_0^1 |R_3(t)|t^{-2}dt \leqq C\int_0^1 a_1 * (dl_\alpha + dk_\alpha/2)*(dl_\alpha + dk_\alpha/2)*b_1 * 1 \cdot t^{-2}dt$$

$$\leqq C\int_0^1 a * (dr_\alpha + dk_\alpha)*(dr_\alpha + dk_\alpha)*b \cdot t^{-2}dt.$$

Taking the Laplace transform of the last integrand we obtain

$$\hat{a}(\widehat{dr}_\alpha + \widehat{dk}_\alpha)^2\hat{b} = \hat{a}^{3/2}\,[-\sqrt{\lambda}\hat{a}'\hat{a}^{-5/4}/4 + 1/(\sqrt{\lambda}\hat{a}^{1/4})]^2$$

$$= \lambda(\hat{a}')^2\hat{a}^{-1}/16 - \hat{a}'/2 + \hat{a}/\lambda;$$

on the other hand

$$\sqrt{\lambda}\,\hat{a}'/(2\hat{a}^{1/2}) = \sqrt{\lambda}(\hat{a}^{1/2})' = [\lambda\hat{a})^{1/2}]' - \hat{a}^{1/2}/(2\sqrt{\lambda}) = (\hat{a}_1^{1/2})' - \hat{a}_1^{1/2}/(2\lambda),$$

and so

$$\hat{a}(\widehat{dr}_\alpha + \widehat{dk}_\alpha)^2\,\hat{b} = [-(\hat{a}_1^{1/2})']^2/4 - \frac{5}{8}\hat{a}' + \frac{15}{16}\hat{a}/\lambda.$$

Let $dc \geqq 0$ denote the measure such that $\widehat{dc}(\lambda) = [\hat{a}_1(\lambda)]^{1/2}$; existence of dc follows from another application of Bernstein's Theroem since

$$\hat{a}_1^{1/2} = \pi^{-1}\int_0^\infty \hat{a}_1(\hat{a}_1 + r)^{-1}dr/\sqrt{r}$$

is completely monotonic by log-convexity of $a_1(t)$. We finally have

$$\int_0^1 |R_3(t)|t^{-2}dt \leqq C \int_0^1 [tdc * tdc/4 + 5ta(t)/8 + (15/16) \int_0^t a(s)ds]t^{-2}dt$$

$$\leqq C \int_0^1 [dc * dc/4 + (25/16)a(t)/t]dt < \infty .$$

(g) $G_4(\lambda)$ can be treated in the same way as $G_3(\lambda)$; we only have to observe that $a_1 * S$ has to be replaced by $R * R$ which can be estimated by $Ca(t)$. The proof of the first three assertions for the case $\mu = \infty$ is complete.

(h) Assume in addition A invertible, $1/\hat{a}(\lambda) \in \rho(A)$ for all $\text{Re}\,\lambda \geqq 0$, $\lambda \neq 0$, if $\omega_0(A) > 0$, let $a_0 = 0$, $\mu = \infty$ and (3.9) holds. Then by Theorem 3.1 we have $S \in L^1(\mathbb{R}_+, \mathcal{B}(X))$ and by Theorem 3.2 also $S'A^{-1} \in L^1(\mathbb{R}_+, \mathcal{B}(X))$; since $S''A^{-1} \in L^1_{loc}(\mathbb{R}_+, \mathcal{B}(X))$ we have

$$(tS''A^{-1})\hat{}(\lambda) = - G'(\lambda)A^{-1},\ (S'A^{-1})\hat{}(\lambda) = (1/\hat{a}-A)^{-1},\ \lambda > 0. \tag{6.10}$$

The relation

$$-\lambda^2\hat{a}'(\lambda) = \lambda^2(ta)\hat{}(\lambda) = (t\dot{a}_1)\hat{} + 2\hat{a}_1(\lambda) = (t\dot{a}_1)\hat{}(\lambda) + 2\lambda\hat{a}(\lambda)$$

with (6.2) then yields

$$(tS''A^{-1})\hat{}(\lambda) = -(S'A^{-1})\hat{}(\lambda) + 2[(S'A^{-1})\hat{}(\lambda)]^2+(t\dot{a}_1)\hat{}(\lambda)\hat{S}(\lambda)(S'A^{-1})\hat{}(\lambda)$$

i.e.

$$tS''(t)A^{-1} = - S'(t)A^{-1} + 2(S'*S')A^{-2} + (t\dot{a}_1) * S * S'A^{-1}(t),\quad t > 0.$$

The first two terms belong to $L^1(\mathbb{R}_+, \mathcal{B}(X))$ as does $S_2 = S * S'A^{-1}$; therefore we obtain the estimate

$$\int_1^\infty |S''(t)A^{-1}\ dt = \int_1^\infty |tS''(t)A^{-1}|t^{-1}dt$$

$$\leqq |S'A^{-1}|_{L^1} + 2|S'A^{-1}|^2_{L^1} + \int_1^\infty (-(t\dot{a}_1) * |S_2|)t^{-1}dt < \infty$$

since

$$\int_1^\infty (-(t\dot{a}_1)*|S_2|)t^{-1}dt \leq |S_2|_{L^1}[\int_1^\infty (-t\dot{a}_1(t))t^{-1}dt + \int_0^1 - t\dot{a}_1(t)dt]$$
$$\leq |S|_{L^1} \cdot |S'A^{-1}|_{L^1} \cdot \int_0^1 a_1(\tau)d\tau < \infty .$$

Thus $S''A^{-1}$ belongs to $L^1(R_+, B(X))$ and the proof is complete. □

References

[1] Bland, D.R., The Theory of Linear Viscoelasticity. Pure and Applied Mathematics 10, Pergamon Press, New York 1960.

[2] Carr, R.W., Hannsgen, K.B., A nonhomogeneous integrodifferential equation in Hilbert space. SIAM J. Math. Anal. 10, 961-984 (1979).

[3] Carr, R.W., Hannsgen, K.B., Resolvent formulas for a Volterra equation in Hilbert space. SIAM J. Math. Anal. 13, 453-483 (1982).

[4] Christensen, R.M., Theory of Viscoelasticity. Academic Press, New York 1971.

[5] Clement, P., Nohel, J.A., Abstract linear and nonlinear Volterra equations preserving positivity. SIAM J. Math. Anal. 10, 365-388 (1979).

[6] Clement, P., Nohel, J.A., Asymptotic behavior of solutions of Volterra equations with completely positive kernels, SIAM J. Math. Anal. 12, 514-535 (1981).

[7] Dafermos, C.R., Asymptotic stability in viscoelasticity. Arch. Rat. Mech. Anal. 37, 297-308 (1970).

[8] Dafermos, C.R., An abstract Volterra equation with applications to linear viscoelasticity. J. Diff. Equs. 7, 554-569 (1970).

[9] Desch, W., Grimmer, R., Smoothing properties of linear Volterra integrodifferential equations. Preprint.

[10] Desch, W., Grimmer, R., Schappacher, W., Some considerations for linear integrodifferential equations. J. Math. Anal. 104, 219-234 (1984).

[11] Friedman, A., Shinbrot, M., Volterra integral equations in Banach space. Trans. Amer. Math. Soc. 126, 131-179 (1967).

[12] Goldstein, J., Semigroups of Linear Operators and Applications. Oxford Math. Monographs, Clarendon Press, 1985.

[13] Grimmer, R., Pritchard, A.J., Analytic resolvent operators for integral equations in Banach spaces. J. Diff. Equs. 50, 234-259 (1983).

[14] Grimmer, R., Prüss, J., On linear Volterra equations in Banach spaces. Comp. & Math. with Appls. 11, 189-205 (1985).

[15] Hrusa, W.J., Renardy, M., On wave propagation in linear viscoelasticity. Quart. Applied Math. 43, 237-253 (1985).

[16] Miller, R.K., Wheeler, R.L., Asymptotic behavior for a linear Volterra integral equation in Hilbert space. J. Diff. Equs. 23, 270-284 (1977).

[17] Noren, R., Uniform L^1-behavior for the solution of a Volterra equation with a parameter. Preprint.

[18] Pipkin, A.C., Lectures on Viscoelasticity Theory. Springer-Verlag Berlin 1972.

[19] Da Prato, G., Iannelli, M., Linear abstract integrodifferential equations of hyperbolic type in Hilbert spaces. Rend. Sem. Math. Padova 62, 191-206 (1980).

[20] Da Prato, G., Iannelli, M. Linear integro-differential equations in Banach space. Rend. Sem. Math. Padova 62, 207-219 (1980).

[21] Reuter, G., Über eine Volterrasche Integralgleichung mit totalmonotonem Kern. Arch. Math. 7., 59-66 (1956).

[22] Prüss, J., On linear Volterra equations of parabolic type in Banach spaces. Trans. Amer. Math. Soc. 301, 691-721 (1987).

[23] Prüss, J., Bounded solutions of Volterra equations. SIAM J. Math. Anal. 19, 133-149 (1988).

[24] Prüss, J., Positivity and regularity of hyperbolic Volterra equations in Banach spaces, Math. Ann. 279, 317-344 (1987).

[25] Renardy, M., Some remarks on the propagation and non-propagation of discontinuities in linearly viscoelastic liquids. Rheol. Acta. 21, 251-254 (1982).

[26] Renardy, M., Hrusa, W.J., Nohel, J.A., Mathematical Problems in Viscoelasticity. Longman Scientific and Technical (1987).

Jan Prüss
FB 17 Mathematik
Universität Paderborn
Warburger Strasse 100
D-4790 Paderborn
West Germany.

D. SFORZA

Regularity results for an integrodifferential equation of convolution type in Hilbert space

0. INTRODUCTION

This paper deals with a class of linear Volterra integrodifferential equations of convolution type in a Hilbert space H:

$$\begin{aligned} u'(t) &= -\int_0^t k(t-s)Au(s)ds + f(t), \quad t \geqq 0 \\ u(0) &= x \end{aligned} \tag{0.1}$$

where k is a scalar function such that $t \to e^{-\omega t}k(t)$ is a positive kernel for some $\omega \in \mathbb{R}$, and $A:D(A) \subset H \to H$ is a linear, closed, positive operator with domain D(A) dense in H.

We will solve equation (0.1) by means of a resolvent operator U(t), representing the solution by the formula

$$u(t) = U(t)x + \int_0^t U(t-s)f(s)ds, \quad t \geqq 0.$$

We will then be able to analyze the regularity of the solution u. Problem (0.1) has been studied by several authors, using various methods (see, for instance, [1-6, 9, 10, 12, 15, 17, 20]).

Our work proceeds as follows. In Section 1 we introduce the standard notions of positive and strongly positive kernel, and we obtain some preliminary results. Section 2 is devoted to the construction of the resolvent operator U and to some estimates and properties concerning its regularity.

Section 3 deals with the nonhomogeneous problem. Finally, in Section 4 we study a second order problem, as an application of our results.

We list now some notations.

- H will denote a complex Hilbert space with inner product $\langle .,. \rangle$ and norm $\|\cdot\|$.
- By $L^p_{loc}([0, +\infty[;H)$, $1 \leqq p \leqq +\infty$, we denote the space of all strongly measurable functions $u:[0, +\infty[\to H$ with the property that $u \in L^p([0,T];H)$

for every $0 < T < +\infty$. If $H = \mathbb{R}$, we will simply write $L^p_{loc}([0, +\infty[)$.

- If X and Y are two Banach spaces with norm $\|\cdot\|_X$ and $\|\cdot\|_Y$ respectively, $L(X,Y)$ is the Banach space of all bounded operators $T:X \to Y$, endowed with the norm $\|T\| = \sup\{\|T(x)\|_Y : \|x\|_X \leq 1\}$. We set $L(X) = L(X,X)$.
- If $A:D(A) \subset H \to H$ is a linear, closed operator, we shall endow the domain $D(A)$ with the graph norm of A; in addition, if A is positive, we shall denote by $R(n,A) = (n+A)^{-1}$, $n \in \mathbb{N}$, the resolvent of A.

1. POSITIVE KERNELS

For the sake of completeness, we recall here the well-known notion of positive kernel. A function $k \in L^1_{loc}([0, +\infty[)$ is called a positive kernel if the mapping

$$F: L^2_{loc}([0, +\infty[;H) \to L^2_{loc}([0, +\infty[;H)$$

$$F(v)(t) = \int_0^t k(t-s)v(s)ds$$

is positive on $L^2([0,T];H)$ for all $T > 0$, that is

$$\mathrm{Re} \int_0^T \langle F(v)(t),v(t)\rangle \, dt \geqq 0.$$

Moreover k is called a strongly positive kernel if there exist $\varepsilon > 0$, $a > 0$ such that $k(t)-\varepsilon\, e^{-at}$ is a positive kernel.

It is known that the notion of positive kernel is related with the theory of Laplace transform. We recall that, denoting by

$$\hat{k}(p) = \int_0^{+\infty} e^{-pt}\, k(t)dt$$

the Laplace transform of k, then $k \in L^1([0, +\infty[)$ is a positive kernel if and only if $\mathrm{Re}\, \hat{k}(it) > 0$ for all $t \in \mathbb{R}$ (see [1, 13, 16, 19]).

To prove our existence results, we shall use the following proposition, which seems to be of independent interest.

PROPOSITION 1.1. Let $k \in L^1([0, +\infty[)$ be a positive kernel. Then for all $a > 0$ there exists a constant $\varepsilon > 0$ such that $e^{-at}k(t)-\varepsilon\, e^{-at}$ is a positive kernel (i.e. $e^{-at}k(t)$ is a strongly positive one).

Proof. Since $t \to \operatorname{Re} \hat{k}(it)$ is a continuous and bounded function on $\mathbb{R}$, one may apply Poisson's formula in the half-plane Re z 0 and obtain

$$\operatorname{Re} \hat{k}(a+ib) = \frac{1}{\pi} \int_{-\infty}^{+\infty} \operatorname{Re} \hat{k}(it) \frac{a}{a^2+(t+b)^2} \, dt =$$

$$= \frac{a}{a^2+b^2} \frac{1}{\pi} \int_{-\infty}^{+\infty} \operatorname{Re} \hat{k}(it) \frac{a^2+b^2}{a^2+(t+b)^2} \, dt, \; b \in \mathbb{R}, \; a > 0 \tag{1.1}$$

Then it is easy to check that

$$\frac{a^2+b^2}{a^2+(t+b)^2} \geqq \frac{4a^2}{4a^2+(t+(t^2+4a^2)^{1/2})^2}, \text{ for all } b \in \mathbb{R}, \; t \geqq 0;$$

therefore, since $\operatorname{Re} \hat{k}(it) > 0$ for all $t \in \mathbb{R}$, from (1.1) one gets

$$\operatorname{Re} \hat{k}(a+ib) \geqq \frac{a}{a^2+b^2} \frac{1}{\pi} \int_{0}^{+\infty} \operatorname{Re} \hat{k}(it) \frac{4a^2}{4a^2+(t+(t^2+4a^2)^{1/2})^2} \, dt,$$

so that our statement is proved.

COROLLARY 1.2. Let $k \in L^1_{loc}([0, +\infty[)$ verify the following properties:

(i) k is absolutely Laplace transformable in a half-plane, i.e. there exist α, $M \geqq 0$ such that, for all $t \geqq 0$, $|k(t)| \leq Me^{\alpha t}$;

(ii) there exist a constant $\omega > \alpha$ such that $e^{-\omega t}k(t)$ is a positive kernel.

Then for all $a > 0$ there exists $\varepsilon > 0$ such that $e^{-(a+\omega)t}k(t) - \varepsilon\, e^{-at}$ is a positive kernel (i.e. $e^{-(a+\omega)t}k(t)$ is strongly positive).

Proof. It is sufficient to remark that $e^{-\omega t}k(t) \in L^1([0,+\infty[)$, so that the statement follows directly from Proposition 1.1.

2. EXISTENCE, UNIQUENESS AND REGULARITY OF THE RESOLVENT OPERATOR FOR THE HOMOGENEOUS PROBLEM

In this section we shall prove existence, uniqueness and regularity of the resolvent operator for the problem

$$u'(t) = -\int_0^t k(t-s)Au(s)ds, \quad t \geq 0$$
$$u(0) = x \tag{2.1}$$

Throughout the paper, the following properties of the function k and of the operator A are assumed:

(2.2) $k \in L^1_{loc}([0,+\infty[)$ is absolutely Laplace transformable in a half-plane, i.e. there exist α, $M \geq 0$ such that for all $t \geq 0$ we have $|k(t)| \leq Me^{\alpha t}$. Moreover, there exists a constant $\omega > \alpha$ such that $e^{-\omega t}k(t)$ is a positive kernel;

(2.3) $A : D(A) \subset H \to H$ is a linear, closed, self-adjoint and positive operator (with domain D(A) dense in H).

We now introduce the notions of strict solution and resolvent operator.

DEFINITION 2.1. Let x belong to D(A); we say that u is a strict solution of (2.1) if $u \in C^1([0, +\infty[;H) \cap C([0, +\infty[; D(A))$ and fulfills (2.1).

DEFINITION 2.2. Let $\{U(t)\}_{t\geq 0}$ be a family of linear operators in L(H). We say that $\{U(t)\}_{t\geq 0}$ is a (N,β)-resolvent operator for (2.1) ($N > 0$, $\beta \in \mathbb{R}$) if the following properties are satisfied:

i) for every $x \in H$, $t \to U(t)x \in C([0, +\infty[;H)$;

ii) for every $x \in D(A)$, $t \to U(t)x$ is a strict solution of (2.1) (see Def. 2.1);

iii) $\|U(t)\| \leq N e^{\beta t}$, $t \geq 0$.

Our first result concerns existence and uniqueness of the resolvent operator for problem (2.1).

THEOREM 2.3. The following assertions are true:

1) for all $p \in \mathbb{C}$, Re $p > \omega$ there exists $(p + \hat{k}(p)A)^{-1} \in L(H)$ and $\|(p + \hat{k}(p)A)^{-1}\| \leq 1/(\text{Re } p - \omega)$;

2) there exists a unique $(1,\omega)$-resolvent operator for problem (2.1).

Proof. 1) Let $x \in D(A)$ and $p \in \mathbb{C}$, $\mathrm{Re}\, p > \omega$. We have:

$$\|px + \hat{k}(p)Ax\|^2 = (\mathrm{Re}\, p)^2 \|x\|^2 + 2\, \mathrm{Re}\, p\, \mathrm{Re}\, \hat{k}(p) \langle Ax,x\rangle +$$

$$+ (\mathrm{Re}\, \hat{k}(p))^2 \|Ax\|^2 + (\mathrm{Im}\, p)^2 \|x\|^2 + 2\, \mathrm{Im}\, p\, \mathrm{Im}\, \hat{k}(p) \langle Ax,x\rangle +$$

$$+ (\mathrm{Im}\, \hat{k}(p))^2 \|Ax\|^2 \tag{2.4}$$

We observe that

$$(\mathrm{Im}\, p)^2 \|x\|^2 + 2\mathrm{Im}\, p\, \mathrm{Im}\, \hat{k}(p) \langle Ax,x\rangle + (\mathrm{Im}\, \hat{k}(p))^2 \|Ax\|^2 \geqq 0 \tag{2.5}$$

Actually, if $\mathrm{Im}\, p\, \mathrm{Im}\, \hat{k}(p) \geqq 0$, (2.5) is true, because A is a positive operator, while if $\mathrm{Im}\, p\, \mathrm{Im}\, \hat{k}(p) < 0$, we have

$$(\mathrm{Im}\, p)^2 \|x\|^2 + 2\mathrm{Im}\, p\, \mathrm{Im}\, \hat{k}(p)\langle Ax,x\rangle + (\mathrm{Im}\, \hat{k}(p))^2 \|Ax\|^2 \geqq$$

$$(\mathrm{Im}\, p)^2 \|x\|^2 + 2\mathrm{Im}\, p\, \mathrm{Im}\, \hat{k}(p) \|x\| \|Ax\| + (\mathrm{Im}\, \hat{k}(p))^2 \|Ax\|^2 \geqq 0.$$

Recall now that $\mathrm{Re}\, \hat{k}(q) > 0$ for all $q \in \mathbb{C}$, $\mathrm{Re}\, q > \omega$, because $e^{-\omega t}k(t)$ is a positive kernel; by (2.4) and (2.5) we get

$$\|px+\hat{k}(p)Ax\|^2 \geqq (\mathrm{Re}\, p)^2 \|x\|^2 \geqq (\mathrm{Re}\, p - \omega)^2 \|x\|^2.$$

A similar inequality holds for the adjoint operator

$$\|p^*x+\hat{k}(p)^*Ax\|^2 \geqq (\mathrm{Re}\, p - \omega)^2 \|x\|^2$$

and assertion 1) follows.

2) By virtue of 1), for every $x \in D(A^3)$ there exists a unique absolutely Laplace transformable strict solution u of (2.1) (see [20] Th. 3). If we prove that every strict solution u of (2.1) with initial value $x \in D(A^3)$ verifies the estimate

$$\|u(t)\| \leqq e^{\omega t} \|x\|, \quad t \geqq 0 \tag{2.6}$$

then, again by Th. 3 of [20], there exists a unique $(1,\omega)$-resolvent operator for problem (2.1). To prove (2.6), we observe that

$$\frac{d}{dr}\|e^{-\omega r}\ u(r)\|^2 = -2\omega\ \|e^{-\omega r}\ u(r)\|^2 -$$

$$- 2\mathrm{Re} < \int_0^r k(r-s)e^{-\omega(r-s)}A(e^{-\omega s}u(s))ds,\ e^{-\omega r}\ u(r)>,\ 0 \leqq r \leqq t.$$

Integrating this identity over $[0,t]$ and recalling that $e^{-\omega t}k(t)$ is a positive kernel, we have:

$$\|e^{-\omega t}u(t)\|^2 = \|x\|^2 - 2\omega \int_0^t \|e^{-\omega r}u(r)\|^2 dr -$$

$$-2\mathrm{Re}\int_0^t < \int_0^r k(r-s)e^{-\omega(r-s)}\ \sqrt{A}\ (e^{-\omega s}u(s))\ ds,\ \sqrt{A}\ (e^{-\omega r}\ u(r)) > dr$$

$$\leqq \|x\|^2,$$

that is (2.6).

We shall use in the sequel the following regularity result for the $(1,\omega)$-resolvent operator $\{U(t)\}_{t\geqq 0}$ of (2.1).

PROPOSITION 2.4. The following assertions are true:

i) for all $t \geqq 0$, $\int_0^t U(s)\ ds$ belongs to $L(H,D(\sqrt{A}))$ and $\int_0^t ds \int_0^s U(r)dr$ belongs to $L(H,D(A))$;

ii) for all $x \in H$ the mappings $t \to \sqrt{A}\int_0^t U(s)x\ ds$ and $t \to A\int_0^t ds \int_0^s U(r)x\ dr$ are continuous on $[0, +\infty[$;

iii) for all $t \geqq 0$, $\int_0^t U(s)\ ds$ belongs to $L(D(\sqrt{A});D(A))$;

iv) for all $x \in D(\sqrt{A})$ the mapping $t \to A\int_0^t U(s)x\ ds$ is continuous on $[0,+\infty[$;

v) for all $x \in D(\sqrt{A})$ and $t \geqq 0$ we have

$$U(t)x = x - \int_0^t k(t-s)A(\int_0^s U(r)x\ dr)ds.$$

In addition, if $k' \in L^1_{loc}([0,+\infty[)$, then

vi) for all $t \geq 0$, $\int_0^t k(t-s)U(s)ds$ belongs to $L(H,D(\sqrt{A}))$;

vii) for all $x \in H$ the mapping $t \to \sqrt{A} \int_0^t k(t-s)U(s)xds$ is continuous on $[0, +\infty[$;

viii) for all $t \geq 0$, $\int_0^t k(t-s)U(s)ds$ belongs to $L(D(A),D(A^{3/2}))$;

ix) for all $x \in D(A)$ the mapping $t \to A^{3/2} \int_0^t k(t-s)U(s)xds$ is continuous on $[0,+\infty[$.

Proof. In view of Corollary 1.2, for $a > 0$ there exists $\varepsilon > 0$ such that $\bar{k}(t) = e^{-(a+\omega)t}k(t) - \varepsilon e^{-at}$ is a positive kernel.

Let $x \in D(A)$, $t \geq 0$ and $0 \leq \tau \leq r \leq t$; then $w(\tau) = e^{-(a+\omega)\tau}U(\tau)x$ satisfies

$$w'(\tau) = -(a+\omega)\ w(\tau) - \int_0^\tau e^{-(a+\omega)(\tau-s)}k(\tau-s)Aw(s)ds$$

$$w(0) = x.$$

Integrating over $[0,r]$, we get

$$w(r) = x - (a+\omega) \int_0^r w(s)ds - \int_0^r e^{-(a+\omega)(r-s)}k(r-s)A(\int_0^s w(\tau)d\tau)ds,$$

from which it follows

$$\frac{d}{dr}||\sqrt{A} \int_0^r w(s)ds||^2 = 2\mathrm{Re} \langle x, A\int_0^r w(s)ds\rangle - 2(a+\omega) ||\sqrt{A} \int_0^r w(s)ds||^2 -$$
$$- 2\mathrm{Re} \langle\int_0^r e^{-(a+\omega)(r-s)}k(r-s)A(\int_0^s w(\tau)d\tau)ds,\ A \int_0^r w(s)ds\rangle.$$

Integrating this identity over $[0,t]$ and taking into account that $\bar{k}$ is a positive kernel, we obtain

$$\| \sqrt{A} \int_0^t w(s)ds||^2 \leq 2\mathrm{Re} \langle x, A\int_0^t dr \int_0^r w(s)ds\rangle -$$
$$-2\varepsilon\mathrm{Re} \int_0^t \langle\int_0^r e^{-a(r-s)}A(\int_0^s w(\tau)d\tau)ds,\ A\int_0^r w(s)ds\rangle\ dr. \tag{2.7}$$

Setting $v(r) = \int_0^r e^{-a(r-s)}(\int_0^s w(\tau)d\tau)ds$, $0 \leqq r \leqq t$, and using the identity

$$v'(r) = -av(r) + \int_0^r w(\tau)d\tau, \tag{2.8}$$

from (2.7) we get

$$\| \sqrt{A} \int_0^t w(s)ds \|^2 + \varepsilon \| Av(t) \|^2 + 2 \frac{\varepsilon a}{t} \| \int_0^t Av(\tau)d\tau \|^2 \leqq$$

$$2Re < x, Av(t)> + 2a \; Re < x, A\int_0^t v(\tau)d\tau > \leqq$$

$$\frac{2}{\varepsilon}(1 + ta) \|x\|^2 + \frac{\varepsilon}{2} \|Av(t)\|^2 + \frac{\varepsilon a}{2t} \| \int_0^t Av(\tau)d\tau \|^2 . \tag{2.9}$$

From the latter estimate we obtain

$$\| \sqrt{A} \int_0^t e^{-(a+\omega)r} U(r)x \, dr \| \leqq (\frac{2}{\varepsilon}(1+ta))^{1/2} \|x\|$$

On the other hand, integrating by parts we have

$$\| \sqrt{A} \int_0^t e^{-(a+\omega)r} U(r)x \, dr \| \geqq e^{-(a+\omega)t} \| \sqrt{A} \int_0^t U(r)x \, dr \| -$$

$$- (a+\omega) \| \int_0^t e^{-(a+\omega)r} (\sqrt{A} \int_0^r U(s)x \, ds) \, dr \| .$$

Comparing the two last estimates, one gets

$$e^{-(a+\omega)t} \| \sqrt{A} \int_0^t U(r)x \, dr \| \leqq (\frac{2}{\varepsilon}(1+ta))^{1/2} \|x\| +$$

$$+ (a+\omega) \int_0^t e^{-(a+\omega)r} \| \sqrt{A} \int_0^r U(s)xds \| dr,$$

and now a straightforward calculation involving Gronwall's lemma gives

$$\| \sqrt{A} \int_0^t U(r)x \, dr \| \leqq (\frac{2}{\varepsilon}(1+ta))^{1/2} e^{2(a+\omega)t} \|x\| . \tag{2.10}$$

In addition, (2.9) yields

$$\frac{\varepsilon}{2} \|Av(t)\|^2 + \frac{3\varepsilon}{2t} a \|A \int_0^t v(s)ds \|^2 \leqq \frac{2}{\varepsilon}(1+ta) \|x\|^2,$$

from which, taking into account (2.8), we have

$$\| A \int_0^t dr \int_0^r e^{-(a+\omega)s} U(s)x ds\|^2 \leq \frac{8}{\varepsilon^2} (1+ta)(1+t/3a) \|x\|^2.$$

Repeating the same steps that led us to (2.10), we get

$$\|A \int_0^t dr \int_0^r U(s)x\, ds\| \leq \frac{2}{\varepsilon} \sqrt{2(1+ta)(1+t/3a)} e^{3(a+\omega)t} \|x\| .$$

Therefore, as D(A) is dense in H, i) and ii) are completely proved.
iii) and iv) follow directly by (2.10) and ii), because of the identity

$$A \int_0^t U(s)x\, ds = \sqrt{A} \int_0^t U(s) \sqrt{A}x\, ds, \; t \geq 0$$

which holds for $x \in D(\sqrt{A})$.

Finally, it is easy to check v) when $x \in D(A)$; as D(A) is dense in $D(\sqrt{A})$ and iii) holds, one obtains v).

Suppose now $k' \in L^1_{loc}([0,+\infty[)$. Let $x \in D(A)$ and $t \geq 0$; integrating by parts and using (2.10), one gets

$$\| \sqrt{A} \int_0^t k(t-s)U(s)x\, ds\| = \| \int_0^t k(t-s)\frac{d}{ds} (\sqrt{A} \int_0^s U(r)x\, dr)ds\| \leq$$

$$|k(0)|\ \|\sqrt{A} \int_0^t U(r)x\, dr\| + \int_0^t |k'(t-s)|\ \|\sqrt{A} \int_0^s U(r)x\, dr\|\, ds \leq$$

$$(|k(0)| + \int_0^t |k'(s)|ds)\ (\frac{2}{\varepsilon} (1+ta))^{1/2} e^{2(a+\omega)t} \|x\| . \tag{2.11}$$

Therefore, as D(A) is dense in H, vi) and vii) are proved.

Finally, as $D(A^{3/2})$ is dense in D(A), by using (2.11) we get viii) and ix).

To end this section, we give a regularity result for the strict solution of (2.1).

COROLLARY 2.5. If $k' \in L^1_{loc}([0,\infty[)$, then for $x \in D(A)$ the strict solution U(t)x of (2.1) satisfies:

$$U'(t)x \in C([0,+\infty[;\ D(\sqrt{A})).$$

Proof. The assertion follows obviously from Proposition 2.4, ix).

3. THE NONHOMOGENEOUS PROBLEM

In this section we are concerned with the initial value problem

$$\begin{aligned} u'(t) &= -\int_0^t k(t-s)\, Au(s)\, ds + f(t), \quad t \geqq 0 \\ u(0) &= x \end{aligned} \tag{3.1}$$

where k and A satisfy (2.2) and (2.3), and $f\colon [0, +\infty[\to H$ verifies suitable assumptions.

In particular we shall check that the variation of parameters formula

$$u(t) = U(t)x + \int_0^t U(t-s)f(s)ds, \; t \geqq 0 \tag{3.2}$$

gives the solution of (3.1), where $\{U(t)\}_{t \geqq 0}$ is the resolvent operator for the homogeneous problem (2.1) (see Section 2).

Before proceeding with existence theorems, we introduce some preliminary notations and an auxiliary result.

In the sequel, for $n \in \mathbb{N}$, $A_n = n - n^2R(n,A) = nAR(n,A) \in L(H)$ will denote the Yosida approximation of A. It is known that A_n is a self-adjoint and positive operator. We consider problem (2.1) with A replaced by A_n; in view of Theorem 2.3 there exists a unique $(1,\omega)$-resolvent operator $\{U_n(t)\}_{t \geqq 0}$.

We begin proving a lemma which will be used in the following.

LEMMA 3.1. For all $x \in H$ we have:

$$\lim_{n \to +\infty} U_n(t)x = U(t)x$$

uniformly on every compact set of $[0, +\infty[$.

Proof. Let $x \in D(A^2)$ and $\gamma > \omega$; we have (see [10] Remark 7)

$$U_n(t)x - U(t)x = \frac{1}{2\pi i} \int_{\gamma - i\infty}^{\gamma + i\infty} e^{pt}((p+\hat{k}(p)A_n)^{-1} - (p+\hat{k}(p)A)^{-1})x \; dp =$$

$$= \frac{1}{2\pi i}\int_{\gamma - i\infty}^{\gamma + i\infty} e^{pt}\,\hat{k}(p)\,(p+\hat{k}(p)A_n)^{-1}(p+\hat{k}(p)A)^{-1}dp\,(A_n x - Ax) =$$

$$= \int_0^t k(t-s)\Big(\int_0^s U_n(s-r)U(r)(A_n x - Ax)dr\Big)ds.$$

Therefore,

$$\|U_n(t)x - U(t)x\| \leq \int_0^t ds|k(t-s)|\int_0^s \|U_n(s-r)U(r)(A_n x - Ax)\|\, dr \leq$$

$$\int_0^t |k(t-s)| \int_0^s e^{\omega(s-r)}e^{\omega r}\, dr\, ds\|A_n x - Ax\| \leq$$

$$(te^{\omega t}\int_0^t |k(s)|ds)\, \|A_n x - Ax\|\ ;$$

hence $\lim_{n\to+\infty} U_n(t)x = U(t)x$ uniformly on every compact set of $[0, +\infty[$.
Finally, as $D(A^2)$ is dense in H (see [10], Lemma 5), the proof is complete.

In addition, for $T > 0$, $f \in L^1([0,T];H)$ and $x \in H$, the solutions of the nonhomogeneous problems

$$\begin{aligned} u_n'(t) &= -\int_0^t k(t-s)A_n u_n(s)ds + f(t), \quad t \in [0,T] \\ u_n(0) &= x \end{aligned} \qquad (3.3)_n$$

are given by

$$u_n(t) = U_n(t)x + \int_0^t U_n(t-s)f(s)ds,\ t \in [0,T]. \qquad (3.4)_n$$

We say that u is a strict solution of (3.1) if $u \in C^1([0, +\infty[;H) \cap C([0,+\infty[;D(A))$ and verifies (3.1).

Let us define the strong solution of problem (3.1).

<u>DEFINITION 3.2.</u> Let $T > 0$, $f \in L^1([0,T];H)$ and $x \in H$. We say that $v: [0,T] \to H$ is a strong solution of (3.1) in $L^1([0,T];H)$, if there exists a sequence $\{v_n\}_{n\in\mathbb{N}}$ in $W^{1,1}([0,T];H) \cap L^1([0,T];D(A)$ such that

i) $v_n \to v$ in $L^1([0,T];H)$;

ii) $v_n'(t) + \int_0^t k(t-s)Av_n(s)\,ds \to f(t)$ in $L^1([0,T];H)$;

iii) $v_n(0) \to x$ in H.

<u>THEOREM 3.3.</u> If $T > 0$, f belongs to $L^1([0,T];H)$ and x belongs to H, then

1) the function u defined by (3.2) is the unique strong solution of (3.1) in $L^1([0,T];H)$;
2) $u \in C([0,T];H)$, $u(0) = x$;
3) $u_n \to u$ in $C([0,T];H)$, where $\{u_n\}_{n\in\mathbb{N}}$ is the sequence of Yosida approximations given by $(3.4)_n$.

<u>Proof.</u> 1) Setting, for every $n \in \mathbb{N}$,

$$v_n(t) = nR(n,A)u(t),\ f_n(t) = nR(n,A)f(t),\ x_n = nR(n,A)x,$$

it is clear that $v_n \in W^{1,1}([0,T];H) \cap L^1([0,T];D(A))$, $v_n \to u$ in $L^1([0,T];H)$ and $v_n(0) = x_n \to x$ in H.

Furthermore, one may easily check the identity

$$v_n(t) = U(t)x_n + \int_0^t U(t-s)f_n(s)ds,\ n \in \mathbb{N},\ t \in [0,T].$$

As x_n and f_n belong to $D(A)$, v_n is differentiable, and

$$v_n'(t) = -\int_0^t k(t-s)Av_n(s)ds + f_n(t),\ t \in [0,T].$$

Therefore $v_n'(t) + \int_0^t k(t-s)Av_n(s)ds = f_n(t) \to f(t)$ in $L^1([0,T];H)$, so that u is a strong solution of (3.1).

We now prove the uniqueness of the strong solution. Let v be a strong solution of (3.1) and let $\{v_n\}_{n\in\mathbb{N}}$ be a sequence satisfying i), ii), iii) of Definition 3.2. Setting $f_n(t) = v_n'(t) + \int_0^t k(t-s)Av_n(s)ds$ and $x_n = v_n(0)$, we have, for $t \geq 0$ and $s \in [0,t]$,

$$\frac{d}{ds}(U(t-s)v_n(s)) = \int_0^{t-s} k(t-s-r)AU(r)v_n(s)dr -$$

$$- U(t-s)\int_0^s k(s-r)Av_n(r)dr + U(t-s)f_n(s).$$

Integrating over $[0,t]$ and then letting $n \to +\infty$, one obtains $v = u$.

2) Observe that for $t \in [0,T]$ and $n, m \in \mathbb{N}$ we have

$$\|v_n(t) - v_m(t)\| \leq e^{\omega t}\|x_n - x_m\| + \int_0^t e^{\omega(t-s)}\|f_n(s)-f_m(s)\|\,ds;$$

therefore $\{v_n\}_{n\in\mathbb{N}}$ is a Cauchy sequence in $C([0,T];H)$, which implies $u \in C([0,T];H)$. In addition, $v_n(0) \to u(0)$, and hence $u(0) = x$.

3) For $n \in \mathbb{N}$ and $t \in [0,T]$ it holds

$$\|u_n(t)-u(t)\| \leq \|U_n(t)x-U(t)x\| + \int_0^t \|U_n(t-s)f(s)-U(t-s)f(s)\|\,ds,$$

and by Lemma 3.1 the assertion follows.

THEOREM 3.4. If $f \in W^{1,1}_{loc}([0, +\infty[;H)$ and $x \in D(A)$, then u defined by (3.2) is the unique function satisfying

1) $u \in C([0, +\infty[;D(\sqrt{A})) \cap C^1([0,+\infty[;H)$;

2) $t \to \int_0^t k(t-s)u(s)ds \in C([0,+\infty[;D(A))$;

3) $$\begin{cases} u'(t) = -A\left(\int_0^t k(t-s)u(s)ds\right) + f(t), & t \geq 0 \\ u(0) = x. \end{cases}$$

Proof. It suffices to prove that u satisfies 1), 2), 3) on $[0,T]$ for all $T > 0$. By Theorem 3.3, u is the unique strong solution of (3.1) in $L^1([0,T];H)$. First of all we observe that

$$u(t) = U(t)x + \int_0^t U(s)f(0)ds + \int_0^t \left(\int_0^{t-s} U(r)f'(s)dr\right)ds,$$

hence by Proposition 2.4,ii) u belongs to $C([0,T];D(\sqrt{A}))$.

Let now w be the strong solution of the problem

$$w'(t) = -\int_0^t k(t-s)Aw(s)\,ds + f'(t) - k(t)Ax,\ t \in [0,T]$$

$$w(0) = f(0)$$

(see Theorem 3.3), and let w_n be the solution of the problem

$$w_n'(t) = -\int_0^t k(t-s)A_n w_n(s)ds + f'(t) - k(t)A_n x,\quad t \in [0,T]$$

$$w_n(0) = f(0).$$

In view of $(3.4)_n$ we have

$$w_n(t) = U_n(t)f(0) + \int_0^t U_n(t-s)f'(s)ds - \int_0^t U_n(t-s)k(s)A_n x\,ds,$$

and by a straightforward calculation, one has $w_n(t) = u_n'(t)$, where u_n is the Yosida approximation of u. By 3) of Theorem 3.3, $u_n \to u$ and $w_n \to w$ in $C([0,T];H)$; so that u belongs to $C^1([0,T];H)$ and $u_n' = w_n \to u'$ in $C([0,T];H)$. In addition,

$$\int_0^t k(t-s)nR(n,A)u_n(s)ds \to \int_0^t k(t-s)u(s)ds \text{ in } C([0,T];H),$$

$$-A\left(\int_0^t k(t-s)nR(n,A)u_n(s)ds\right) = -\int_0^t k(t-s)A_n u_n(s)ds =$$

$$= u_n'(t) - f(t) \to u'(t) - f(t) \text{ in } C([0,T];H),$$

and, since A is closed, the function $t \to \int_0^t k(t-s)u(s)ds$ belongs to $C([0,T];D(A))$ and $-A(\int_0^t k(t-s)u(s)ds) = u'(t)-f(t)$.

Finally, uniqueness of the solution satisfying 1), 2), 3) follows from uniqueness of the strong solution.

<u>THEOREM 3.5.</u> If f belongs to $W^{1,1}_{loc}([0,+\infty[;D(\sqrt{A}))$ and x belongs to D(A), then u given by (3.2) is the strict solution of (3.1). Moreover, if k' is in $L^1_{loc}([0,+\infty[)$, then u' is in $C([0,+\infty[;D(\sqrt{A}))$.

<u>Proof.</u> It is easy to check that for $t \geq 0$

$$u(t) = U(t)x + \int_0^t U(s)f(0)ds + \int_0^t ds \int_0^{t-s} U(r)f'(s)dr, \tag{3.5}$$

hence, by Proposition 2.4, iv), $u \in C([0,+\infty[;D(A))$. Therefore the first statement of the theorem follows from Theorem 3.4.

In addition, from (3.5) we get

$$u'(t) = U'(t)x + U(t)f(0) + \int_0^t U(t-s)f'(s)ds, \quad t \geq 0;$$

so that, if k' is in $L^1_{loc}([0,+\infty[)$, then u' belongs to $C([0,+\infty[;D(\sqrt{A}))$ due to Corollary 2.5.

<u>THEOREM 3.6</u>. If f belongs to $L^1_{loc}([0,+\infty[;D(A))$ and x belongs to $D(A)$, u given by (3.2) is in $C([0,+\infty[;D(A)) \cap W^{1,1}_{loc}([0,+\infty[;H)$ and verifies (3.1). Moreover, if k' is in $L^1_{loc}([0,+\infty[)$, then u' is in $L^1_{loc}([0,+\infty[;D(\sqrt{A}))$.

<u>Proof</u>. u obviously belongs to $C([0,+\infty[;D(A))$; in addition, u is differentiable and

$$u'(t) = -\int_0^t k(t-s)AU(s)x\,ds - \int_0^t ds \int_s^t k(t-r)AU(r-s)f(s)dr + f(t) =$$

$$= -\int_0^t k(t-s)AU(s)x\,ds - \int_0^t k(t-r)A(\int_0^r U(r-s)f(s)ds)dr + f(t) =$$

$$= -\int_0^t k(t-s)Au(s)ds + f(t), \quad t \geq 0. \tag{3.6}$$

Suppose now $k' \in L^1_{loc}([0,+\infty[)$; as (3.6) yields

$$u'(t) = -\int_0^t k(t-s)AU(s)x\,ds - \int_0^t ds \int_0^{t-s} k(t-s-r)U(r)Af(s)dr + f(t),$$

then, due to Corollary 2.5 and Proposition 2.4, ix) u' is in $L^1_{loc}([0,+\infty[;D(\sqrt{A}))$.

<u>THEOREM 3.7</u>. If x belongs to $D(A)$, f belongs to $W^{2,1}_{loc}([0,+\infty[;H)$ and $f(0) \in D(\sqrt{A})$, then u given by (3.2) is the strict solution of (3.1). Moreover, if k' is in $L^1_{loc}([0,+\infty[)$, then u' is in $C([0,+\infty[;D(\sqrt{A}))$.

<u>Proof</u>. It is easy to check that

$$u(t) = U(t)x + \int_0^t U(r)f(0)dr + \int_0^t dr \int_0^r U(s)f'(0)\,ds +$$

$$+ \int_0^t ds \int_0^{t-s} d\tau \int_0^\tau U(r)f''(s)dr, \quad t \geq 0 \tag{3.7}$$

hence by Proposition 2.4,ii) $u \in C([0,+\infty[;D(A))$. Therefore the first statement of the theorem follows from Theorem 3.4. In addition, from (3.7) one gets

$$u'(t) = U'(t)x + U(t)f(0) + \int_0^t U(r)f'(0)dr + \int_0^t ds \int_0^{t-s} U(r)f''(s)dr;$$

so that, if k' is in $L^1_{loc}([0,+\infty[)$, then due to Corollary 2.5 and Proposition 2.4, ii), we have $u' \in C([0, +\infty[;D(\sqrt{A}))$.

4. THE SECOND ORDER PROBLEM

We will study in this section an integrodifferential equation of the second order:

$$\begin{aligned} u''(t) &= - k(0)Au(t) - \int_0^t k'(t-s)Au(s)ds + g(t), \quad t \geq 0 \\ u(0) &= x \\ u'(0) &= y \end{aligned} \tag{4.1}$$

where k and A satisfy (2.2), (2.3) and in addition $k' \in L^1_{loc}([0,+\infty[)$. Suitable assumptions will be made on $g: [0,+\infty[\to H$.

First we prove an existence result for the homogeneous problem.

THEOREM 4.1. If x belongs to $D(A)$ and y belongs to $D(\sqrt{A})$, then there exists a unique function $u \in C([0,+\infty[;D(A)) \cap C^1([0,+\infty[;D(\sqrt{A})) \cap C^2([0,+\infty[;H)$ that fulfills

$$\begin{aligned} u''(t) &= - k(0)Au(t) - \int_0^t k'(t-s)Au(s)ds, \quad t \geq 0 \\ u(0) &= x \\ u'(0) &= y \end{aligned} \tag{4.2}$$

Proof. We observe that a solution u of (4.2) satisfies

$$\begin{aligned} u'(t) &= - \int_0^t k(t-s)Au(s)ds + y, \quad t \geq 0 \\ u(0) &= x \end{aligned} \tag{4.3}$$

By Theorem 3.5 there exists a unique solution $u \in C([0,+\infty[;D(A)) \cap C^1([0,+\infty[;D(\sqrt{A}))$ of (4.3). In addition, from (4.3) it follows that u' is differentiable and

$$u''(t) = - k(0)Au(t) - \int_0^t k'(t-s)Au(s)ds \in C([0,+\infty[;H);$$

so the proof is complete.

On the other hand, the following existence theorem holds for problem (4.1).

THEOREM 4.2. Let $x \in D(A)$, $y \in D(\sqrt{A})$ and let g be in $L^1_{loc}([0,+\infty[;D(\sqrt{A}))$. Then there exists a unique function $u \in C([0,+\infty[;D(A)) \cap C^1([0,+\infty[;D(\sqrt{A})) \cap W^{2,1}_{loc}([0,+\infty[;H)$, that satisfies (4.1). If, in addition, g belongs to $C([0,+\infty[;H)$, then also u" belongs to $C([0,+\infty[;H)$.

Proof. It is clear that a solution u of (4.1) satisfies

$$\begin{aligned} u'(t) &= - \int_0^t k(t-s)Au(s)ds + y + \int_0^t g(s)ds, \quad t \geqq 0 \\ u(0) &= x \end{aligned} \tag{4.4}$$

Thanks again to Theorem 3.5, there exists a unique solution u of (4.4), which belongs to $C([0,+\infty[;D(A)) \cap C^1([0,+\infty[;D(\sqrt{A}))$. Moreover, from (4.4) it follows that u' is differentiable and

$$u''(t) = - k(0)Au(t) - \int_0^t k'(t-s)Au(s)ds + g(t) \in L^1_{loc}([0,+\infty[;H).$$

The last statement of the theorem obviously follows from last equality.

THEOREM 4.3. Let $x \in D(A)$, $y \in D(\sqrt{A})$ and $g \in W^{1,1}_{loc}([0,+\infty[;H)$. Then there exists a unique function $u \in C([0,+\infty[;D(A)) \cap C^1([0,+\infty[;D(\sqrt{A})) \cap C^2([0,+\infty[;H)$ that satisfies (4.1).

Proof. A solution u of (4.1) satisfies (4.4). Thanks to Theorem 3.7, there exists a unique solution $u \in C([0,+\infty[;D(A)) \cap C^1([0,+\infty[;D(\sqrt{A}))$ of (4.4). Moreover, from (4.4) it follows that u' is differentiable and

$$u''(t) = -k(0)Au(t) - \int_0^t k'(t-s)Au(s)ds + g(t) \in C([0,+\infty[;H).$$

References

[1] V. Barbu, Nonlinear Semigroups and Differential Equations in Banach Spaces, Noordhoff Ed. Academiei, Leyden Bucharest (1976).

[2] V. Barbu, Nonlinear Volterra Integrodifferential Equations in Hilbert Spaces, in Conferenze Sem. Matematico Univ. di Bari (1976).

[3] V. Barbu - M.A. Malik, Semilinear Integrodifferential Equations in Hilbert Space, J. Math. Anal. Appl. 67 (1979), 452-475.

[4] R.W. Carr - K.B. Hannsgen, A nonhomogeneous integrodifferential equation in Hilbert space, SIAM J. Math. Anal. 10 (1979), 961-984.

[5] M.G. Crandall - S.O. Londen - J.A. Nohel, An Abstract Nonlinear Volterra Integrodifferential Equation, J. Math. Anal. Appl. 64 (1978), 701-735.

[6] C. Corduneanu, Integral Equations and Stability of Feedback Systems, Academic Press, New York (1973).

[7] C.M. Dafermos, An abstract Volterra equation with applications to linear viscoelasticity, J. Diff. Eq. 7 (1970), 554-569.

[8] C.M. Dafermos, Asymptotic Stability in Viscoelasticity, Arch. Rat. Mech. Anal. 37 (1970), 297-308.

[9] C.M. Dafermos - J.A. Nohel, Energy methods for nonlinear hyperbolic Volterra integrodifferential equations, Comm. in Partial Diff. Eq. 4(3) (1979), 219-278.

[10] G. Da Prato - M. Iannelli, Linear Integrodifferential Equations in Banach Spaces, Rend. Sem. Mat. Padova, 62 (1980), 207-219.

[11] H. Engler, A version of the chain rule and integrodifferential equations in Hilbert spaces, SIAM J. Math. Anal. 13 (1982), 801-810.

[12] R.C. Grimmer - F. Kappel, Series expansions for resolvents of Volterra integrodifferential equations in Banach spaces, J. Math. Anal. Appl. 15 (1984), 595-604.

[13] A. Halanay, On the Asymptotic Behaviour of the Solutions of an Integrodifferential Equation, J. Math. Anal. Appl. 10 (1965), 319-324.

[14] J.J. Levin, The asymptotic behaviour of the solution of a Volterra equation, Proc. Amer. Math. Soc. 14 (1963), 534-541.

[15] S.O. Londen, An existence result on a Volterra equation in a Banach space, Trans. Amer. Math. Soc. 235 (1978), 285-304.

[16] R.C. MacCamy - J.S.W. Wong, Stability theorems for some functional equations, Trans. Amer. Math. Soc. 164 (1972), 1-37.

[17] R.C. MacCamy, An integrodifferential equation with applications in heat flow, Q. Appl. Math. 35 (1977), 1-19.

[18] R.C. MacCamy, A model for one-dimensional, nonlinear viscoelasticity, Q. Appl. Math. 35 (1977), 21-33.

[19] J.A. Nohel - D.F. Shea, Frequency domain methods for Volterra equations, Advances in Math. 22 (1976), 278-304.

[20] D. Sforza, On the Hille-Yosida theorem for integrodifferential equations in Banach spaces, Boll. U.M.I. Anal. Funz. Appl. Serie VI, Vol. V-C,N.1-1986, 169-173.

Daniela Sforza
Dipartimento di Matematica
Università di Pisa
via Buonarroti 2
I-56100 Pisa
Italy.

V. VESPRI

Linear stochastic integrodifferential equations

ABSTRACT

We study existence and regularity of solutions to a class of abstract stochastic integrodifferential parabolic equations in suitable interpolation spaces. Under different hypotheses, "mild" and strict solutions are obtained. The method consists in considering the stochastic and integro-differential terms as perturbations of an abstract parabolic equation.

1. INTRODUCTION

The purpose of this paper is to study the existence, uniqueness and regularity of the solutions of the following class of stochastic integro-differential equations

$$\begin{aligned} dy &= \left(Ay + \int_0^t K(s,t)y(s)ds + f\right)dt + \\ &+ \sum_{i=1}^{n} \left(B_i y + \int_0^t T_i(s,t)y(s)ds + f_i\right)dW_i(t) \\ y(0) &= x \end{aligned} \tag{1.1}$$

where $A, B_1, \ldots, B_n$ are linear operators (possibly unbounded) in a Hilbert space H; $K(s,t)_{0\leq s\leq t\leq T}$ and $T_i(s,t)_{0\leq s<t\leq T}$ $i = 1,\ldots,n$ are closed linear operators in H and $w_1,\ldots,w_n$ are real mutually independent Brownian motions in a probability space (Ω, E, P). Stochastic evolution equations with bounded operators in Hilbert space have been extensively considered by several authors (see, e.g., [2], [6], [7], [9], [10], [15], [18], [19], [24], [27], [32]). Also, deterministic integrodifferential equations (which describe many meaningful physical situations) have been widely studied (see, e.g. [13]). However, only recently, the importance of the stochastic integro-differential equations has been recognized (for physical motivations, see [3]) : Kannan studied random integrodifferential evolution equations in the finite dimensional case [25], and Kannan and Barucha-Reid [3] analyzed a

class of stochastic integrodifferential equation of Volterra type with unbounded coefficients in Hilbert space. In this paper stochastic integro-differential Equation (1.1) is studied and the existence and uniqueness of suitable solutions is proved. Assume that A is an infinitesimal generator of an analytic semigroup of operators and write Equation (1.1) in the following mild form:

$$y(t) = e^{tA}x + \int_0^t e^{(t-s)A}\left[\int_0^s K(\rho,s)y(\rho)d\rho + f(s)\right]ds + \sum_{i=1}^n \int_0^t e^{(t-s)A}\left[B_i y(s) + \int_0^s T_i(\rho,s)y(\rho)d\rho + f_i(s)\right]dW_i(s) \quad (1.2)$$

By using some results for the stochastic convolution (see [8]) it is possible to apply the contraction principle in $M_W^2(0,T;X)$ and prove that existence and uniqueness of the solution of the Eq. (1.2), where X is a suitable interpolation space and $M_W^2(0,T;X)$ is the set of all square integrable H valued stochastic processes adapted to $W = (w_1,\ldots,w_n)$. Since the natural spaces in which it is convenient to study Equation (1.2) are the extra-polation and interpolation ones, in Section 2 the definition and properties of these functional spaces are recalled. In Section 3 the existence and uniqueness of mild solution of (1.2) in $M_W^2(0,T;D_A(\ ,2))$ $\frac{1}{2} < \theta < 1$ is proved. In Section 4 we study the regularity in $M_W^2(0,T;D_A(\frac{1}{2},2))$ under more restrictive assumptions. Strict solutions are analyzed in Section 5, while Section 6 is devoted to a special class of integrodifferential stochastic equations that can be reduced into determinsitic ones by a suitable trans-formation.

2. INTERPOLATION AND EXTRAPOLATION SPACES

SPACE $D_A(\theta,2)$

Let A be a linear operator in H. Suppose that

$$\text{A is the infinitesimal generator of an analytic semigroup } e^{tA} \text{ such that } |e^{tA}| \leq Me^{-\ t} \text{ for some } M > 0,\ \omega > 0 \quad (2.1)$$

Denote by

$$D_A(\theta,2) = \{x \in H: \int_0^{+\infty} h^{1-2\theta}|Ae^{hA}x|^2 dh = |x|^2_{D_A(\theta,2)} < +\infty\}.$$

We note that $D_A(\theta,2)$ is a Banach space with norm $|\cdot|_{D_A(\theta,2)}$. The main results about these interpolation spaces were obtained by Lions-Peetre [30], Grisvard [23], Butzer and Berens [4]. Set

$$D_A(\theta+1,2) = \{x \in D(A) : Ax \in D_A(\theta,2)\}$$

In the sequel we recall some properties of these spaces, for the proof of which we refer the reader to the Appendix of [9].

PROPOSITION 2.1. Let $x \in D_A(\theta,2)$ and $u(t) = e^{tA}x$. If $\theta = \frac{1}{2}$ we have $u \in L^2(0,T;D(A))$ and if $\theta \in]0,\frac{1}{2}[$ then $u \in L^2(0,T;D_A(\theta + \frac{1}{2},2))$.

PROPOSITION 2.2. Let A be a negative self-adjoint operator. Then for any $\theta \in]0,1[$ we have $D_A(\theta,2) = D((-A)^\theta)$ and

$$|x|^2_{D_A(\theta,2)} = 2^{2\theta-1}\,\Gamma(2-2\theta)\,|(-A)^\theta x|^2, \quad \forall x \in D_A(\theta,2)$$

where $(-A)^\theta = \int_{-\infty}^{0} (-\lambda)^\theta \, d\,E_\lambda(x)$ and E_λ is the spectral family projectors associated with A (see [21]).

PROPOSITION 2.3. (see [26] and [28]). If A is dissipative, $D_A(\theta,2) = D(-A)^\theta$ $\theta \in]0,1[$ where $-A^\theta$ is the fractional power of order θ of $-A$.

EXTRAPOLATION SPACES

The extrapolation spaces were introduced and studied by Da Prato-Grisvard (see [12] and [9]). We refer to these works for the main definitions and properties. Set

$$G_A = \{(x,y) : x \in D(A),\ y = Ax\}$$

$$\Gamma = (H \times H)/G_A$$

If j is the embedding of $H \to \Gamma$ defined by

$$j : H \to \Gamma\ ,\ x \to (0,x)^\sim$$

(where $(0,x)^{\sim}$ is the coset of $(0,x)$), we define a linear operator $\tilde{A}$ on Γ by

$$D(\tilde{A}) = j(H)$$

$$\tilde{A}(0,x) = -(x,0)^{\sim}$$

The following properties hold:

PROPOSITION 2.4.

1) $\tilde{A}$ is an extension of A. In fact $\tilde{A}(J(x)) = j(A(x))$ $\forall x \in D(A)$

2) $\tilde{A}$ is a generator of an analytic semigroup on Γ. In addition $D(\tilde{A}^2) = J(D(A))$

3) $\forall\theta \in]0,1[$ the following equality holds

$$D_{\tilde{A}}(1 + \theta,2) = j(D_A(\theta,2))$$

We further denote

$$D_A(\theta-1,2) = D_{\tilde{A}}(\theta,2) \qquad \theta \in]0,1[,$$

identify $j(H)$ with Γ and call $D_A(\theta-1,2)$ the extrapolation space of order $\theta-1$. If A is a self-adjoint operator the extrapolation space $D_A(\theta-1,2)$ coincides with the closure of the space H in the seminorm $[(-A)^{\theta-1}x]$.

EXAMPLE

If A is a second order elliptic operator defined on a bounded domain $\Omega \in \mathbb{R}^N$ under Dirichlet boundary conditions, then [29]

$$D_A(\theta,2) = \begin{cases} H^{2\theta}(\Omega) & \text{if } 0 < \theta < \frac{1}{4} \\ H_0^{2\theta}(\Omega) & \text{if } \frac{1}{4} < \theta \leq \frac{1}{2} \\ H_0^1 \cap H^2 \ (\Omega) & \text{if } \frac{1}{2} < \theta < 1 \\ H^{-2\theta}(\Omega) & \text{if } -1 < \theta < 0 \end{cases}$$

3. MILD SOLUTIONS

Now we can state the main result of this paper.

THEOREM 3.1.

Let $\theta \in]0,\frac{1}{2}[$ and assume that A fulfills condition (2.1). Also suppose that

$$x \in D_A(\theta,2) \tag{3.1}$$

$$f \in M_W^2([0,T];D_A(-\frac{1}{2}+\theta,2)) \tag{3.2}$$

$$f_i \in M_W^2([0,T];D_A(\theta,2)) \tag{3.3}$$

$B_i \in L(D_A(\theta+\frac{1}{2},2);D_A(\theta,2))$ and there exists (3.4)

$\eta \in]0,1-2\theta[$ such that $\sum_{i=1}^{n} |B_i x|_\theta^2 \leq \eta |x|^2_{\theta+\frac{1}{2}}$

There exists an extension of K, $\tilde{K}(s,t) \in L(D_A(\frac{1}{2}+\theta,2);$ $D_A(\theta-\frac{1}{2},2))$. Furthermore

$\int_{\Delta(T)} \|\hat{K}(s,t)\|^2 dsdt < \infty$ where $\Delta(r) = \{(s,t) \in \mathbb{R}^2 : 0 \leq s < t \leq r\}$

$T_i(s,t) \in L(D_A(\frac{1}{2}+\theta,2);D_A(\theta,2))$. Moreover (3.6)

$$\int_{\Delta(T)} \|T_i(s,t)\|^2 dsdt < +\infty$$

Then there exists a unique $v \in M_W^2([0,T];D_A(\frac{1}{2}+\theta,2))$ mild solution of (1.2) with continuous paths in x.

Proof. Our goal is to prove that for small t_0 the linear endomorphism T of $M_W^2([0,t_0];D_A(\frac{1}{2}+\theta,2))$ is a contraction, where

$$\begin{aligned} T_u = e^{tA}x &+ \int_0^t e^{(t-s)A}[\int_0^s K(\rho,s)u(\rho)d\rho + f(s)]ds \; + \\ &+ \sum_{i=1}^{n} \int_0^t e^{(t-s)A}[\int_0^s T_i(\rho,s)u(\rho)d\rho + B_i u(s) + f_i(s)]d\,W_i(s) \end{aligned} \tag{3.7}$$

Observe first that, by stochastic convolution, the map

$$v \to \sum_{i=1}^{n} \int_0^t e^{(t-s)A} B_i v(s) d\, W_i(s) = X(t)$$

satisfies the following estimate ([8])

$$|x(t)|_{\theta + \frac{1}{2}} \leq \frac{\eta}{1-2\theta} |v|_{\theta + \frac{1}{2}} \tag{3.8}$$

We also need an estimate of the map

$$v \to \int_0^t e^{(t-s)A} \int_0^s K(\rho,s)v(\rho)d\rho ds = Y(t)$$

Set $F(s) = \int_0^s K(\rho,s)v(\rho)d\rho \in M_W^2([0,T];D_A(-\frac{1}{2} + \theta,2))$

therefore

$$|F(s)|^2_{\theta - \frac{1}{2}} \leq \left(\int_{\Delta(t_0)} \|k(s,t)\|^2 dsdt\right)|v|^2_{\theta + \frac{1}{2}}$$

But y(t) is the solution of the equation

$$y'(t) = \tilde{A}y(t) + F(t)$$

$$y(0) = 0$$

where $\tilde{A}$ is the realization of A in the extrapolation spaces. Therefore by the classical evolution theory, we derive

$$|y(t)|^2_{\theta + \frac{1}{2}} \leq \left(\int_{\Delta(t_0)} \|K(s,t)\|^2 dsdt\right)|v|^2_{\theta + \frac{1}{2}} \tag{3.9}$$

With the same argument, it is possible to prove that also the following map

$$v(t) \to \sum_{i=1}^{n} \int_0^t e^{(t-s)A} \int_0^s T_i(\rho,s)v(\rho)d\rho\, d\, W_i(s) = Z(t)$$

satisfies a similar estimate

$$|z(t)|^2_{\theta + \frac{1}{2}} \leqq \sum_{i=1}^{n} \int_{\Delta(t_0)} \|T_i(s,t)\|^2 dsdt) |v(t)|^2_{\theta + \frac{1}{2}} \tag{3.10}$$

Therefore, from (3.8)-(3.10), it easily follows that the map T defined in (3.7) is a contraction for small t_0.

In fact we have that for every $u,v \in M^2_W([0,t_0];D_A(\theta + \frac{1}{2}))$

$$|T(u-v)|^2_{\theta + \frac{1}{2}} \leqq (\frac{n^2}{(1-2\theta)^2} + \int_{\Delta(t_0)} [\ \|K(s,t)\|^2 + \sum_{i=1}^{n} \|T_i(s,t)\|^2]ds \cdot dt) |u-v|^2_{\theta + \frac{1}{2}}$$

Now, we have proved the existence and uniqueness of a solution $\bar{v}$ of (1.2) in $[0,t_0]$. In order to extend such solution in $[t_0,2t_0]$, we consider the following map T_1 from $M^2_W([t_0,2t_0],D_A(\theta + \frac{1}{2}))$ into itself.

$$\begin{aligned} T_1 v = e^{tA}x &+ \int_0^t e^{(t-s)A} \int_0^{s \wedge t_0} K(\rho,s)\bar{v}(\rho)d\rho + \\ &+ \int_0^t e^{(t-s)A} \int_{t_0}^{t_0 \vee s} K(\rho,s)v(\rho)d\rho + \int_0^t e^{(t-s)A} f(s)ds + \\ &+ \sum_{i=1}^{n} \int_0^t e^{(t-s)A} \int_0^{s \wedge t_0} T_i(\rho,s)\bar{v}(\rho)d\rho \; d\,W_i(s) + \\ &+ \sum_{i=1}^{n} \int_0^t e^{(t-s)A} \int_{t_0}^{t_0 \vee s} T_i(\rho,s)v(\rho)d\rho \; d\rho W_i(s) + \\ &+ \sum_{i=1}^{n} \int_0^{t_0 \wedge t} e^{(t-s)A}[B_i\bar{v}+f_i]d\,W_i(s) + \sum_{i=1}^{n} \int_{t_0}^{t_0 \vee t} e^{(t-s)A}[B_i v+f_i]d\,W_i(s). \end{aligned}$$

By arguing exactly as before, it is easily checked that T_1 is a contraction. Repeating the same contraction argument and jointing the solutions obtained step by step, one easily proves the existence and uniqueness of a solution of (1.2). The continuity of solutions derives by the continuity of the stochastic convolution ([8], [27]) and by standard interpolation theory ([20]).

EXAMPLE

Assume $A = \sum_{i,j=1}^{N} a_{ij}D_{ij} + \sum_{i=1}^{N} b_i D_i + c$ is an elliptic operator with continuous coefficients defined on a domain of $\mathbb{R}^n$ under Dirichlet boundary conditions;

$$B_i = \sum_{i=1}^{N} \alpha_i D_i \text{ where } \alpha_i \in C^1(\bar{\Omega})$$

$$T_i(s,t) = \sum_{i=1}^{N} \beta_i (t-s)^{-\delta} D_i \text{ where } \beta_i \in C^1(\bar{\Omega}) \text{ and } \delta < \frac{3}{4} \tag{3.11}$$

$$K(s,t) = \sum_{i,j=1}^{N} \gamma_{ij}(t-s)^{-\delta} D_{ij} + \sum_{i=1}^{N} \gamma_i D_i + \gamma \text{ where } \gamma_{ij}, \gamma_i, \gamma \in C^1(\bar{\Omega}).$$

For small $\varepsilon > 0$ (see [11]) the integrodifferential equation

$$du = (Au + \int_0^t K(s,t)u\, ds + f)dt +$$

$$+ \sum_{i=1}^{n} (\varepsilon B_i u + \int_0^t T_i(s,t)u\, ds + f_i) d W_i(t)$$

$$u(0) = x$$

has a unique "mild" solution $u \in M_W^2([0,T]; H^{2\theta} \cap H_0^1)$.

REMARK 3.2.

The condition that A is an infinitesimal generator of an analytic semigroup of negative type can be easily removed. In fact if A is a generator of positive type, then by the transformation $y(t) = e^{\alpha t} z(t)$, one obtains that $z(t)$ solves the integrodifferential stochastic equation

$$dz = (A_1 z + \int_0^t k(s,t) e^{(s-t)\alpha} z + e^{-\alpha t} f)dt +$$

$$+ \sum_{i=1}^{n} (B_i z + \int_0^t T_i(s,t) e^{(s-t)\alpha} z\, ds + e^{-\alpha t} f_i) d W_i(t)$$

$$z(0) = x$$

where $A_1 = A - \alpha$ is a generator of negative type.

REMARK 3.3

It is possible to assume that for every $\omega \in \Omega$ there exists a linear operator $A(\omega):D(A(\omega)) \subset H \to H$. In fact the same result holds ([11]) supposing that

- the operators $A(\omega)$ have the same domain
- $\forall\omega \in \Omega$, $A(\omega)$ is an infinitesimal generator of analytic semigroups of negative type. Moreover there exists $\eta < 0$ and $M > 0$ such that for every $\omega \in \Omega$ $|A-\lambda I|^{-1} \leq M|\lambda-\eta|^{-1}$ for every $\lambda \in C$ such that Re $\lambda > \eta$
- $\forall x \in H$ $e^{tA}x$ is a measurable function. In addition it is independent of W_t $\forall t > 0$
- on $M_W^2([0,\infty[;D_A(\theta,2))$ $\theta > 0$, the following norm have to be considered

$$|||g|||_\theta^2 = \int_\Omega P(\delta\omega) \int_0^\infty dt \int_0^\infty \xi^{1-2\theta} |A(\omega)e^{\xi A(\omega)}g(t,\omega)|^2 d\xi$$

$\forall g \in M_W^2([0,T];D_A(\theta,2))$.

REMARK 3.4.

With the same argument used in [9], it is possible to study the non-autonomous equation (obtained assuming the operators A, B_i in equation (1.1) are functions of t). The thesis of Theorem 3.1 holds if we impose the below conditions on the operators A(t) and B(t).

There exists a linear operator S in H, fulfilling hypothesis (2.1), such that $D(A(t)) = D(S)$ $\forall t \in [0,T]$, and $A^{-1}(t)S$, $S^{-1}A(t)$ are continuous in H $\forall t \in [0,T]$. (3.14)

There exists $\theta \in]0, \frac{1}{2}[$ such that (3.15)

$$A \in C([0,T];L(D_S(\theta + \tfrac{1}{2},2); D_S(\theta - \tfrac{1}{2},2))$$

$$B \in C([0,T];L(D_S(\theta + \tfrac{1}{2},2); D_S(\theta,2))$$

If $\eta(t) = \sum_{i=1}^{n} |B_i(t)|^2$, then (3.16)

$$\sup_{t\in[0,1]} \eta(t) = \eta < (1-2\theta).$$

Actually the thesis of Theorem 3.1 holds for more general equations:

$$dy = (A(t)y + C(t)y + \int_0^t K(s,t)y(s)ds + f)dt +$$

$$+ \sum_{i=1}^{n} (B_i(t)y + \int_0^t T_i(s,t)y(s)ds + f_i)d\,W_i(t)$$

$$y(0) = x$$

under the following assumptions

- conditions (3.14) and (3.15) hold
- $A(t)$ is negative self-adjoint $\forall t \in [0,T]$
- $c \in C([0,T];L(D_S(\theta + \frac{1}{2},2);D_S(\theta,2))$
- There exists $\eta \in [0,1[$ such that $\forall x \in D_S(\theta + \frac{1}{2},2)$

$$\frac{1}{2}\sum_{i=1}^{n} |((-A(t))^{\theta})B_i(t)x|^2 \leq \eta|((-A(t))^{\theta+\frac{1}{2}})x|^2.$$

REMARK 3.5.

All the results of this section can be easily generalized to the case of a Brownian motion W that takes its values in another Hilbert space K (see [8]).

4. THE CASE $\theta = 0$

THEOREM 4.1. Assume that (3.2) and (3.5) hold for $\theta = 0$. In addition suppose that

A satisfies condition (2.1) (4.1)

$D(A) = D(A^*)$ (4.2)

$A + A^*$ is self-adjoint and negative (4.3)

$x \in H$ (4.4)

$f_i \in M_W^2([0,T];H)$ (4.5)

$B_i \in L(D_A(\frac{1}{2},2);H)$ and there exists $\eta \in]0,1[$ such that (4.6)

$$\frac{1}{2}\sum_{i=1}^{n}|B_i(x)|^2 + \eta \langle Ax,x\rangle \leq 0 \quad \forall x \in D(A)$$

$$T_i(s,t) \in L(D_A(\tfrac{1}{2},2);H) \text{ and in addition} \tag{4.7}$$

$$\int_{\Delta(T)} \|T_i(s,t)\|^2 dsdt < +\infty$$

Then there exists a unique $v \in M_W^2([0,T];D_A(\frac{1}{2},2)$ mild solution of (1.1).

SKETCH OF THE PROOF

Let L be a negative self-adjoint operator satisfying the following property $D(L) = D(A) = D(A^*)$ (for the existence of such an operator see [29]). Denote with Z the extrapolation space $D_{\tilde{L}}(1-\eta,2) = D_L(-\eta,2) = D_A(-\eta,2)$. Defining $D(L_1) = D(A_1) = D_A(1-\eta,2)$ and $x \quad D(A_1)\ A_1(x) = \tilde{A}(x)$ and $L_1(x) = \tilde{L}(x)$, we have $D_{A_1}(\eta,2) = H$ and $D_{A_1}(\eta + \frac{1}{2},2) = D_L(\eta + \frac{1}{2},2) = D(-L)^{1/2}) = D_A(\frac{1}{2},2)$. The thesis easily follows applying Theorem 3.1 to the Hilbert space Z.

REMARK 4.2.

It is possible to prove Theorem 4.1 under weaker assumptions than (4.3) and (4.6):

$A + A^*$ is a self-adjoint operator and there exists $\mu > 0$ such that $A + A^* - 2\mu$ is a negative operator (4.8)

$B_i \in L(D_A(\frac{1}{2},2),H)$ and there exist $\eta \in]0,1[$ and $\lambda > 0$ such that (4.9)

$$\frac{1}{2}\sum_{i=1}^{n}|B_i x|^2 + \eta \langle Ax,x\rangle \leq \lambda\, |x|^2 \quad \forall x \in D(A)$$

In fact if $y(t) = e^{\alpha t}z(t)$ $\alpha > \mu$, then $z(t)$ solves the integrodifferential equation (3.13) where $A_1 = A-\alpha$ satisfies condition (4.3). Furthermore

$$\frac{1}{2}\sum_{i=1}^{n}|B_i x|^2 + \eta \langle Ax,x\rangle \leq (\lambda-\eta\alpha)|x|^2$$

Therefore if $\alpha > \frac{\lambda}{\eta}$, also condition (4.6) holds.

REMARK 4.3.

As shown by Da Prato (Remark 3.7 of [11]) condition (4.9) is essential.

EXAMPLE

Let A be a variational elliptic operator defined on a smoothly bounded domain of $\mathbb{R}^N$. Set

$$A = \sum_{i,j=1}^{N} \frac{\partial}{\partial x_j} (a_{ij} \frac{\partial u}{\partial x_i}),\ H = L^2(\Omega),\quad D_A = H^2(\Omega) \cap H_0^1(\Omega),$$

$$Bu = \sum_{i=1}^{N} \frac{\partial}{\partial x_i} (b_i u),\ D(B) = H_0^1(\Omega) \text{ and } D_A(\tfrac{1}{2}, 2) = H_0^1(\Omega).$$

Suppose also that conditions (3.11) and (3.12) are fulfilled, then (4.6) holds if and only if

$$\sum_{i,j=1}^{N} (a_{ij} - \frac{1}{2} b_i b_j)\xi_i\xi_j \geqq \nu|\xi|^2 \text{ where } \nu > 0 \text{ and } \xi = (\xi_1,\ldots,\xi_N) \in \mathbb{R}^N.$$

We obtain the same condition applying the variational approach (see [32]).

5. THE STRICT SOLUTIONS

To have strict solutions of equation (1.1), it is necessary to make more restrictive assumptions on the operators A and B_i. Because the demonstration of the following theorem is quite similar to the previous ones, the proof is omitted.

THEOREM 5.1. Assume that $\theta \in [0,1]$ and that conditions (3.1)-(3.3) and (3.5)-(3.6) hold. Suppose in addition that

$$A \text{ is self-adjoint negative} \tag{5.1}$$

$B_i \in L(D(-A)^{\theta + \frac{1}{2}}, D((-A)^{\theta}))$ and there exists $\eta \in [0,1[$ such that

$$\frac{1}{2}\sum_{i=1}^{n} |(-A)^{\theta}B_i x|^2 \leq \eta|(-A)^{\theta + \frac{1}{2}} x|^2 \quad \forall x \in D_A(\theta + \tfrac{1}{2},2)$$

The equation (1.1) has a unique solution $y \in M_W^2([0,T];D((-A)^{\theta + \frac{1}{2}})$.

Furthermore if $\theta \geqq \frac{1}{2}$ y is a strict solution.

6. A SPECIAL CASE. REDUCTION TO A DETERMINISTIC INTEGRODIFFERENTIAL EQUATION

Consider this special case of equation (1.1)

$$dy = (Ay + \int_0^t K(s,t)y(s)ds + f)dt + BydW_t$$
$$y(0) = x \tag{6.1}$$

Where W_t is a one-dimensional Brownian motion.

As shown in [11], [16] and [31] using the Itô formula, $v = e^{-BW_t}y$ formally satisfies

$$v_t(t,\omega) = [e^{-BW_t}(A - \frac{B^2}{2})e^{BW_t}\; v + e^{-BW_t}\; f +$$
$$+ \int_0^t e^{-BW_t}\; K(s,t)e^{BW_s}\; v\; ds]dt \tag{6.2}$$
$$v(0,\omega) = x.$$

This determinisitic equation is called the robust equation associated to the equation (6.1). In the stochastic case the existence and regularity of the solutions of (6.2) has been widely studied ([5], [16], [17], [22], [33]) in the finite dimensional case and ([15], [18]) in the finite dimensional case. Suppose that

there exists a Banach space $E \subset H$ such that (6.3)

$C = A\text{-}B^2/2$ is the infinitesimal generator of an analytic semigroup in H and E (6.4)

$D(A) \subset D(B^2)$ and B is the infinitesimal generator of a strongly continuous group in H and E (6.5)

There exists an operator k such that $\forall s \in \mathbb{R}^+$ (6.6)

$$e^{sB}C\; e^{-sB}x = (C+K)x = \tilde{C}x$$

and $\tilde{C}$ is the generator of an analytic semigroup in H and E with constant domains.

$$\|1-\tilde{C}(\tau)C^{-1}(t)\|_{L(E)} \leq L|t-\tau|^{\alpha} \qquad \forall t,\tau \in [0,T] \tag{6.7}$$

$$\forall (t,s) \in \Delta(T) \quad e^{-BWt} K(s,t)e^{BWs} K(s,t)e^{BWs} = R(t,s):D_{R(t,s)} \subseteq E \to E \tag{6.8}$$

is a closed linear operator and $D_{R(t,s)} \subseteq D_{\tilde{C}}$

$$\|R(t,s)\|_{L(D_{\tilde{C}},E)} \leq \frac{k}{(t-s)^{1-\beta}}, t,s \in \Delta(t) \text{ for some } k > 0 \ \beta \in]0,1[. \tag{6.9}$$

$\forall \varepsilon > 0$ there exists $H_{\varepsilon} > 0$ such that (6.10)

$$\|R(t,s)-R(r,s)\|_{L(D_{\tilde{C}},E)} \leq H_{\varepsilon} \frac{(t-r)^{\varepsilon}}{(r-s)^{1+\varepsilon-\beta}}$$

Under assumption (6.3)-(6.10) we can now apply the result of [1].

<u>THEOREM 6.1</u>. Let $\delta \in [0,\alpha]$, $x \in D_{\tilde{C}}$, $f \in C^{\delta}([0,T];E)$ and $A(0)x + f(0) \in D(\tilde{C})$. Then

$$\|v^t\|_{C([0,T];E)} + \|\tilde{C}v\|_{C([0,T];E)} + \|Sv\|_{C([0,T];E)} \leq$$

$$\leq c\{\|C(0)x\|_E + \|f\|_{C^{\delta}([0,T];E)}$$

where $Sv(t) = \int_0^t e^{-BWt} K(s,t)e^{BWs} v\, ds$

v^t, $\tilde{C}v \in C^{\delta\wedge\varepsilon}(]0,T],E)$ $\forall \varepsilon \in]0,\beta]$

If $\tilde{C}(0)x + f(0) \in D_{\tilde{C}}(\delta\varepsilon,\infty)$ and $\varepsilon \in]0,\beta[$

then

$$\|v^t\|_{C^{\delta\wedge\varepsilon}([0,T];E)} + \|\tilde{C}v\|_{C^{\delta\wedge\varepsilon}([0,T];E)} + \|Sv\|_{C^{\varepsilon}([0,T];E)} \leq$$

$$\leq C(\varepsilon)\{\|\tilde{C}(0)x\|_E + \|f\|_{C^{\delta}([0,T];E)} + \|\tilde{C}(0)x+f(0)\|_{D_{\tilde{C}}(\delta\wedge\varepsilon,\infty)}.$$

Because the solution of the integrodifferential stochastic equation is $y = e^{BW_t}v$, the regularity of y easily follows from that of v.

EXAMPLE

Let us consider this integrodifferential stochastic partial differential equation

$$dy = [Ay + \int_0^t K(s,t)y(s)ds]dt + h(x)y \, d\, W(t)$$

$$(t,x) \in]0,T] \times \mathbb{R}^N$$

where

$$A = \sum_{i,j=1}^{N} a_{ij}D_{ij} + \sum_{j=1}^{N} b_jD_j + c$$

is a strongly elliptic symmetric operator and

$$a_{ij}, b_j, c \in C_b^{0,\alpha}(\mathbb{R}^N),\ h \in C_b^{2,\alpha}(\mathbb{R}^N) \text{ and } K(s,t) = \sum_{j=1}^{N} \alpha_jD_j$$

and $\alpha_j \in C_b^{0,\alpha}(\mathbb{R}^N)$. Under such assumptions the robust equation is

$$dv_t = (e^{-hW_t}(A - \frac{h^2}{2})e^{hW_t} v + \int_0^t \sum_{j=1}^{N} e^{-hW_t} \alpha_jD_j(e^{hW_s} v)ds)dt$$

$$v(0) = x$$

and after an easy calculation

$$dv_t = [(A - \frac{h^2}{2})v + 2W(t) \sum_{i,j=1}^{N} a_{ij}D_ih\, D_jv +$$

$$+ W(t)^2 \sum_{i,j=1}^{N} a_{ij}D_{ij}\, h\cdot v + W(t) \sum_{j=1}^{N} b_jD_jh\cdot v +$$

$$+ \int_0^t e^{-h(W(t)-W(s))} \cdot (\sum_{j=1}^{N} \alpha_jD_jv)ds +$$

$$+ \int_0^t e^{-h(W(t)-W(s))}W(s)\cdot(\sum_{j=1}^{N} D_jh\, \alpha_jv)ds]dt$$

$$v(0) = x.$$

Obviously the above equation satisfies the condition of Theorem 6.1 if

$$E = C_b^{0,\alpha}(\mathbb{R}^N) \text{ and } H = \{u \in L^2(\mathbb{R}^N) : \frac{u}{(1+|x|^2)^N} \in L^2(\mathbb{R}^N)\}$$

(see also [5], [16] where more general stochastic equations are analyzed).

References

[1] P. Acquistapace, "Existence and Maximal Time Regularity for Linear Parabolic Integrodifferential Equations". J. Integral Equations (1985), Special Conference Issue: Trento, Italy, 1984, pp. 5-43.

[2] A. Balakrishnan, "Stochastic Bilinear Partial Differential Equations". U.S.A. Italy Conference on Variable Structure Systems, Oregon (1974).

[3] A.J. Bharucha-Reid and D. Kannan, "On a Stochastic Integrodifferential Evolution Equation of Volterra Type" J. Integral Equations, (1985) Special Conference Issue: Trento, Italy, 1984, pp. 351-379.

[4] P. Butzer, H. Berens, "Semigroup of Operators and Approximation", Springer-Verlag (1967).

[5] P. Cannarsa, V. Vespri, "Existence and Uniqueness of Solutions to a Class of Stochastic Partial Differential Equations", Stoch. Anal. Appl. 3 (1985), pp. 315-339.

[6] A. Chojnowska-Michalik, "Stochastic Differential Equations in Hilbert Spaces and some of their Applications", Thesis, Institute of Mathematics, Polish Academy of Sciences, 1977.

[7] R.F. Curtain, "Stochastic Evolution Equations with general white Noise Disturbance", J. Math. Anal. Appl. 60 (1977), pp. 570-595.

[8] G. Da Prato, "Regularity Results of a Convolution Stochastic Integral and Applications to Parabolic Stochastic Equations in a Hilbert Space", Conferenze del Seminario di Matematica dell'Università di Bari (1982).

[9] G. Da Prato, "Some Results on Linear Stochastic Evolution Equations in Hilbert Spaces by the Semigroups Methods". Stoch. Anal. Appl. 1 (1983), pp. 57-88.

[10] G. Da Prato, "Maximal Regularity for Stochastic Convolutions and Applications to Stochastic Evolutions Equations in Hilbert Spaces", Stoch. Space Time Models and Limit Theorems, L. Arnold, P. Kotelenez (eds.) (1985) D. Reidel Publishing Company pp. 41-52.

[11] G. Da Prato, "Equations aux Derivees Partielles Stochastiques et Applications" Centre de Mathematiques Appliquees, R.I. n. 148 (1986).

[12] G. Da Prato, P. Grisvard, "Maximal Regularity for Evolution Equations by Interpolation and Extrapolation", J. Funct. Anal. (1984), pp. 107-124.

[13] G. Da Prato, M. Iannelli, eds. of Journal of Integral Equations, 1985, Special Conference Issue; Trento, Italy, 1984, Integrodifferential Evolutions equations and Applications.

[14] G. Da Prato, M. Iannelli, L. Tubaro, "Semilinear Stochastic Differential Equations in Hilbert Spaces", Boll. U.M.I., (1979), pp. 168-185.

[15] G. Da Prato, M. Iannelli, L. Tubaro, "Some Results on Linear Stochastic Differential Equations in Hilbert Spaces", Stochastic (1982), pp. 105-116.

[16] G. Da Prato, M. Iannelli, L. Tubaro, "On the Path Regularity of a Stochastic Process in a Hilbert Space defined by the Ito Integral", Stochastic (1982), pp. 315-322.

[17] G. Da Prato, M. Iannelli, L. Tubaro, "An Existence Theorem for a Stochastic Partial Differential Equation arising from Filtering Theory", Rend. Sem. Mat. Univ. Padova (1983), pp. 217-222.

[18] G. Da Prato, L. Tubaro, "Some Results on Semilinear Stochastic Differential Equations in Hilbert Spaces", Stochastic 15 (1985), pp. 271-281.

[19] D.A. Dawson, "Stochastic Evolution Equations", Math. Biosciences (1972), pp. 287-316.

[20] G. Di Blasio, "Linear Parabolic Evolution Equations in L^p-spaces", Ann. Mat. Pura Appl. (1984), pp. 55-104.

[21] N. Dundfors, J. Schwartz, Linear Operators II, Interscience (1963).

[22] W.H. Fleming, S.K. Mitter, "Optimal Control and Nonlinear Filtering for Nondegenerate Diffusion Processes", Stochastic 8, (1982), pp. 63-77.

[23] P. Grisvard, "Commutativite de deux foncteurs d'interpolation and applications", J. Math. Pures Appl. (1966), pp. 143-290.

[24] A. Ichikawa, "Linear Stochastic Evolution Equations in Hilbert Spaces", J. Diff. Equs. (1978), pp. 266-283.

[25] D. Kannan, "Random Integrodifferential Equations", pp. 87-167 in A.T. Bhapucha-Reid ed., Probabilistic Analysis and Related Topics, Vol. 1, (Academic, New York, 1978).

[26] T. Kato, "Fractional Powers of Dissipative Operators II", J. Math. Soc. Japan (1962).

[27] P. Kotelenez, "A Submartingale type Inequality with Applications to Stochastic Evolution Equations", Stochastic 8 (1982), pp. 139-151.

[28] J.L. Lions, "Espaces d'Interpolation et Domaines de puissances fractionaires d'operateurs", J. Math. Soc. Japan (1962), pp. 234-241.

[29] J.L. Lions, E. Magenes, Problemes aux limites non-homogenes et applications, Dunod, Paris, (1968).

[30] J.L. Lions, J. Peetre, "Sur une classe d'espaces d'interpolation", Publ. I.H.E.S. (1964), pp. 5-68.

[31] M. Metivier, J. Pellaumail, "Stochastic Integration", Academic Press, New York, 1980.

[32] E. Pardoux, Equations aux dérivées partielle stochastiques non linéaires monotones, These, Paris XI (1975).

[33] E. Pardoux, "Stochastic Partial Differential Equations and Filtering of Diffusion Processes", Stochastic 3, (1979), pp. 127-167.

Vincenzo Vespri
II[a] Università di Roma
Dipartimento di Matematica
Via Orazio Raimondo
00173 Rome
Italy.